T0191467

The Hunterian Lectures
in
Comparative Anatomy
May–June, 1837

The Hunterian Lectures
in
Comparative Anatomy
May-June, 1837

Richard Owen

*Edited, and with an Introductory Essay
and Commentary, by*
PHILLIP REID SLOAN

The University of Chicago Press

Sir Richard Owen (1804–1892), comparative anatomist, associate and later antagonist of Darwin, and founder of the British Museum (Natural History), was a major figure in Victorian science. His contemporaries— Darwin, Lyell, Grant, Huxley, and others—knew his ideas and agreed or argued with him while developing their own views.

Phillip Reid Sloan is professor in and chair of the Program of Liberal Studies at the University of Notre Dame, and a faculty member in the Notre Dame Program in History and Philosophy of Science. He is coauthor, with John Lyon, of *From Natural History to the History of Nature: Readings from Buffon and His Critics.*

This book has been printed from material set in type by the editor.

The University of Chicago Press, Chicago 60637
Natural History Museum Publications, London

© Introductory essay and commentary, Phillip Reid Sloan, 1992
All rights reserved. Published 1992
Printed in the United States of America
99 98 97 96 95 94 93 92   5 4 3 2 1

ISBN 0-226-64189-9 (cloth)
ISBN 0-226-64190-2 (pbk.)

Library of Congress Cataloging-in-Publication Data

Owen, Richard, 1804–1892.
    The Hunterian lectures in comparative anatomy, May and June 1837 /
    Richard Owen ; edited and with an introductory essay and
commentary by Phillip Reid Sloan.
        p.   cm.
Includes bibliographical references and index.
1. Anatomy, Comparative.   I. Sloan, Phillip R.   II. Title.
QL808.O84   1992   91-21017
591.4—dc20   CIP

∞ The paper used in this publication meets
the minimum requirements of the American National
Standard for Information Sciences—Permanence of
Paper for Printed Library Materials, ANSI Z39.48-1984.

To my parents, family, and especially to Sharon,
who have all inspired much more than this work

Table of Contents

List of Illustrations

List of Tables

Preface

As historians of science are shifting their attention to the issues of the social construction of knowledge, patronage, disciplinary formation, and micro-contextual studies of historical episodes, production of an edited text serves a function beyond that of making available a document from the past. It also opens a window through which the historian can explore the micro-features of the intellectual and social landscape inaccessible in any other way. An edition of a set of public lectures delivered to a large and illustrious metropolitan audience also provides insight into the specific community of discourse uniting a speaker with his or her auditors. The presumptions made upon the audience, the cryptic and implicit references to authorities presumably understood by lecturer and listener, the often abbreviated allusions to contemporary issues, the specific context in which the speaker wishes to situate their materials—all these serve to illuminate the nature of the common reference frameworks mediating the dialogue of the public lecture.

The following lectures display many of these features. Delivered during a remarkable period of the growth of British and Continental science, the 1830s, they situate themselves within a discussion of issues transpiring in the London biomedical community. Their reference audience is urban and professional rather than academic; they are part of an effort of an institution, the Royal College of Surgeons, to emphasize the professional status of the surgeon; they are taking place within the context of the recent reopening of a grand and newly-remodelled facility, redesigned by one of London's finest architects. This was grand scientific spectacle in an age which felt the heady power of Bacon's promises now seemingly realized with the railroad and steam engine. At the same time, we see specific and detailed scientific doctrine expounded by a young anatomist who was to become one of the authoritative spokespersons for science in the Victorian era that effectively began with the death of King William during the course of these lectures.

In carrying out this project, many of my own perceptions of this period were altered. Two issues were most important for a revision in prior assumptions about the theoretical content of the biomedical sciences in this era: first, the lectures have revealed some of the subtle ways in which German biomedical thought and philosophy entered metropolitan London science in the 1820s and 30s. Prior to this research, I was much more prepared than I am at present to associate German biology in England with variants of Romanticism, even though we find in the person of Joseph Henry Green a subtle point of connection between the Coleridgeans and workers like Owen. But foremost is the impact of German and French functional physiologists, physiological chemists, and developmental anatomists on the scientific discourse. Under the rhetorical sanction provided by appeals to the illustrious John Hunter, the auditor at these lectures was often receiving the theories of the German physiologist Johannes Müller, the chemistry of Berzelius, the anatomy of Cuvier, and the embryology of Von Baer. But they were not listening to reworkings of the views of Schelling, Oken, Spix, Carus or Ritter. If there are ideological components to Owen's lectures, they cannot be assimilated to a framework of Coleridgean Romanticism.

The second insight I achieved was a new understanding of the early scientific thought of Richard Owen as he began his career as a member of the London scientific elite. For too long, Owen has been read in reverse, his scientific thought refracted through his later *On the Archetype of the Vertebrate Skeleton* and *On the Nature of the Limbs*. In the following lectures we encounter a young anatomist, formed jointly by Scottish, French, German and London traditions, concerned to ground comparative anatomy on medical physiology, who sees himself standing at a juncture between contemporary French, British and German inquiries into life and organization. We also perceive the importance of material conditions in the definition of the content of these lectures. Owen was lecturing to a specific target audience, in a precise physical environment, with constant need to make reference to a museum collection which he was charged to expound. But this was neither a display he had collected, nor did he have the freedom to alter its arrangement to suit his own theoretical interests. As a result, certain theoretical turns in Owen's thought, which in abstraction may appear as free developments of ideas, or as ad-hoc manoeuvres to satisfy Oxbridge and metropolitan constituencies, can better be related to the specifics of Owen's professional location. In this, I feel, lies some solution to the elusive problem of the unity and coherence of Owen's theoretical thought. The Owen which emerges from this study is an unexpected Owen. His theoretical positions develop over several years in conjunction with a complex museum collection. His eventual differences from Darwin relate as much to the specific differences in their respective conditions of theoretical work as they do to personal rivalry between them. Owen has emerged for me in this study a more interesting scientific thinker than I have encountered in the literature. Perhaps this renders Richard Owen the highest service that can be given as we approach the centenary of his death.

This project arose in an almost accidental way. While investigating Darwin's contacts with Owen when he returned from the *Beagle* years, I came across the early undated drafts of these lectures at the College of Surgeons, catalogued simply as "MS. of Some of his Lectures Delivered at the Royal College of Surgeons. n.d." Others, most notably the late Dov Ospovat, had seen these materials before, and it was his pioneering work which first led me to examine the Owen archive at the RCS. But none of the extant scholarship was able to illuminate the location or importance of these lectures in Owen's body of work.

Exploring this archive also led me to the discovery of the curious diagram in Owen's hand, reproduced as figure 6 below, on the back of a memorandum dated 20 January 1837. The presence of such a diagram in Owen's materials, apparently dating from early 1837, was understandably of interest to one concerned with the Owen-Darwin relationship and the genesis of Darwin's earliest transformist reflections in the early months of 1837. Exploring the background of this tree diagram of animal group relationships in Owen's earlier manuscripts, with the help of a footnote from Trevore Levere, led me to the remarkable manuscript lectures of Joseph Henry Green and Owen's notes on these, wherein I found similar representations in Green's Hunterian lectures for 1827–28.

The distance from Owen's initial drafts of the 1837 Hunterian lectures to the delivered text reproduced below was greater than anticipated. Determining the sequence of the lectures and the relation of these to the recopied documents at the RCS and in the Owen archives at the Natural History Museum meant resolving several uncertainties which remained even after discovering the final dated versions. The principal difficulty lay in the small degree of correlation between the dates and contents of these lectures and the fragmentary description of Owen's inaugural series in the *Life of Richard Owen* by his grandson Richard Starton Owen. R. S. Owen's account had the advantage of access to the now-missing daily diary of Caroline Clift Owen. The discovery of a crucial misdating in the *Life* of the letter of 6 May 1837 reproduced below (page 66) was finally able to resolve several chronological anomalies.

The appearance of Adrian Desmond's remarkable book, *The Politics of Evolution*, at a time when the preparation of these lectures was far advanced proved particularly beneficial. Independently we had come to focus on the importance of the London biomedical community as the context in which to understand metropolitan transformism. Desmond's masterful research and synthesis of a vast range of primary sources in this same period added greatly to my understanding of this wider context as I completed this project. Although we have come to disagree on some issues, particularly on the character of the German tradition expounded in the lectures at the College, on one critical point we agree completely, namely on the need for a shift of attention from the natural history community to the metropolitan medical and anatomical context as the framework for further understanding of British transformism.

Essential to the completion of a project of this nature was the continued assistance of the staff of the Royal College of Surgeons in London, which has been patiently cooperative during several years of work in obtaining materials from storage, making photocopies, answering correspondence, and generally making the library of the College an unusually congenial place to work. I particularly thank Mrs Michele Gunning and Mr Matthew Derrick for their regular assistance. The generous help of the librarians of the College, Mr E. H. Cornelius, and his successor, Mr Ian Lyle, and the current Quist Curator of the Hunterian Museum, Miss Elizabeth Allen and her assistant Mrs Muriel Gibbs, has been especially valuable in tracing down materials in the College archives used in the illustrations.

The staff of the British Museum of Natural History have also been extremely helpful, and without their willingness to supply photocopy materials this project could not have been completed. My editor at the Natural History Museum, Mrs Myra Givans, and her staff have well deserved the praises given by my colleagues in the Darwin industry. She has been a rigorous and helpful editor as well as a congenial hostess, and has been most patient with the unforeseen delays in the completion of this project. Her service in obtaining photographic materials at the end of the project was particularly appreciated. Ms Lynda Stratford of the Natural History Museum rendered valuable assistance at a crucial point in the project. An anonymous reader for the Natural History Museum provided very valuable comments on an early draft of the Introduction.

My editor at the University of Chicago Press, Susan Abrams and the senior manuscript editor, Jennie Lightner, were invaluable in the final stages of this project as rigorous copyeditors and critics. I owe a particular debt to their work and to the staff at the University of Chicago Press generally for their services in this phase of the project. Two reviewers selected by the University of Chicago provided particularly helpful comments at a critical point in the preparation of this manuscript.

I also have been assisted in several ways in this task by the library staffs of Cambridge University, the American Philosophical Library, and Temple University. Professor Jacob Gruber of Temple kindly made accessible to me his calendar of the Owen letters and shared with me a very useful essay based on his vast knowledge of Owen's life and work. The staff of the Victoria College Library, University of Toronto, should also be acknowledged for their timely assistance with Coleridge and Joseph Henry Green materials.

At Notre Dame, I have been assisted by my secretaries, Mrs Mary Etta Rees and Mrs Debra Kabzinski. The work of my student assistants over the years— Thomas Berry, Susan Prahinski, Mary Gipping, Kevin Crooks, Margaret Bilson, Samuel Nigro, Treven Santicola and Christopher Hatty—is appreciated.

Financial support for this project has been given by two Faculty Support Grants from the Institute for Scholarship in the Liberal Arts of the University of Notre Dame, a Travel-to-Collections grant from the National Endowment for the Humanities, a grant from the American Philosophical Society and a travel grant for final phases of this work from the Wellcome Institute for the History of Medicine in London. My initial interest in this project was generated by the research on Darwin's early contacts with German science made possible by National Science Foundation Grant SES—7925113.

I must also acknowledge the importance of comments from members of the Wellcome Seminar, the History and Philosophy of Science Department at the University of Leeds and the Centre Alexandre Koyré in Paris.

I also thank my colleagues in the Program of Liberal Studies who have provided a unique and supportive teaching and learning community. My ability to draw upon their wide range of learning in several disciplines has been very helpful on several issues. Much of my original interest in Owen was inspired by discussions with the late Stephen Rogers of the Program who first enabled me to see the relevance of my interest in Owen to his own interest in British Romanticism. Michael Crowe has been a continual support. His rigorous scholarship, encouragement and critical eye have been invaluable in aspects of this project.

Notre Dame, Indiana
June, 1992

PART ONE

Introductory Essay

On the Edge of Evolution

Introductory Essay
On the Edge of Evolution

This first publication of Richard Owen's inaugural series of Hunterian lectures in comparative anatomy is intended to serve a two-fold purpose. Its primary aim is to extend the understanding of the foundations of Owen's biological thought as it developed within a context of London's scientific and medical circles during a highly creative period of British science. Only through the deeper understanding of Owen's larger research project can the basis of his eventual opposition to Darwinism, an opposition which has too often served to define his importance in the history of science, be fully understood.

Secondly, these lectures provide an insight into the development of Darwin's transformism itself by revealing the specific theoretical context of London biomedical science exactly in the critical months when Darwin had moved his residence to London and was formulating the first version of his theory on the transformation of species. In this aspect the publication of these lectures is intended as a contribution to Darwin studies.

Determining the full context surrounding the origins of Darwin's evolutionary theory has involved recent investigators in close exploration of the latter period of the *Beagle* voyage and the first ten months after Darwin's return to England on 24 October 1836.[1] The traditional image of Darwin's "eureka" experience on the Galapagos archipelago has dissolved under the scrutiny of this scholarship, and a set of new issues has occupied Darwin scholars. Had Darwin talked about the species question with John Herschel when they met at the Cape in April 1836? To what degree was he synthesizing issues during the final leg of the *Beagle* voyage as he organized his notes and collections? What impact did the analysis of his collections by individuals like John Gould have on his thinking? Can more be determined about the events of March of 1837, the month he retrospectively identified as the period of the initial genesis of his theory?[2] If his ideas were not

---

1   F. J. Sulloway, "Darwin's Conversion: the *Beagle* Voyage and Its Aftermath," *Journal of the History of Biology* 15 (1982): 325–96; idem, "Darwin and His Finches: the Evolution of a Legend," *Journal of the History of Biology* 15 (1982): 1–53; Sandra Herbert, "The Place of Man in the Development of Darwin's Theory of Transmutation, Part I. To July 1837," *Journal of the History of Biology* 7 (1974): 217–58; and M. J. S. Hodge, "Darwin Studies at Work: A Re-examination of Three Decisive Years (1835–7)," in: *Nature, Experiment, and the Sciences: Essays on Galileo and the History of Science in Honour of Stillman Drake*, ed. T. H. Levere and W. R. Shea (Dordrecht: Kluwer, 1990), 249–74.

2   Darwin's "Journal" as published in F. Burkhardt, S. Smith et al., eds. *The Correspondence of Charles Darwin*, 5 vols. (Cambridge: Cambridge University Press, 1985-90), 3: "Appendix." Hereafter cited as *Darwin Correspondence*.

inductively derived from his empirical investigations, what causes led him to strike out in new territory in the early notebooks of 1837 and 1838? The period of time in which Richard Owen delivered the following lectures was one in which British intellectual, technological, social, political and scientific life were undergoing a remarkable series of developments. King William IV, whose reign as monarch was followed by that of Victoria, died during the course of these lectures, and his death resulted in the cancellation of one of them. In 1837, British citizens were experiencing the ability to travel at breathtaking speeds from city to city; a new, self-sustaining industrialism was evident in the countryside; and the social-political structure of England had been altered with power in the hands of the new mercantile and manufacturing classes as a result of the Reform Bill of 1832.

The social and intellectual context of science was also in dramatic development. The "society" model of British science had been transformed by a new style of national organization, the British Association for the Advancement of Science; an aspiring scientific elite, self-consciously designating themselves "scientists," had emerged to dominate the British Association and the other professional societies. Owen was one of these individuals. The days of the amateur and autodidact were quickly waning, although in some respects Darwin always had one foot in this older tradition. Traditional scientific organizations, such as the Linnean Society of London, were fragmenting by 1837 into smaller specialist organizations such as the Zoological and the Entomological societies, to enable the development of a more professional line of inquiry into specialized topics.

During the years of 1831–36, when Darwin had been at sea on the circumglobal voyage of *HMS Beagle*, Owen had been living in the metropolis of London. He was present as the authority of the scientific leaders of the 1820s—Cuvier, Bichat, Lamarck, E. G. St Hilaire, Virey—was being replaced by a new list of scientific authorities, most of them German or writing in German. In the life sciences, K. E. Von Baer, J. Müller, G. R. Treviranus, T. Schwann, C. Ehrenberg, C. G. Carus, F. Tiedemann, F. Wöhler, and J. J. Berzelius had in many respects supplanted the French authors.

Owen was delivering these lectures at a time in London when the controversy between French and German traditions was often reflected in institutional strife. Adrian Desmond has illuminated in detail the struggles between competing factions among British, and in particular London, comparative anatomists occurring in these years. Lamarckianism and the unity-of-type transformism of Geoffroy St Hilaire, of primarily French origin, were in conflict with German comparative anatomy and medical physiology, moving into London particularly through the College of Surgeons.[3]

It was this social world, which surrounded Richard Owen as he prepared to deliver these lectures in the summer and autumn of 1836, that Charles Darwin re-entered on the return of the *Beagle* to England in October. The immediate impact of

---

3   Adrian Desmond, *The Politics of Evolution: Morphology, Medicine, and Reform in Radical London* (Chicago: University of Chicago Press, 1990), chp. 6.

this changed environment on Darwin's thinking is not easy to assess. Cryptic notes and early reading lists compiled after his return suggest that Darwin made an attempt to bridge this gap in his knowledge.[4] We find him trying to learn German with the help of his elder brother Erasmus, who had been in Germany and was in contact with the Germanophiles associating with Thomas Carlyle.[5] His daily association with the London scientific and biomedical community after his move to the city in March of 1837 brought him into immediate proximity with this environment.

Four individuals—Charles Lyell, Robert Brown, John Stevens Henslow and Richard Owen—formed the immediate scientific contacts established by Darwin on his return to England. Others, among them William Sharp MacLeay, Andrew Smith, William Whewell, George Waterhouse and Charles Stokes, were in contact with Darwin, but evidence of their importance for his initial theoretical reflections is circumstantial.[6] Brown and Henslow were well known to Darwin personally before the cruise.[7] The new acquaintance with Lyell has received the greatest attention in the literature in view of Darwin's personal conversion to Lyellian geology during the *Beagle* years. Robert Edmond Grant, the London Lamarckian, Professor of Zoology and Comparative Anatomy at London University, and Darwin's mentor in marine biology during his last year as a medical student at the University of Edinburgh, is surprisingly absent from Darwin's list of new or renewed scientific associates. As an authority in matters of comparative anatomy and functional biology, he seems to have been quickly supplanted in Darwin's circle of associates by the young comparative anatomist at the College of Surgeons in Lincoln's Inn Fields, Richard Owen.

Little is actually known about the contacts between Owen and Darwin in the early months after Darwin's return. Traditionally it has been assumed that their interchange was occasional and limited to specific issues related to the *Beagle* fossils. But much of this interpretation of their encounter has been conditioned by a lack of understanding of Owen's own scientific thinking in this period. The close geographical proximity of Darwin and Owen during the 1837–42 period is itself responsible for the scarcity of documents detailing their relationship.

---

4   These initial efforts, predating the systematic reading lists maintained in the Darwin manuscripts (DAR 119, DAR 120), would include the list in the Edinburgh Zoology Notebook (DAR 118), covering the period from 1826 to sometime after the return; the lists at the end of the "B" and "C" Notebooks; and fragmentary lists such as that at DAR 91: 70 that seem to predate the "C" Notebook list.

5   Letter of S. E. Wedgwood to H. Wedgwood, 16 November 1836, in *Darwin Correspondence* 1: 519–20.

6   William Sharp MacLeay, the founder of quinirianism, had returned to London in 1836, and was in some direct contact with Darwin in the spring of 1837. Andrew Smith, with whom Darwin had conversed and travelled at the Cape, was in London briefly in the same period, and exhibited specimens at the Zoological Society meeting of 27 June 1837. See *Proceedings of the Zoological Society of London* 4 (1837), entry for 27 June.

7   These consultations are outlined in a pre-*Beagle* manuscript in the Darwin Archives, DAR 29.3: fol. 78.

Informal meetings at sessions of the Zoological and Geological societies, at dinners with mutual friends and over the fossil materials at the College of Surgeons provided numerous opportunities to discuss questions of mutual interest. The difficulty until now has been to discern what could have been the content of any conversations between them in this period other than those relating to the taxonomy of fossil mammals, a subject about which Darwin would seem to have had little to contribute.

The following set of lectures, written during the initial period of the Owen-Darwin encounters, and delivered publicly in the months just before Darwin penned his first reflections on the transformism of species in July 1837, provides the missing context within which we can situate the initial Owen-Darwin interactions. From these lectures, we can determine that in the period between October 1836 and July 1837 a remarkable convergence of interests was occurring between the two men on several issues, providing a new context for understanding the origins of Darwinism. In Owen, Darwin had direct contact with an emerging savant of the London biomedical community who was engaged in the synthesis of several different traditions in comparative anatomy and medical physiology and was preparing to present this synthesis to an illustrious audience composed of leaders of the London scientific and medical community.

The importance of these lectures is not solely resident in the light they might shed on the Owen-Darwin relationship. They also provide us with the first clear insight into the foundations of the scientific thought of one of the major scientific figures of Victorian Britain, whose varied scientific output has seemed previously to defy systematic analysis.

A. The Formation of a Comparative Anatomist

Owen was born in the north-west of England in the family house on Thurnham Street, Lancaster, on the twentieth of July 1804, the son of Richard and Catherine Parrin Owen. The family names denote both the Lancashire and French Huguenot background of his parentage.[8] Owen's lean appearance and dark hair are said to

---

8   Details on Owen's life have been constructed from the often unreliable, but essential, Richard S. Owen, *Life of Richard Owen*, 2 vols. (London: Murray, 1894; reprinted Gregg: Westmead, 1970), cited subsequently as *Life*. Unfortunately, this biography is often the only source of many documents and letters, including the extracts from Caroline Clift Owen's daily diaries that were kept regularly from 1827 to 1873. These diaries cannot now be located. Use has also been made of the valuable biography by C. W. G. Rohrer, "Sir Richard Owen: His Life and Works," *Bulletin of the Johns Hopkins Hospital* **22** (1911): 132–40. Rohrer is the last individual I can verify to have had personal access to the missing Caroline Owen diaries and other important family materials that are quoted by him without specific reference to the documents he is employing. Extensive use has also been made of correspondence and documents in the Owen papers at the Royal College of Surgeons, the British Museum of Natural History, the American Philosophical Society and Temple University in Philadelphia. Owen's own short autobiographical sketch, drawn up on 25 November 1874, was also

have derived from his mother's French heritage.

Fatherless from the age of five in a family of six children, Owen did not have the financial resources to obtain more than a limited formal education. In 1810 he entered Lancaster Grammar School to follow in the footsteps of his older brother. It was the same grammar school attended by his fellow Lancastrian and close neighbor, William Whewell, ten years his senior.[9] Reminiscent of the paternal denunciation Darwin received at a young age that he was "good for nothing but dogs, shooting and rat-catching," Owen was reported by one of his schoolmasters to be "lazy and impudent," and a bad end was predicted for him.[10] Owen, like the young Darwin, displayed a precocious enthusiasm for hobbies, in his case heraldry, and some of his later interests in animal form and structure can be seen to have their seeds in this early enthusiasm.

The financial condition of the family prevented further education at this point. Owen enlisted in the Royal Navy and served as a midshipman aboard *HMS Tribune*, and it was evidently at this time he became interested in surgery.[11] Returning to civilian life in Lancaster, Owen selected surgery as a career, a vocation almost cut short by his grisly experience with the severed head of a cadaver.[12] He was indentured for a five-year service on 11 August 1820 to a surgeon-apothecary by the name of Leonard Dickson and subsequently to the surgeon Joseph Seed upon Dickson's death in 1822. In 1823 the indenture was transferred to another surgeon-apothecary, James Stockdale Harrison, and was

---

(contd.)

utilized (American Philosophical Library Owen Papers B. OW 2). I have also made use of a valuable introduction by Professor Jacob Gruber of Temple University in Philadelphia to his unpublished calendar of the Owen letters (private communication). The Owen papers at the BMNH include a volume of biographical materials and details collected by Owen himself in the 1880s under the title of "Memoir of Profr Owen, 1868" (BMNH MSS L.O.C. O.o.24). This contains two unsigned biographical sketches, one appearing in *Leisure Hour* **32** (1883): 523–4, and one extracted from *Leaders in Science* (1884). These biographies have been dated personally by Owen, and presumably he has accepted their accuracy. The biographical remarks in the unsigned review by William Broderip of Owen's anatomical work, "Progress of Comparative Anatomy," *Quarterly Review* **90** (1851–2): 362–413, was also helpful. The early naval service on *HMS Tribune* is reported in the Broderip review, and repeated in the biographical sketches in the "Memoir." As the anonymous *Leisure Hour* sketch reports: "In early youth he served as midshipman on H. M. S. *Tribune*, but all prospects of naval advancement were speedily clouded by the peace that ensued at the close of the American war of 1814. The youthful middy returned to school on shore, and subsequently studied medicine under Mr. Baxendale, a surgeon of his native town" (523–4). The ship's log of *HMS Tribune* at the Public Record Office, Kew (ADM 51), verifies that the *Tribune* was in service from 1803 to July 1813, and was then decommissioned until 14 September 1818. It is therefore unlikely that Owen could have been aboard the ship prior to the 1818 period. The service of the *Tribune* for the 1818–22 period was primarily in the Caribbean.

9   On the early Owen-Whewell associations see Stair Douglas, *The Life and Selections from the Correspondence of William Whewell D. D.* (London: Kegan Paul, Trench & Co. 1882) 2–5.

10   R. S. Owen, *Life* 1: 7.

11   See note 8.

12   *Life* 1: 22.

terminated in October of 1824 when Owen was admitted into the University of Edinburgh medical school.[13] Owen thus commenced his studies at this center of English-speaking medical and anatomical science exactly one year before Charles Darwin would pursue the same path. Owen's course of studies and even his reaction to it were similar to those of his illustrious successor. Enrolling, as would Darwin, in Thomas Hope's outstanding chemistry lecture series, Andrew Duncan's *Materia Medica* course, Robert Jameson's lectures on natural history and W. P. Alison's lectures on theoretical and applied medical physiology, Owen took the standard curriculum for a medical student entering Edinburgh's influential school.[14] Darwin's autobiographical remarks have made famous the apparently dreadful lectures of Alexander Monro (tertius) in anatomy, and Owen's reaction to this portion of the curriculum was similar, even though the practical results were different in important ways from those of Darwin. Disgusted with Monro's teaching, Owen enrolled in the private anatomical school of John Barclay at No. 10 Surgeon's Square in the final year of Barclay's active directorship.[15] Such private schools flourished in Edinburgh in the 1820s, and these private schools, out of control of the university, were also the main routes by which French comparative anatomy was to enter the British Isles.[16]

At Barclay's school, Owen apparently made initial contact with Robert Edmond Grant, a part-time comparative anatomy instructor working for Barclay, charged with offering invertebrate anatomy lectures at the school;[17] it was the same Grant who two years later was to introduce Charles Darwin to the mysteries of marine invertebrate zoology.[18] At this point, the comparative careers of Owen and Darwin diverged markedly. Whereas Grant would most deeply influence Darwin in his Edinburgh years, it was Barclay who would leave the deepest mark upon Owen's thought, while Grant subsequently became his rival in London.

John Barclay's impact on Owen was not confined to the area of medical anatomy, but also extended to more philosophical issues. In the early 1820s Edinburgh was the scene of important theoretical debates over the possibility of reducing life and mind to states of matter. Medical theorists were engaged in the

---

13   Rohrer, "Richard Owen," 134 (source unspecified).
14   The syllabi and detailed descriptions of these courses in the 1820s are given in *Report of the Parliamentary Commission on Evidence, Oral and Documentary, Taken and Received by the Commissioners Appointed . . . for Visiting the Universities of Scotland*, vol. 1 (London: Clowes and Sons, 1837), "Appendix."
15   On Barclay's school see J. D. Comrie, *History of Scottish Medicine* 2nd ed., 2 vols. (London: Baillière, Tindall and Cox, 1932), 2: 495; see also P. H. Rehbock, *The Philosophical Anatomists* (Madison: U. Wisconsin Press, 1983), chp. 1.
16   Desmond, *Politics of Evolution*, chp. 2.
17   Documentation of contact between Grant and Owen in these years has been impossible to obtain for the period when Owen was at Barclay's school. The congenial relations between them are evident at least until the late 1830s. See note 111 below and the original ending of Lecture Four of 9 May.
18   P. R. Sloan, "Darwin's Invertebrate Program, 1826–1836, Preconditions for Transformism," in: D. Kohn, ed., *The Darwinian Heritage: A Centennial Retrospect.* (Princeton: Princeton U. Press, 1985), 73–86.

debates over immanentist and transcendentalist theories of mind and body, which had recently been transported to Edinburgh from the London debates between William Lawrence and John Abernethy.[19] Controversies over materialism were also generated in Edinburgh circles by disciples of Jeffrey Combe, who had formed the Edinburgh Phrenological Society to promulgate his theories.

Directly responding to these challenges, Barclay had published in 1822 an important defense of body-mind dualism, to defeat this trend to materialism. Examining the long history of the mind-matter problem, he drew from historical data evidence for the claim that mind and matter needed to be distinguished and could not be reduced one to the other.[20]

Barclay's philosophical anti-materialism marks one important point of divergence between Owen's and Darwin's intellectual development in their respective Edinburgh periods. Darwin's association with Robert Grant drew him directly into the circle of Barclay's opponents. From his Edinburgh years date the first beginnings of Darwin's concern to explain mind in terms of states of matter. Owen, however, decisively adopted the side of Barclay and Abernethy. As he recalled the situation years later in a letter to a friend:

> The present phase of physiology substitutes 'brain' for 'mind'. The latter signifies the sum of the acts and powers of its organ, as 'Life' is the sum of those of the organism. When I was an Edinburgh Student <in 1822,>[sic] our best Anatomist, Dr. Barclay, put forth his ~~best~~ work in defence of a 'Vital Principle'(see his 'Life and Organization'—still worth a perusal.) What 'Soul' or 'Mind' ~~is~~ <was> to 'Brain', 'Vital Principle' ~~is~~ <was> to 'Frame'. ~~Vital~~ <That>'Principle' & 'Soul' were held to be Entities, not mere ~~are~~ 'abstract Ideas' and ~~are~~ <to be> alike 'imperishable'. In the other view 'Vital Principle' and 'Soul' end with death. But only as regards this small globe, one of millions of such 'worlds'. . . .[21]

This divergence was manifest in the different social circles with which Darwin and Owen would respectively affiliate as Edinburgh medical students. Unique to the Edinburgh environment in these years was the formation of small student societies, meeting with university sanction, in which various topics were debated by students and faculty. Darwin's affiliation with the student Plinian Natural History Society is well known, and it is evident that the discussions of this group very often converged on the issues of materialism, phrenology and reductive biology.[22] The Barclay disciples, however, formed a rival society, the Hunterian

---

19 L. C. Jacyna, "Immanence or Transcendence: Theories of Life and Organization in Britain, 1790–1835," *Isis* **74** (1983): 311–29.

20 John Barclay, *An Inquiry into the Opinions Ancient and Modern Concerning Life and Organization* (Edinburgh, 1822).

21 Draft letter of Owen to Author of "Creeds of the Day," RCS Owen MSS File L1 Misc. No date, no wm. Angle brackets denote interlineations.

22 The Darwin letters and his other unpublished material from 1825–27 suggest that he was drawn increasingly into association with phrenological materialists and zoological reductionists through his association with Grant and the student Plinian Society. He was

Society, which Owen, along with fellow student Gavin Milroy, helped found.[23] It is difficult to determine, in the absence of crucial documents, the content of the student-society meetings of this Hunterian Society. It seems safe to conclude that its agenda was probably in opposition to that of the Plinian.

Leaving Edinburgh without a degree in April 1825, but with Barclay's personal recommendation to support him, Owen moved to London to begin an apprenticeship at St Bartholomew's Hospital with John Abernethy himself, the main figure in the anti-materialist debates that had generated the Edinburgh controversies.[24]

Offered the post of Prosecutor for Abernethy's surgical lectures at St Bartholomew's, Owen received his specialized training in pathological anatomy. As current president of the Royal College of Surgeons, John Abernethy was also in a position to encourage Owen's application for membership and licensure by the now-prestigious College, a distinction that was granted on 18 August 1826. Licencing by the College had become a crucial certification of professional stature in a highly polarized London medical context. It set apart those recognized as properly trained professionals from the large numbers of graduates of the popular proprietary anatomy and surgery schools that had extended from Edinburgh to London by the middle decade.[25] Richard Owen now had appropriate certification to set up his own private practice in London as a fee-collecting surgeon. However, his career was to undergo some surprising changes of direction due to the intrusion of external events.

A.1 The Hunterian Museum as Material Context

Although, as its name suggests, it was principally oriented to the training and licencing of surgeons, the Royal College of Surgeons was also to play a major role in the theoretical development of aspects of life science—principally comparative anatomy, physiology and paleontology—in London not unlike that played by the *Muséum national d'histoire naturelle* in Paris. But the origins of this function of the College were very different from the complex history of its French counterpart.

In 1793, John Hunter, a prominent surgeon and the keeper of a large anatomical

---

(contd.)
> recommended for membership in the Plinian Society by W. A. F. Browne and Robert Grant, both members of the Edinburgh Phrenological Society (Minutes Book of the Phrenological Society, University of Edinburgh archives). Browne was a dues paying member, and Browne and Grant were in regular attendance at the closed meetings of the Society (ibid.). Browne's advocacy of a materialist theory of mind at a meeting of the Plinian Society attended by Darwin in November 1826 is carefully excised from the *Minutes Book of the Plinian Society*, entry for 21 Nov. 1826 (University of Edinburgh Archives DC 2 2: 53–4). For discussion see H. E. Gruber and P. H. Barrett, *Darwin on Man* (London: Wildwood, 1974), 478, and Desmond, *Politics*, 67–9.

23   *Life*, 1: 27.
24   Rohrer ("Richard Owen,"135) gives the date of Barclay's letter of introduction as 25 April 1825.
25   Desmond, *Politics*, chp. 4.

and natural curiosity collection, had died suddenly. The collection of over thirteen thousand specimens was bequeathed in Hunter's will to his son-in-law, Everard Home, and his nephew, Matthew Baillie, with the specification that it should first be offered for sale to the Crown.[26] Purchase of this collection from John Hunter's estate was agreed to and made part of the establishment of the Royal College of Surgeons of London in 1799, and by this act the new College became the proprietor of one of the largest museums of its kind in the world.[27] In 1806, this collection had been moved from its Leicester Square residence to its permanent location on the premises of the Royal College of Surgeons in Lincoln's Inn Fields, as part of a national memorial to Hunter's achievements. The College became the possessor of one of the finest comparative anatomical museums in the world.

The great issue facing the College curators and lecturers upon acquiring this large collection, which included anatomical preparations, monsters, fossils, bones and preserved vertebrates and invertebrates, was that its plan of organization and intentions were deeply obscure. John Hunter's death left the collection without an extended catalogue. He also did not have a professional disciple—an Achilles Valenciennes or a John Pentland—who knew the intimacies of the collector's theoretical intentions. Hunter's organizational principles underlying the arrangement of his collection were in manuscript form. No guide to the collection existed except two small published guides, and in manuscript a series of several small booklets giving short listings of specimens in numerical order.[28] Hunter's heir-apparent and first Curator, his son-in-law Sir Everard Home, was to prove an inefficient and, it appears, thoroughly dishonest individual who seems to have used the manuscripts primarily to plagiarize and publish as works under his own name.

A condition of the purchase and grant of this collection to the College of Surgeons by the Crown in 1799 was that it be a national resource, accessible to the medical public. The move of the museum to the College premises promised to fulfill this condition. But by the 1820s, the College was also involved in a political struggle with rival London medical establishments centering around its claims to control the licencing of surgeons. Not only membership in the College, but also admission to the College premises, including the Museum and Library, was restricted to recognized professionals, admission being only by ticket or with

---

26　"John Hunter's Will," John Hunter Papers, RCS H. CR. HUN: J 49.3.73. This is a later recopy by Clift of the copy held in the Court of Canterbury dated 26 Feb. 1824.

27　Relevant documents are reprinted as an appendix to the "Report from the Select Committee on Medical Education, with the Minutes of Evidence, Appendix and Index" [Warburton Committee], *Reports from the Parliamentary Committees 1834, Part II: Royal College of Surgeons, London* (London, 1834), 59–68. Hereafter cited as *RPC 1834* .

28　The published guides were [E. Home and W. Blizard] *Summary of the Arrangement of the Hunterian Collection of the Royal College of Surgeons* (London, 1813) and [Wm Blizard] *Synopsis of the Arrangement of the Preparations, in the Gallery of the Museum, of the Royal College of Surgeons* (London: Carpenter and Son, 1818). A summary of all the guides before 1910 is given in Arthur Keith's *Guide to the Museum of the Royal College of Surgeons* (London: Taylor, 1910).

the personal recommendation and signature of a recognized member of the College. Restricting entry to the Museum seems to have been due more to the embarrassing lack of a guide to the collection than to an explicit desire   to restrict outsiders.[29] However, to medical   professionals or other interested persons unable   to gain access to this supposedly national collection,   these restrictions were understandably seen as conscious attempts by the College to exclude non-members.[30]

By the mid 1820s, clamor over these practices of the College had reached the level of Parliamentary debate. The College was under severe external pressure to make its Museum collection generally accessible to all interested medical personnel and students without such restrictions.   But those within the College were fully aware that the collection was usable only with some coherent explanation of its arrangements.   Visitors to the Paris Museum, or to museums in Berlin, Breslau and other European centers of comparative anatomy, had examples of the way in which an anatomical museum could be utilized as a teaching collection with specific theoretical goals intended in its viewing. The Paris *Muséum national,* and specifically its *Cabinet d' anatomie comparée,* for example, was organized in accord with Cuvier's *Leçons d' anatomie comparée* and this work could function almost as a guide to the collection.[31] The best the Hunterian Collection could offer as guides for the visitor were the sketchy outlines of 1813 and 1818 by Everard Home and William Blizard (see Table One). Visitors to the Museum were simply conducted around it by William Clift in large groups "peeping over each other's shoulders" as he explained the baffling collection.[32]

But the preparation of an adequate guide to the Hunterian Collection was bedeviled by a series of adversities. Through an almost comical series of events that included the near burning down of Everard Home's house, most of John Hunter's manuscripts, including those illuminating the main theoretical framework for the organization of this collection, had been intentionally destroyed by Sir Everard Home in 1823 before a catalogue had been worked out. To Clift's horror, Home casually mentioned the dastardly deed during a carriage ride in 1823. Recovery of

---

29   The minutes of the meetings of the Trustees and those of the Museum Curators suggest that the state of the museum, rather than a strong political motive, was directly responsible for this restrictive policy. There is evident frustration within the Board of Trustees with the failure of the Conservator William Clift to solve this problem. See letter of 14 June 1826 of John Elliotson of the Board of Trustees in *Lancet* **10** (1826): 409–12. Elliotson is replying to Thomas Wakley's charges of strong political motivation that Wakley published in a distorted report of a meeting of the Trustees in ibid., 340–1.

30   For a moderate expression of the "smothered indignation" felt by professional physicians at these restrictive policies, see *Medical-Chirurgical Review* (N. S. Jan. 1827): 283–9.

31   Details on the arrangement of Cuvier's *Cabinet d' anatomie comparée* in the early 1830s can be determined from the posthumous *Catalogue des préparations anatomiques laissées dans le Cabinet d' anatomie comparée de Muséum d' histoire naturelle,* prepared by Achilles Valenciennes and John B. Pentland (Paris: privately published, [1832]).

32   See description by John Elliotson in his letter to the editor in *Lancet,* above note 29.

Table One:
The Hunterian Collection, 1818[33]

PART I: Organs in Plants and Animals for Special Purposes	PART 2: Organs, in Plants and Animals, for Propagation of the Species
Series I: Parts Employed in Progressive Motion	Series 13: Organs, in Plants and Animals, Which Are Double in the Unimpregnated State
Series 2: Organs of Digestion; and Instruments of Preparing the Food	Series 14: Male Organs, in Plants and Animals, of Distinct Sexes
Series 3: Intestinal Canal; and Glands, Connected with It	Series 15: Female Organs, in Plants and Animals, of Distinct Sexes, Unimpregnated
Series 4: Absorbent Vessels	Series 16: Coitus
Series 5: Heart; and Blood-Vessels	Series 17: Production of Young, in Plants, and Animals, with Double Organs
Series 6: Organs for Aeration of the Blood	Series 18: Production of Young, in Animals of Distinct Sexes, from Ova
Series 7: Organs for the Secretion of Urine	Series 19: Production of Young, in Animals Which Have a Uterus
Series 8: Brain and Spinal Marrow	Series 20: Foetal Peculiarities
Series 9: Organs of Sense	Series 21: Growth in Young, in Plants, and Animals
Series 10: Cellular Membrane; and Animal Oils	Series 22: Nourishment, and Protection, Afforded by the Mother to Her Young
Series 11: Cuticle: Its Different Forms	
Series 12: Peculiarities in Vegetables, and Animals	

33 Adopted from the analysis in [Blizard] *Synopsis*; see also [R. Owen & W. Clift] *Descriptive and Illustrated Catalogue of the Physiological Series of Comparative Anatomy Contained in the Museum of the Royal College of Surgeons, London*, 5 vols. (London: Taylor, 1833–40) 5, ii–viii. Hereafter cited as *DIC*.

some remains of the Hunterian papers in the public water closet on Sackville Street by the ever-thorough Clift could not repair the loss.[34] A crucial body of manuscripts needed for completing an adequate catalogue was no longer extant, and one would have to be painfully reconstructed.

The disaster this disclosure implied for the College ironically played an important role in Richard Owen's career and in the subsequent history of biology. In the opinion of the Council and Trustees of the Museum, it was important that William Clift's full attention be directed to the construction of a catalogue to cover this embarrassing event. As Clift was often distracted by menial duties as well as the requirement that he conduct visitors through the Hunterian Collection, competent assistance was required.[35] It was under these conditions that Richard Owen was appointed on 6 March 1827 as a third assistant to William Clift.[36] His immediate task was to assist Clift and his assistants Samuel Stutchbury and William Home Clift in the construction of a new catalogue. The results of this labor were to issue in William Clift's brief *Catalogue of the Contents of the Museum*,[37] and finally in the five-volume *Descriptive and Illustrated Catalogue of the Physiological Series* that was published in 1833.[38]

One of the conditions of the purchase of the Hunterian Collection by the Crown in 1799 was the stipulation "that a course of lectures, not less than twelve in number, upon comparative anatomy, illustrated by the preparations, shall be given

34   Details of this are supplied in the transcription of the testimony by Clift before the Parliamentary inquiry of 1834 ([Warburton], *RPC 1834*, 59–68). Selections from Clift's testimony, with much silent editorializing, were republished in *Lancet* (1834–5), 471–6. Other selections were also reprinted more faithfully in "Appendix A" to J. Hunter, *Essays and Observations on Natural History, Anatomy, Physiology, Psychology, and Geology*, 2 vols., ed. Richard Owen (London: J. Van Voorst, 1861). See also Clift's comment below, Lecture Four, fol. 80. The issue has been discussed by Jessie Dobson, *William Clift* (London: Heinemann, 1954), chp. 6. Home's disservice to the history of science was not limited to this deed. He is also the last known person to have had in his possession in 1820 the priceless collection of Leeuwenhoek microscopes belonging to the Royal Society of London. See Brian Ford, *Single Lens: The Story of the Simple Microscope* (New York: Harper and Row, 1985), 74–75.

35   Clift had claimed to the Trustees in 1826 that eight years would be needed to complete a catalogue in view of the destruction of the manuscripts. See *Lancet* **10** (1826): 341.

36   Minutes of Museum Board of Curators, vol. 5, entries for 9 February and 6 March 1827, RCS Archives. Owen was appointed as an assistant to work with Samuel Stutchbury. William Home Clift, Clift's son, had been appointed in 1824 to assist WC in the task of preparing the catalogue (Minutes of Curators, 5 January 1824), and served as the Assistant Conservator until his accidental death in 1832. Owen accepted the offer on 30 March. Clift had proposed the project of a complete catalogue of the collection to the College sometime in the late 1820s (Hunter Papers, RCS H.CR.HUN: J 49.3.67, n.d.) that would reconstruct the collection according to the system "Hunter himself has given of the preparations he has made," determined by collecting all the remaining manuscripts he could assemble.

37   [W. Clift & R. Owen] *Catalogue of the Contents of the Museum of the Royal College of Surgeons in London*, 6 parts (London: Taylor, 1830–31).

38   See note 33.

twice a year by some member of the Surgeons Company."[39] These lectures had been a regular feature of the spring curriculum at the College of Surgeons since their establishment in 1810. Initiated by twelve lectures on comparative anatomy by Everard Home, paired with William Blizard's thirteen lectures on surgery and pathology, the comparative anatomy lectureship had been occupied successively by Astley Cooper (1814–15), William Lawrence (1816–19), Benjamin Brodie (1820–21, 1823), again by Everard Home (1822), Joseph Henry Green (1824–28), Herbert Mayo (1829–30), and Charles Bell (1832–33).[40] John Abernethy and William Lawrence, the two great protagonists in the controversy over materialism, which had rocked British and Scottish medical circles, had engaged in this controversy through immediately sequential College lectureships, with Lawrence holding that in comparative anatomy and Abernethy the appointment in surgery (1814–17). Lawrence's Hunterian lectures of 1817 on human comparative anatomy were to form an important part of his controversial publication *The Anatomy and Physiology of Man*, the work responsible for the debate with Abernethy.[41]

A more relevant theoretical dimension of the Hunterian lecture series for Owen's development dates from the appointment, in July 1823, of Joseph Henry Green to the Hunterian lectureship in comparative anatomy. When Owen joined the College staff in early 1827, Green was still the occupant of this position, and he opened his 1827 series on the comparative anatomy of birds on 27 March with Owen in attendance. Understanding Green's larger importance for Owen's thought requires a select examination of the developments taking place in German biology and anatomy in the generation following the philosophical revolution instituted by Immanuel Kant, and the way in which these formed an important component of Green's lectures as Owen encountered them.

A.2 Intellectual Context: German Biology at the College

Since the pioneering studies of Réné Wellek, the influx of German ideas into England in the early nineteenth century has been of general interest to literary scholars and intellectual historians.[42] The extension of these inquiries into the analysis of nineteenth century British science by Elinor Shaffer and Trevor Levere

---

39 Letter of Charles Long to the Court of Assistants, College of Surgeons, 7 December 1799, *RSC 1834*, 62. This stipulation was revised in Long's letter of 7 January 1800 to read "one course of lectures, not less than 24 in number, on comparative anatomy and other subjects, illustrated by the preparations" (ibid.). Until 1837, the twenty-four lectures were divided nearly equally between surgical and comparative anatomical topics.

40 The lectures were suspended in 1831 due to the public disruption of the series by Thomas Wakley and Thomas King on 8 March. See below, note 120. For an approximately accurate chronology of the lectures see V. G. Plarr, *List of Lecturers and Lectures at the Royal College of Surgeons of England 1810–1900* (London: Taylor and Francis, 1900).

41 W. Lawrence, *Lectures on Physiology, Zoology, and the Natural History of Man*, 2 vols. (London: Benbow, 1822).

42 R. Wellek, *Kant in England*, rev. ed. (Princeton: Princeton University Press, 1972); see also R. Ashton, *The German Idea* (Cambridge: Cambridge University Press, 1981).

has also demonstrated the interaction of scientific and literary traditions.[43]

Intellectual contacts between British and German scientific, literary and philo-sophical circles had surely begun before the nineteenth century. The Hanoverian connection of the British Royalty gave a strong basis for this in the eighteenth cen-tury, and the political alliance between England, Austria and Prussia during the Revolutionary and Napoleonic wars increased these general contacts. But in spite of these wider connections, there was still a sense of novelty expressed in the early decades of the nineteenth century by a circle of writers who managed to travel and study in the German states. German science and philosophy promised to many of the young Romantics a great alternative to the materialism of late-eighteenth century Newtonianism and the empiricist epistemology of Locke, Hartley, Hume, Bentham and Priestley. Principally, but not exclusively, through the activities of the poet-philosopher Coleridge—that meteoritic intellectual of the early decades of the cen-tury—this Germanophilia received its great popular advocacy. Coleridge had stud-ied at German universities, attending the lectures of the renowned anatomist and theoretical physiologist J. F. Blumenbach. From this experience, Coleridge had returned to London concerned to generate broader contacts among his fellow countrymen with the ideas of Kant, Schelling, Heinrich Steffens, Carl Gustav Carus and J. H. Ritter. In the early decades of the century a wave of translations and secondary reports on German science, literature and philosophy poured into London and Scottish intellectual circles.

But German thought was novel, complex and often difficult for its new English-speaking audience to comprehend. Its early dissemination relied heavily on a small group of individuals who had travelled to Germany and carried out the early translation, explication and often misinterpretation of complex issues in the German intellectual world. Letters from travellers proclaiming their escape from the stifling world of empiricist epistemology and etherealist Newtonianism were means by which English readers were often exposed to German thought.[44] Literary journals such as the *Foreign Quarterly Review* and scientific journals such as the *Edinburgh New Philosophical Journal* and the *British Foreign and Medical Review* provided a fertile source of translated or summarized Germanic literature, philosophy and science for those interested in pursuing matters further.

Among Coleridge's close circle of friends was none other than a young surgeon from St Thomas' Hospital who was to become the Hunterian lecturer in compara-

43   Trevor Levere, *Poetry Realized in Nature: Samuel Taylor Coleridge and Early Nineteenth-Century Science* (Cambridge: Cambridge University Press, 1981); Elinor S. Shaffer, "Romantic Philosophy and the Organization of Disciplines," in: A. Cunningham and N. Jardine, eds., *Romanticism and the Sciences* (Cambridge: Cambridge University Press, 1990), 38–54; *idem.,* "Coleridge and Natural Science: A Review of Recent Literary and Historical Research," *British Journal for the History of Science* **12** (1974): 284–98. For several relevant discussions on the complexities of these interactions see Cunningham and Jardine (op. cit.), especially essays by Rehbock, Dietrich von Englehardt and Trevor Levere.

44   See especially Henry Crabb Robinson's "Letters on Kant and German Literature," *The Monthly Register and Encyclopedian Magazine* (1802–3): 6–12; 205–208; 294–8; 397–403; 411–16; 485–8; 492–3. I am indebted to Gregory Maertz for this valuable reference.

tive anatomy at the College, Joseph Henry Green. A new intellectual component was now added to Owen's Edinburgh schooling and London hospital apprenticeship. Through Green, Owen first encountered German philosophical anatomy and biology. But to understand the significance of this requires analysis of important distinctions between the different intellectual traditions within German theoretical science of the early nineteenth century. The misleading polarity of speculative *Naturphilosophie,* represented in English dress by Coleridge, and native English and Scottish empiricism, articulated by Bentham, Thomas Reid and the Millses, has tended to distort some important issues in the analysis of the German impact on British science in these decades.[45] A more accurate picture of the complexity of this German influence in specific detail can be gained by an examination of Joseph Henry Green's unique blend of philosophical comparative anatomy and German philosophy.

Joseph Henry Green was born in London in November 1791 to a wealthy merchant family, the maternal nephew of Henry Cline, surgeon and officer of the Royal College of Surgeons. This relationship to Cline would prove valuable for Green's own later prominence at the College. In 1806, at the age of fifteen, Green was sent, in the company of his mother, to the German states for continued education, an unusual choice for his parents to have made at this turbulent era in European history. For three years he studied in Germany, chiefly in Hanover. His biographer, John Simon, reports that this early German education gave him "those habits of methodical industry and deliberate reflection and conscientiousness that marked him till the end of his career."[46]

After his return to England in late 1809, Green was apprenticed to his uncle, Henry Cline, at St Thomas' Hospital, London, beginning an association that would later culminate in his directorship of the institution. In 1815, Green obtained his surgeon's licence from the College of Surgeons and opened a practice in Lincoln's Inn Fields. His association with St Thomas' continued during the next two years, and he delivered, in his unpaid post of Demonstrator of Anatomy, a series of lectures on anatomy and surgery. Green's serious involvement with German biology and speculative philosophy was occasioned by contact he made with Johann Ludwig Tieck (1773–1833), the German Romantic poet, translator and literary critic, who had travelled to London in 1817 to make use of the British Library, visit Samuel Taylor Coleridge and meet scholars at Oxford.[47] This visit

---

45   John Stuart Mill's 1834 characterization of British intellectual traditions as derivative of either Bentham or Coleridge, while useful for some issues, confuses many others.

46   J. Simon, "Memoir" prefacing J. H. Green, *Spiritual Philosophy, Founded on the Teaching of the Late Samuel Taylor Coleridge*, ed. J. Simon, 2 vols. (London: MacMillan, 1865), 1: iii.   The most recent and thorough account of Green's biography and relationship to Coleridge is Heather Jackson, "Coleridge's Collaborator, Joseph Henry Green," *Studies in Romanticism* 21 (1982): 1611–79.

47   Samuel Taylor Coleridge to Thomas Boosey, 14 June 1817, *The Collected Letters of Samuel Taylor Coleridge,* ed. E. L. Griggs (Oxford: Clarendon, 1959), 4: 738. Green seems to have known Tieck before meeting Coleridge, and the first contact of Coleridge and Green was evidently through a reception for Tieck at Green's apartments in Lincoln's Inn Fields.

seems also to have been the occasion on which Green and Coleridge first met, possibly drawn together by James Gillman, a member of the College of Surgeons and a close collaborator with Coleridge in the writing of his *Essay on Scrofula* and his *Theory of Life*.[48] As a consequence of this visit, Tieck personally arranged for Green to travel to Berlin in the summer of 1817 to undertake a private reading course in contemporary German philosophy from Tieck's close friend, the German philosopher Karl Wilhelm Ferdinand Solger (1780–1819).[49] As Tieck described the circumstances, Green had for some time been particularly eager

> to go to Germany especially to be instructed by Schelling, whom he knows and
> reveres best, and particularly to be able to learn about the history of the new philo-
> sophy. But he had to postpone the fulfillment of this hope for several years,
> since he was employed in London as a teacher of anatomy. I do not believe that
> Schelling is the one who can enlighten the young man in this, since he is wholly
> incompetent in conversation. [50]

Solger seemed to promise Tieck more patience with Green's conversational deficiencies, and arrangements were made for a private tutorial in Berlin.

The intensive study of German philosophy during what seems to have been a two-week tutorial with Solger confirmed Green in his conclusion that in German philosophy was to be found the theoretical basis for unifying natural philosophy, anatomy and aesthetics.[51] Kant's own combination of aesthetics with reflections on organic teleology and speculations on comparative anatomy in his *Kritik der Urteilskraft* of 1790 provided Green with a model for this unification, and Green's general enterprise represents a conscious attempt to realize this synthesis in practical work.

Following this tutorial, Green and his wife remained until November in the

---

(contd.)
> Coleridge, Green and Tieck seem to have had common interests in the theory of animal
> magnetism at this time. See letter of Coleridge to Frere, 27 June 1817, ibid., 744–45.

48   Levere, *Poetry,* 42–5.
49   L. Tieck to F. Solger, 26 July 1817, *Nachgelassene Schriften und Briefwechsel;
     herausgegeben von L. Tieck und F. v. Raumer* (Leipzig: Brockhaus, 1826; reprinted
     Schneider: Heidelberg, 1973), 1: 550–2. This letter is reprinted in P. Matenko, ed., *Tieck and
     Solger: The Complete Correspondence* (New York: Westermann, 1933), 370–3. Additional
     information on Green's visit is given in J. Simon, "Memoir," vi–vii. Green travelled after
     this to the northern German states in company with his wife, and returned to England in
     November. See letter of Coleridge to Green, 14 November 1817, *Collected Letters,* 4: 783.
     Green's knowledge of German and German philosophy was thereafter often relied upon by
     Coleridge. See letters of 13 December 1817 (ibid., 791) and S.T. to W. H. Coleridge, 1
     April 1818, *Unpublished Letters of Samuel Taylor Coleridge,* ed. E. L. Griggs (New Haven:
     Yale University Press, 1933), 2: 238.
50   Ibid., 370–1.
51   Solger's complex philosophical positions and his relations to German Idealism are spelled out
     by Percy Matenko in the introduction to *Tieck and Solger,* 38–63. Solger's aesthetic theory is
     developed in his *Vorlesungen über Ästhetik,* ed. Karl Wilhelm Ludwig Heyse (1829, reprinted
     Darmstadt: Wissenschaftliche Buchgesellschaft, 1980).

German states, but his contacts in this period are uncertain and can only be inferred. Whatever these might have been, in the period following his 1817 travels, Green displays clear familiarity with many issues in German biomedical science and philosophy.

The importance of Green's direct, rather than indirect, contact with German philosophy and, it seems, with German biomedical science in this period should be underlined. He was much more than a mere disciple of his friend Coleridge, and on many issues there is a clear difference between them. Most significant is the evidence that Green seems to have been deeply conversant with Immanuel Kant's technical philosophy in a way not displayed by Coleridge. The importance of this point will emerge subsequently.

In 1820 on the sudden death of his uncle, Henry Cline, Green was elected to the Surgeonship of St Thomas' Hospital and became closely associated with Sir Astley Cooper, a former Hunterian lecturer in comparative anatomy at the College of Surgeons and later to be President of the College. In 1831 Green was to collaborate with Cooper in the publication of the popular *Manual of Surgery*. The appointment of Green to the Hunterian comparative anatomy lectureship at the College followed in 1823, and in 1825 he was made Professor of Anatomy at the Royal Academy of Art. Subsequently he was selected to deliver an annual series of six lectures to art students at Somerset House, lectures he continued until 1852. In 1830 Green resigned his Chair of Surgery at St Thomas' to inaugurate the prestigious new Professorship of Surgery at the new King's College, London, lecturing there from 1831 to 1836.

With this background, the appointment of Green in 1823 as Hunterian lecturer in comparative anatomy rather than to the position of surgical lecturer is somewhat surprising. Nothing in his prior work would have immediately suggested his suitability for this position that required him to lecture on the various invertebrates and vertebrates in the massive Hunterian Collection. His selection probably had much to do with his friendship with Astley Cooper and his reputation as a fine lecturer. Although it has proven impossible to document the point, Green must have displayed some interests in the speculative theories of the biologists of his day, making his appointment acceptable.[52]

---

52 Green's intellectual and scientific development before 1824 has proven difficult to document directly. I have been unsuccessful, in spite of repeated searches, in locating any substantial body of manuscript or correspondence surviving for Green beyond the notes on his Hunterian lectures at the College of Surgeons, one set of surgical lecture notes taken by a student (St. Thomas' Hospital Archives, M 68) and a series of fourteen letters of Green to Coleridge in the Coleridge papers at Victoria College, University of Toronto, dated between 27 December 1817 and 16 April 1833. Neither correspondence nor notebooks appear to have survived from his Continental travels. The student notes on his surgical lectures were donated to St. Thomas' hospital by a party in South Africa in 1953, and it is possible that additional materials are located in that country. Claims of a strong ideological motivation behind Green's appointment are difficult to substantiate. Green was elected by ballot with an ambiguous plurality at the meeting of the Council of the College in July 1823 to replace Benjamin Collins Brodie, who had resigned on 3 May (Minutes of the Council of the College of

As Hunterian lecturer, Green was considered to be an invigorating speaker and, initially at least, attracted wide attention from a portion of the London medical community. The *Medico-Chirurgical Review* speaks of Green's inaugural 1824 series of lectures in glowing terms, describing it as expounding "the whole of animated nature, from the minutest animalcula up to man himself," and commented that "nothing could exceed the *matter* thus brought forward, except the *manner* in which it was delivered. The most abstruse points of physiology and anatomy were descanted on in an easy and flowing oratory that would have done honour to the senate or the bar."[53]

Green opened his first course of lectures on 30 March 1824 by stating that his plan was to "give a general view of animal Nature—commencing with the simplest and tracing forwards."[54] In accord with the formal obligation of the lecturer to expound the views on comparative anatomy of John Hunter and the materials in the Hunterian Collection, the lectures paid due respect to their titular inspirer. But the incomplete character of Hunter's printed thought, exacerbated by the destruction of the Hunterian manuscripts by Everard Home, meant that Green, like all his predecessors, had little constraint on the content from this Hunterian presence.[55]

This very inadequacy provided the crucial social and institutional conditions necessary to render these lectures an ideal vehicle for the importation of novel ideas into an established British institution under the sanction of a revered member of the scientific establishment. Hunter functioned more as an occasion than a source for the exposition of any number of biological topics by the lecturers, all of whom

---

(contd.)

   Surgeons, 4 (1820–27), entry for 11 July 1823, RCS Archives). The minutes of the Council discussion of his appointment display no concern with his intellectual positions. It is probable that his skill as a lecturer had much to do with this appointment. His lectures were to be sandwiched between the surgical lectures of the dying Thomas Chevalier, who was universally acknowledged to be a dull and unimaginative speaker. Green had, by contrast, the reputation as a stimulating lecturer in his post at St Thomas'. For a gentle review of Chevalier's lectures of 1823–24 see *London Medical and Physical Journal* 52 (Jan.–June 1824), 93. For a vituperative analysis see *Lancet* 2 (1824): 383–4. Chevalier died on 9 June 1824. See also Desmond, *Politics*, 260–75.

53   Anon, *Medico-Chirurgical Review* 1 (June, 1824): 251. Green's abilities as a thinker and lecturer were predictably savaged by the *Lancet*. See *Lancet* 3 (19 June 1824): 370–1.

54   Green 1824 Lectures (in notes by William Clift), Green Papers, RCS 67. b. 11. Hereafter cited as "Green 1824." This is a complete series of detailed notes on the entire series. No manuscripts have been located in Green's own hand on any of the series.

55   The primary published works of Hunter, apart from a few papers in the *Philosophical Transactions*, were his *Treatise on the Blood* (London, 1794); *Observations on Certain Parts of the Animal Oeconomy* (London, 1786); *Treatise on Venereal Disease* (London, 1786); and the *Treatise on the Natural History of the Human Teeth*, 2nd ed. rev. (London, 1778). The collection in four volumes of Hunter's extant writings, *The Works of John Hunter*, ed. J. J. Palmer (London, 1835–7), provided a more coherent framework of Hunter's thought. Only with the publication in 1861 of the large body of recopied Hunterian manuscripts in possession of William Clift and released to Owen in Clift's waning days (Hunter, *Essays and Observations*) was a more complete view of Hunter's biological thought available to the public.

could summon Hunter's ambiguous authority with almost complete philosophical latitude. Abernethy had used the lectures to expound an electrical theory of life, drawing on Humphry Davy's electrical researches, under the name of Hunterian doctrine.[56] William Lawrence utilized the comparative anatomy lectures of 1817 and 1818 to develop the groundwork of his controversial views on life and organization, and their application to the comparative anatomy of man.[57] Green's utilization of these lectures for non-Hunterian ends was no exception. Hunter's authority, summoned at appropriate moments, became the sanction for Green's utilization of these lectures to present his synthesis of German and French biology to his audience of students. It is important to see what this did and did not imply in this context.

Green began each year's series of lectures with a general overview of the aims of the course to be covered, making cross-references, after the first year, to the lectures of the previous years. A characteristic of the opening lecture in each year was the display of and comment upon a series of diagrams presenting a general breakdown of the animal kingdom and its classification. He then proceeded to elaborate on a particular portion of the groups so schematized. The two schema displayed each year at the outset, as taken from Richard Owen's notes on the 1827 series, are illustrated on page 22.

The conjunction of these two diagrams is curious. The first is generally a breakdown of the animal kingdom in conformity with Cuvier's well-known four *embranchements* as set forth in his *Le Règne animale* of 1817. The second, however, is a remarkable diagram in the context of the period.[58] It most closely resembles, but also goes beyond, the diagrams given by Jean Baptiste Lamarck in a set of "Additions" to his *Philosophie zoologique* in 1809 and in his *Histoire naturelle des animaux sans vertèbres,* published in 1815.[59] The oddity in this is that these two authors were presumably in conflict in this period, with Cuvier attacking the transformism implied in Lamarck's ascending and branching series through his comparative anatomical critique. Green has, however, directly conjoined these. This puzzling strategy invites explication.

---

56 J. Abernethy, *Introductory Lectures Exhibiting Some of Mr. Hunter's Opinions Respecting Life & Disease* (London: Longman et al., 1815) (the surgical lectures for 1814–15).

57 On the complex interpretations of Hunter summoned by medical theorists, see L. C. Jacyna, "Images of John Hunter in the Nineteenth Century," *British Journal for the History of Science* 21 (1983): 85–108.

58 On the early history of the use of tree-diagrams, see P. F. Stevens, "Augustin Augier's Arbre botanique (1801), a Remarkable Early Representation of the Natural System," *Taxon* 32 (1983): 209–11.

59 Lamarck first used a branching diagram in the *Philosophie zoologique.* This linked the groups in a branching diagram, but as a descending and not ascending series. The second usage appears at the conclusion of the "Discours préliminaire" to the *Histoire naturelle des animaux sans vertèbres,* 5 vols. (Paris: Verière, 1815-22), 1: 131. This displays several invertebrate groups linked together in *descending* diverging groups, but without a more general connection between them. Lamarck's work on the invertebrates would have been directly relevant to Green's lectures on the invertebrates between 1824 and 1826 .

The bulk of Green's presentation in the lectures that followed these introductory remarks was devoted to an analytical description of the main functional systems—muscles, skeleton, digestion, feeding mechanisms, nervous system and reproduction—by comparisons of the forms with one another. In this, specimens from the Hunterian collections were utilized that were themselves arrayed, as we have seen from the synopsis of the 1818 arrangement, in generally functional categories. This functional presentation resembles, with important differences, Cuvier's practice in his *Leçons d' anatomie comparée.*[60]

As the opening lectures proceeded to develop the specific topics under consideration each year, Green followed a sequence dictated not by Cuvier, but instead by the branching diagram derived from Lamarck. This was not accidental. Green interpreted this ascending series to be a more dynamic means to represent organic relationship. As William Clift reports Green's comments in his notes on the second lecture of the 1824 series:

> Thus said he is all the Animal Kingdom divided—tis said he [it is] (reffering [sic] to these divisions) like the mighty sovereign of the Forest—The mighty oak—which from its huge trunk divides into branches—which is[?] again and again subdivided—into lesser branches and twigs—yet from the most minute twig—all all [sic] arise from the same Vitality—like these ascending gradations— Is as it were a grand march of Nature—each part in succession, shewing what will follow[.][61]

But Green was not following familiar traditions, and placing Cuvier and Lamarck in sharp opposition. Instead he was *combining* the insights of these two authors. There is little evidence to conclude that he was attacking Lamarckianism in the name of an Idealist morphology.[62]

To understand this curious strategy is to see the subtlety of the "German" influence upon Green's College lectures. In *content* there is little about Green's lectures to suggest his endorsement of traditional images of "romantic" or idealist biology. A summary of his lecture course for 1827, which can be reconstructed from

---

60   For important differences between the Hunterian and Cuvierian arrangements, see below, p. 39.
61   "Green 1824," fol. 30.
62   His most explicit comment on Lamarck remains praiseworthy: "I concluded my last lecture. . . with an attempt to form Animated Nature into a genesis—shall to day attempt to give a general view of the gradual ascent of the animal Kingdom from the most simple—an ascending Series—I am exceedingly happy in this to adopt the two-fold style of Lamarck—with some few alterations to my own Ideas. . . .Commence with the lowest and move upwards" ("Green, 1824," fol. 32). Green's more pointed criticisms of Lamarck's transformism seem to enter only in the reprinted "Recapitulation" of the 1828 course published in 1840 in his *Vital Dynamics: the Hunterian Oration Delivered. . .14th February, 1840* (London: Pickering, 1840), 108. The likelihood that this comment represents a later revision of the manuscript is supported by the absence of similar comments in the nearly identical 1827 opening lecture (see Appendix). Record of such negative comments on Lamarck in the 1828 course is also absent from Richard Owen's lecture notes (Owen Papers, RCS 67.b.11).

three different sources, reveals the structure displayed in Table Two. This curious conjunction of Lamarck's notion of an ascending and branching series of groups with the content displayed by this outline, is to be explained not by incoherent eclecticism, but by a complex methodological and philosophical position derived from German sources that underlies his reflections. Because of the importance of this theoretical strategy for the understanding of Green's own lectures, and subsequently for analyzing the complex character of the German tradition Owen initially endorsed in his own, it is necessary to explore some of this philosophical backdrop in German philosophy.

A.3 Green's Lectures and the Kantian Tradition in Natural History

Green's personal study in the German states on two different occasions and his technical study of German philosophy with Solger resulted in his grasp of important details of German philosophy that eluded many of his English contemporaries because their acquaintance with the German tradition came principally through imperfect periodical articles or from intermediary sources. Green's writings display a subtle understanding of Immanuel Kant's philosophical program precisely in those points where it was often badly misunderstood by other British enthusiasts, such as Coleridge, who tended to lump together the intellectual programs of Kant, Schelling and Hegel and the more speculative theories of Henrik Steffens and Lorenz Oken.[63] By contrast, Green is a careful thinker who displays awareness of distinctions in Kant's philosophical program as these had been emphasized in some of Kant's later writings of the 1780s.[64]

Briefly summarized, Kant's larger philosophical project, a project that he envisioned as creating a revolution in philosophy as profound as the Copernican theory had been in astronomy, reoriented the epistemological questions of modern philosophy through his restoration to the mind of the dynamic activity that he felt had been illegitimately extirpated by the "way of ideas" of the British philosophical tradition. Kant's well-known solution to this problem involved the proposal of a constructive power of the mind in terms of a distinction of four levels of mental activity, each of which contributed its own *a priori* to experience. The first level of mental activity, *Sinnlichkeit,* concerned the activity involved in the first-order sensory awareness of objects, and in receiving this sense information the faculty of Sensibility imposed upon the raw material of sensation the pure forms of intuition, space and time. Such temporally and spatially structured sensation then could serve as the material for the second level of mental structuring by the *Verstand* or Understanding, typically used in a technical sense in the Kantian writings to denote the primary structuring faculty of the mind. This faculty imposed on the material

---

63   The full character of Green's contact with technical German philosophy and biology has proven difficult to document directly. See note 52.

64   Green's copy of Kant's multi-volume *Vermischte Schriften* (Halle, 1799), containing several of his late essays, is in the Coleridge collection at the British Library.

Table Two:
Joseph Henry Green's 1827 Lectures
on the Comparative Anatomy of Birds

Lecture	Date of Delivery	Contents of Lecture
1	27 March	Introduction; 3 views of nature; primary divisions and classification of the birds; Cuvier's and Lamarck's arrangements of the animal kingdom diagrammed; summary review of the invertebrate lectures
2	29 March	Cuvier's four classes of animals; diagram of the relations of the birds; distinguishing characters of the birds; distinction from the reptiles; distinctive digestive organs; ordering of birds by degree of irritability; Palmipedes, Gallmaneae
3	31 March	Cuvier's 5 families and various subdivisions; Palmipedes, Water-living, forms of beak and food of main groups
4	2 April	Passeres. Cuvier's two families; plumage, distribution, migration of birds and its causes
5	4 April	Organs of motion: bones & muscles; variation in skeletons; individual bones; proportion if irritability in muscles; individual bones of the skeleton
6	7 April	Muscles of bird described; flight of birds and variations; bills of parrots
7	10 April	Organs for the prehension of food; forms of beaks, digestive organs
8	12 April	Proventriculus and its variations, doudenum; experiments of Spallanzani and Everard Home on digestive organs; Ostrich, Cassowary; large and small intestines
9	14 April	Appendages to alimentary canal; liver, gall bladder, Vauquelin's experiments on the bile; work of Tiedemann, Meckel, Bichat; role of lungs compared with reptiles; lacteals and absorbents; Hunter's experiments on absorption by veins; Magendie's experiments
10	17 April	Lungs and relation to trachea; carotid arteries, subclavian; Jenner's experiments on muscular fibers; organs of voice; Yarrell on the Swan; Monro on decapitated birds; comparison with quadrupeds
11	19 April	Nervous system; comparison of nervous system in vertebrates to invertebrates; relation to cervical vertebrae; brain; Cuvier, Tiedemann and Carus on opthalmic nerves; medulla
12	21 April	The 5 senses; touch, taste, smell and relation to feeding; sight muscles of eye and structure of eye; problem of the first "marsupium" of the eye; lens, humours
13	24 April	Hearing; correspondence to form of voice; tympanic membranes; labyrinth, skin, anatomy of feathers
14	26 April	Urinary organs; organs of generation; Reaumur's experiments; Leeuwenhoek on the semen; kidneys, ureter, testicles, oviduct, mating
15	28 April	Incubation of the egg; Spallanzani, Jenner and Blumenbach on the return of swallows; Blumenbach on the structure of the egg; Jenner on the Cuckoo; Hunter's placement of nests among organs of generation; development of the egg and Hunter's experiments on this; 1st observable parts; the 21 days of development; Harvey's studies on development; arteries, chorion, filament, description of the chalazae

presented by the Sensibility the twelve *a priori* categories or concepts of the pure Understanding. Such "categorized experience," Kant repeatedly would claim, constituted, and even exhausted, the domain of genuine knowledge. Kant's apparent epistemological scepticism emerged from his limitation of knowledge proper to "categorized" experience, and as a result he directly excluded the possibility of supersensible access to reality in the form claimed by transcendental metaphysics. This much constitutes the best known, and most heavily analyzed, aspect of the Kantian philosophy for modern readers. The most fundamental misinterpretations of Kant's philosophy by his nineteenth-century heirs lay in their unwillingness to accept this restriction on the genuinely knowable.[65]

For Kant's successors, however, it was the third faculty of the mind, Reason (*Vernunft*), that opened up new doors to novel areas of exploration. The faculty of Reason, as distinct from the Understanding, constituted a higher power of the mind, functioning to unify the particularized and categorized knowledge of the Understanding, forming this disconnected totality into a whole under unifying Ideas of Reason. Kant's introduction of the concept of Ideas in the second, dialectical section of the *Kritik der reinen Vernunft* was one of the most influential, and also most easily misinterpreted, dimensions of his philosophy. Kant understood by Ideas archetypal principles or ideal maximalizations, arrived at by reflection on the categorised experience. Crucial to Kant's philosophy was the explicit claim that these reflections of Reason could never constitute genuine knowledge, and that any attempt to go beyond their limitations involved Reason in insoluble dialectical antinomies. This important limitation marks the boundary between Kant's own philosophy and that of his Idealist successors.

It would require a history of German philosophy to detail the various ways in which much of what occurred in the decades following Kant's work involved a technical misunderstanding of his philosphical project. His aim had been to limit, rather than encourage, speculative philosophy. But in spite of all his strictures on genuine knowledge, Kant nonetheless had left room for another interpretation by those who wished to explore the supersensible by means of the *Ideen*. Such ideas were, in Kant's own words, something more than purely subjective principles or merely organizing concepts:

> I understand by idea a necessary concept of reason to which no corresponding object can be given in sense-experience. Thus the pure concepts of reason, now under consideration, are *transcendental ideas*. They are concepts of pure reason, in

---

65   The Platonic, or more correctly Neo-Platonic, reading of Kant, while unfamiliar to most Anglo-American interpretations, was not an alien reading to the British tradition. This reading in early nineteenth century England was encouraged by the revival of Neo-Platonism by Thomas Taylor. It is common to see Kantianism linked with Platonism by British authors. Humphry Davy, for example, characterizes Kantianism as "pseudo platonism" (Davy Papers, Royal Institution MS, quoted in Trevor Levere, *Poetry Realized in Nature:. . .,* 2). On Thomas Taylor's importance for the revival of Neo-Platonism in England see K. Raine, "Thomas Taylor in England," in: K. Raine and G.M. Harper, eds., *Thomas Taylor the Platonist: Selected Writings* (Princeton: Princeton University Press, 1969), 3-48. Coleridge seems most responsible for putting Taylor and Kant together.

that they view all knowledge gained in experience as being determined through an absolute totality of conditions. They are not arbitrarily invented; they are imposed by the very nature of reason itself, and therefore stand in necessary relation to the whole employment of understanding. Finally, they are transcendent and overstep the limits of all experience; no object adequate to the transcendental idea can ever be found within experience. [66]

Although such statements were surrounded with explicit limitations on the epistemological status of these ideas, these were strictures that for many were ambiguously intended and could even be ignored or explained away.[67]

The Ideas provided the opening in Kant's philosophy through which Schelling and his disciples could rush with enthusiasm. Particularly if one emphasized certain passages of the *Kritik* that Kant did not seek to revise or discount in subsequent editions, such statements could, and did, suggest to his readers that Kant actually held to the theory of the Divine Ideas of the Neoplatonic tradition, in which these Idealizations served as archetypes of things-in-themselves that the Ideas of the faculty of Reason contacted. In a passage with profound importance for the reading of Kant by subsequent life scientists and comparative anatomists, he wrote:

A plant, an animal, the orderly arrangement of the cosmos—presumably therefore the entire natural world—clearly show that they are possible only according to ideas, and that though no single creature in the conditions of its individual existence coincides with the idea of what is most perfect in its kind. . . , these ideas are none the less completely determined in the Supreme Understanding, each as an individual and each as unchangeable, and are the original causes of things. But only the totality of things, in their interconnection as constituting the universe, is completely adequate to the idea. If we set aside the exaggerations in Plato's methods of expression, the philosopher's spiritual flight from the ectypal mode of reflecting upon the physical world-order to the architectonic ordering of it according to ends, that is, according to ideas, is an enterprise which calls for respect and imitation. [68]

In this passage lies the preformed germ for the whole program of Transcendental Morphology as it was explored by an enthusiastic group of German anatomists. But to pursue these speculations in the imaginative way carried out by

---

66   Kant, *Critique of Pure Reason*, trans. N. K. Smith (London: MacMillan, 1963), 318, B 383-4, hereafter cited as *CPR*.

67   Coleridge justifies this reading by the claim that during his trip to Göttingen in 1798, he encountered two versions of Kant, one a public "sceptical" Kantianism that imposed a large range of restrictions on knowledge, and another considered to be the true but "secret" teaching of Kant being taught at Göttingen by Kant's disciples that acknowledged the access of the mind to things-in-themselves. See S.T. Coleridge, *Biographia Literaria*, in: W. G. Shedd, ed., *Complete Works of Samuel Taylor Coleridge*, vol. 3 (New York: Harper, 1853), 256–60. This displays the importance, even at Göttingen, of Schelling's interpretations of Kant. For evidence see J. C. D. Wildt, "Ideen zur Naturphilosophie," *Magazin für den neusten Zustand der Naturkunde* (=*Voigts Neues Magazin*) 9 (1805): 389–97; idem, "Ueber Naturphilosophie," ibid. 12 (1806): 3–17. Wildt was a professor of philosophy and mathematics at Göttingen.

68   Ibid. 313, B 374-5.

many of these individuals necessarily meant that much of Kant's explicit philosophy had to be discounted or ignored. This is exactly the move that Coleridge chose to make, but Joseph Henry Green seems to have avoided, at least in the 1820s.

As Kant's philosophy was developed in his final writings, its relevance for biomedical problems became more explicit. It was at this time that the importance of a fourth mental power, Judgment (*Urteilskraft*), was emphasized. The *Kritik der Urteilskraft* of 1790 expounded upon the epistemological function of Judgment and its special domain.[69] As Kant discussed this at length in this final *Kritik,* the faculty of *Urteilskraft* had two functions. The first, the *determinant* use of Judgment, was necessarily employed whenever empirical experience was subsumed under the categories. It is distinguished from a *reflexive* employment "which is compelled to ascend from the particular to the universal."[70]

In making this distinction between the two functions of Judgment, Kant seemed to be making the following point. In theory, all experience comes to us as categorized under the laws or categories of the Understanding, subject to determinant judgments. Nevertheless, certain objects, for example, organic beings, as a matter of fact are presented to us empirically in experience as if subject to their own laws, not seeming to derive these from any *a priori* imposition of the knowing mind. They display merely empirical laws that we cannot see immediately connected with more general principles of the Understanding. Such laws are not seen as *a priori* necessary, but are simply empirically given in experience. The task of the reflective judgment is then to seek some higher rule under which these can be subsumed by the determinant judgment acting under the categories. [71]

In practice, this implies that upon encountering many empirical phenomena, the best that can be done is to seek for some kind of tentative unifying principle or empirical law that orders the phenomena in a way that is still to be specified more completely. Such a unity neither is possessed of the *a priori* necessity of the categories nor is properly an Idea of pure reason, although in practice such empirically derived rules can be used as hypotheses for the interrogation of nature and in this respect have an analogical resemblance to the regulative ideas.[72]

Those interested in exploring the constructive and speculative dimensions of Kant's philosophy found fertile soil in Transcendental Dialectic of the *Kritik der*

---

69   The *CPR* primarily speaks of *Urteile,* designating judgment in its "determinant" dimension. It is principally discussed in connection with its role in the necessary subsumption of empirical experience under the categories (B 176 ff.). The *Critique of Judgement* concentrates on the more expanded concept of *Urteilskraft,* with the suffix implying more the sense of an active faculty of the mind.

70   Kant, *Critique of Judgement,* trans. J. C. Meredith (Oxford: Clarendon, 1928; reprinted 1973), Pt. I, 18. Hereafter cited as *CJ.*

71   *CJ,* 18.

72   The critical difference between the maxims of the reflective judgment and the Ideas of Pure Reason lies in their different origins.The latter are demands of Reason imposed on categorized experience itself, and are dependent upon this. For this reason they have greater force and authority than mere principles derived empirically from phenomena that may be due only to an incompleteness in our knowledge.

reinen Vernunft and the extension of these principles into the *Kritik der Urteilskraft*. and in the projects set forth in the *Physicische Geographie* and *Anthropologie*. Kant's influential periodical articles of the 1780s, extending his philosophy into the theory of comparative anatomy, physical anthropology and human, animal and plant geography, also provided important inspirations. These important works displayed for numerous workers in the German scientific tradition the means by which Kant's larger philosophical project could be extended concretely into the domain of empirical biology and natural history.[73] The importance of these various distinctions for the development of life sciences in the early nineteenth century can be elaborated in terms of the different research programs into natural science they made possible for the German tradition. These manifest themselves in the important distinction between the "Description," "History" and "Physiology" of nature.

The differences between the first two enterprises—*Naturbeschreibung* and *Naturgeschichte*—can perhaps best be conveyed by example. This example is provided by the issue that divided the two giants of natural history of Kant's age—Linnaeus and Buffon—over the classification of organisms, the first advocating arrangements by logical distinctions of genera and species, and the other by associations of presumed lineage-relations.[74]

Kant was concerned, from a very early period in his philosophical development, with the possibility that organic beings could be understood as related not simply in terms of formally logical, or "Linnean," relationships in a hierarchy of genera and species, but also in terms of genetic and historical linkages through Buffonian *Stammgattungen*. In 1775, he made this explicit in a paper discussing the unity of the human races:

> The logical division [of Linnaeus] proceeds by classes according to similarities; the natural division considers them according to the stem [*Stämme*], and divides animals according to genealogy, and with reference to reproduction. One produces an arbitrary system for the memory, the other a natural system for the understanding [*Verstand*]. The first has only the intention of bringing creation under titles; the second intends to bring it under laws. [75]

In brief, Kant's claim was that the distinction between descriptive and formal relationships—those determined by the anatomist and taxonomist, and those of

---

73 See especially Christoph Girtanner, *Ueber das Kantischen Prinzip für Naturgeschichte* (Göttingen: Vandenhoek & Ruprecht, 1796; reprinted Brussels: Culture et Civilization, 1962). I have discussed this work briefly in my "Buffon, German Biology, and the Historical Interpretation of Biological Species," *British Journal for the History of Science* **12** (1979): 109–53.

74 See my "The Buffon-Linnaeus Controversy," *Isis* **67**, 356–75.

75 I. Kant, "Von der Verschiedenheit der Rassen der Menschen," *Teutscher Merkur* 1775, reprinted with revisions in J. J. Engel, *Der Philosophie für die Welt* (Leipzig, 1777). See text in *Kants Werke,* vol. 2 (Berlin Akademie; Vorkritische Schriften, Berlin: Reimer, 1912), 429–33. Green's personal copy of this essay, lacking annotations, as it was reprinted in *Kants Vermischte Schriften,* 4 vols. (Halle, 1799) (vol. 2), is in the Coleridge library at the British Library. I have analyzed Kant's views on these issues in my "Buffon, German Biology. . .".

historical and genetic kinship, postulating historical derivation from a common ancestry—fell along the formal division lines between the Understanding and Reason and the mediating function of Judgment.

In empirical experience, one is presented only with discrete species, even, it would seem, only with discrete individuals, and it seems to be Kant's claim in the 1790 treatise that this would give access only to the empirical relationships of the reflective judgment.[76] Reason, however, necessarily imposes a search for the subsumption of these under the necessary laws of the categorized understanding, and finally above this, Reason introduces the principles of the unity of appearances under higher genera, that of the diversity of species within these genera, and finally the unity of all genera in some common principle.[77]

> The first law [imposed by Reason] thus keeps us from resting satisfied with an excessive number of different original genera [*Gattungen*], and bids us pay due regard to homogeneity; the second, in turn, imposes a check upon this tendency towards unity, and insists that before we proceed to apply a universal concept to individuals we distinguish subspecies [*Unterarten*] within it. The third law combines these two laws by prescribing that even amidst the utmost manifoldness we observe homogeneity in the gradual transition from one species [*Species*] to another, and thus recognize a relationship of the different branches, as all springing from the same stem [*aus einem Stamme entsprossen sind*].[78]

Kant applied similar means for adjudicating between alternative theoretical positions in biology in his shorter essays and at length in the *Kritik der Urteilskraft* of 1775.[79] The limitations Kant imposed on the faculty of Reason in the first Critique were now supplemented by the greater freedom allowed the reflective judgment, which could begin its reflections from empirical experience and postulate principles based upon empirical data.

In this same domain of complex epistemological distinctions lies the complex issue of Kant's teleological conception of Nature. Within the larger framework of Kant's philosophy, one finds two clearly distinguishable concepts of nature. These two notions provided the means by which Kant resolved the conflict between teleological and mechanistic interpretations of the world, the one based upon an Aristotelian-Leibnizian conception of nature as a dynamic and unfolding system ordered to self-contained teleological ends, and the other the Cartesian-Newtonian view of nature as a mechanical system of law-governed moving bodies whose teleological ends, if existent at all, are simply imposed from without by creation.[80] Kant resolved this issue, as he did many fundamental oppositions, by a dual per-

---

76   *CPR*, 539 (B 681).
77   *CPR*, 542.
78   *CPR*, 543 (B 688).
79   See also Kant, "Bestimmung des Begriffs einer Menschenrasse," in: *Kant's Gesammelte Schriften*, 23 vols. (Berlin and Leipzig: Walter de Gruyter, 1928), 8: 89–106, first published in *Berlinischer Monatschrift*, 6 (1785); idem, "Ueber den Gebrauch teleologischer Prinzipien in der Philosophie," in Kant, *Werke* 8: 489–516. This first appeared in the *Teutscher Merkur* 61 (1788): 36–52; 123–36.
80   On Kant's conception of nature and its relation to that of his predecessors, see especially D. Kolb, "The Systematic Unity of Kant's Idea of Nature," Unpublished Ph.D. Dissertation, Department of Philosophy, University of Notre Dame, 1983.

spective on the problem. The distinction of these two views again lies along the separation of the roles of the Reason and the Understanding. As an object of knowledge, in Kant's technical sense of the term, nature is to be understood as a system subject to *a priori* mathematical laws, operating according to the principles of Newtonian physics.[81] Conceived empirically, as the "connection of appearances in terms of their existence according to necessary rules, i.e. according to laws,"[82] nature is subject to the restrictions of the categories of the Understanding and is to be understood purely as a causal-mechanical system. But as nature is understood as an object of reason (*Vernunft*), it comprises the "Totality of appearances, insofar as these are entirely connected together by means of an inner principle of causality,"[83] and in this way forms a teleological whole, purposive and developing in history to a goal. This point was expressed by Kant in an influential essay of 1788 that pointed the way for the application of his complex philosophy to concrete problems of natural science. Natural science, in contrast to metaphysics, employs only empirically ascertained purposes or ends, with which it analyzes natural objects:

> If one understands by Nature the totality of everything which exists determined according to laws—the world (as Nature, properly so-called) taken together with its highest cause—then natural inquiry (*Naturforschung*), which is called physics (*Physik*) in the first case, and metaphysics in the second, can attempt to use purposes for its intentions (*Absicht*) in two ways, either the theoretical or the teleological; on the latter path, as Physics, only such purposes as can be known through experience [can be employed]. As metaphysics, on the other hand, corresponding to its vocation, only one purpose can be employed, which is determined by pure reason.[84]

The distinction made in 1775 between a "descriptive" and "historical" science of nature is also given further clarification in this 1788 essay. Kant now introduces an important distinction between these two approaches in view of his mature philosophy of nature. The first, the viewpoint of physics, as well as that of descriptive science, is confined to a description of nature and its laws, a

---

81 See Kant's 1786 *Metaphysiche Anfangsgrunde der Naturwissenschaft,* translated as *Metaphysical Foundations of Natural Science,* trans. J. Ellington (Indianapolis: Bobbs-Merrill, 1970), "Preface," 4-6. This discussion should be closely read in light of the explicit distinction of the two conceptions of nature that opens the treatise. This will avoid the misconception that Kant considered only Newtonian physics as worthy of scientific interest. The often-quoted statement by Kant in this preface that "Ich behaupte aber, dass in jeder besonderen Naturlehre nur so viel eigentliche Wissenschaft angetroffen werden könne, also darin Mathematik anzutreffen ist " (Kant, *Metaphysische Anfangsgründe der Naturwissenschaft* [Erlangen: Harold Fischer, 1984; reprint of original edition of 1786], viii) must be read against his technical distinction between *Naturlehre* and *Wissenschaft* in this text.

82 "Unter Natur (im empirischen Verstande) verstehen wir dem zusammenhang der Erscheinungen ihrem Dasein nach, notwendig en Regeln, d.i. nach Gesetzen. " I have used the German edition of the *KRV,* ed. E. Adickes (Berlin: Mayer &Kunsteler, 1889), B 263, 231.

83 Ibid., B 446.

84 Kant, "Ueber den Gebrauch . . .", 489. Joseph Henry Green's copy of this essay is found in the third volume of Kant's *Vermischte Schriften* in the Coleridge collection at the British Library. This bears Coleridge's, but not Green's, annotations.

Naturbeschreibung or, as he now terms it, a *Physiographie*. Conceived as an inner-connected teleological system according to purposes of reason, it is the subject, however, of a systematic history of nature, a *Naturgeschichte* or, in its new designation, a *Physiogonie*.[85] The latter, subject to the epistemological restrictions always operative on the transcendental uses of reason, can serve to organize, but never explain, natural phenomena.[86] Teleological principles are to be used in science either at a proximate level or in terms of a conception of all of nature as a total system moving toward a teleological goal, that of human consciousness and freedom.[87]

Within the context of this complicated philosophy of nature, Kant also distinguished a third enterprise of relevance to our topic. This is the concept of *Physiologie*. His unusual meaning of this term relates directly to the distinction of the two meanings of "Nature." As this concept was discussed at the conclusion of the *Critique of Pure Reason*, "physiology" is contrasted with transcendental philosophy. The latter deals generally with the conditions of the possibility of objects in general. Physiology deals with nature. This notion is then further discriminated into an inquiry dealing with objects within the empirical conception of nature—i.e. nature as the sum of objects given according to definite laws—and an inquiry dealing with nature in its transcendental meaning as a purposive system with a teleological ordering.[88]

Although the "transcendental" meaning of "Physiology" would have little concrete relation to the kinds of inquiry of interest to medical men and anatomists, the "immanent" meaning could be directly related to these inquiries—as an inquiry concerned with the physiological laws and forces between bodies. It is in this sense that we will find the concept employed by Joseph Henry Green. This definition could be directly related to the analysis of physiological process by teleologically acting physiological forces exemplified by Blumenbach's conception of the *Bildungstrieb*.

The full importance of these somewhat baroque distinctions for the development of German and British science can only be summarized here. They implied, at least for those willing to attend closely to these philosophical issues, that three different, but intimately related, kinds of natural inquiry were possible, albeit pursued on different epistemological footings. Furthermore, these inquiries need not be placed in opposition to one another precisely because they stood on these different levels of philosophical warrant. It is for this reason that we cannot adequately conceive the philosophical relationships within German biology in these early decades of the nineteenth century in terms of competing research programs.

---

85   "Ueber den Gebrauch. . .", 493. The systematization of these distinctions is made even clearer by Christoph Girtanner in his application of Kant's natural philosophy to the issues of natural history and biology: "Die Naturbeschreibung (Physiographie) ist de Kenntniss der natürlichen Dinge, wie sie jetzt sind. . . .Die Naturgeschichte (Physiogonie) ist der Kenntniss von demjenigen, was die natürlichen dinge ehemald gewesen sind, und von der Reihe Veränderungen, durch welche sie gegangen sind, wer on jedem Orte in ihren gegenwärtigen Zustand zu gelangen." Girtanner, *Über das Kantischen Prinzipien für die Naturgeschichte*, 1–2.

86   *CJ*, 53. Kant there speaks of this "as a guide to judgement in its reflection upon the products of nature."

87   Kant, *CJ*, Pt. II, 92.

88   Kant, *CPR*, B 874.

One and the same individual could, and often did, pursue all three inquiries without confusion as long as these distinctions were kept clear. German life scientists of the early nineteenth century had, in these Kantian distinctions, unique resources for pursuing issues on several different levels. Apparently conflicting positions on issues of morphology, historical biology and functional physiology are adopted without apparent deep conflict. Genuine *Wissenschaft* is restricted effectively by Kant's distinctions to *Physiographie* and *Physiologie*. Only inquiries in these domains satisfy the restriction of knowledge to categorized experience.[89] Genuine science (*Wissenschaft*) is restricted to the *physiographic* analysis of phenomena, which interprets nature as a Newtonian system. Historical science, represented by the speculative cosmogony expressed in Kant's day by Buffon and Thomas Wright, even one employing Newtonian mathematical principles, could not qualify as "science" in this strict sense. But at the level of *Naturlehre*, such speculative inquiry formed an important part of a more general inquiry into nature. In immediately relevant terms, this would imply that such inquiries as descriptive anatomy and taxonomy, biogeography, and functional physiology could indeed claim the status of sciences that make genuine knowledge claims about their subject matter.

The status of *Physiogonie* or *Naturgeschichte* constituted a more controversial issue, dependent on one's realist or regulative reading of Kant. Those who made the "Schelling revision," as it might be termed, refused to accept the limitations imposed by Kant on such historical knowledge, and instead granted to their speculations in biology, geology and speculative geochemistry genuine access to an "inner history" of nature.[90] These works constituted the mainstream of what has commonly come to be known as "Romantic" science. Others were willing to accept the restrictions, and granted to their claims about these matters only regulative status. J. F. Meckel, Karl E. Von Baer, even Johannes Müller, all showed some willingness to engage in transcendental speculations and at the same time published works characterized by a rigorous empiricism and philosophical caution.[91] The complexity of Kant's philosophical program in natural science allowed this duality, and by this means one could combine issues and traditions typically placed in opposition by other national traditions—teleological purposiveness of organisms *and* their explanation by rigorous mechanism; Cuvierian functional anatomy *and* speculations about transcendental archetypes; distinctness of species *and* the concept of the unity of forms in historical derivation from a common stem. The reconciliation of these alternatives is made possible through their status as different *interests* of reason that need not, and properly could not,

---

89 This would bear on some of the distinction Kant makes between mathematical physics and other kinds of scientific inquiry in the preface to the *Anfangsgrunde* (see above, note 81). *Naturgeschichte* in Kant's technical meaning of this term falls under *Naturlehre* but cannot constitute a proper *Wissenschaft* for the reasons we have outlined.

90 See for example Henrik Steffens, *Beyträge zur innern Naturgeschichte der Erde* (Freiberg, 1801). Coleridge's annotated copy of this is in the British Library.

91 See for example K. E. Von Baer,"Ueber das äussere und innere Skelet," *Meckels Archive für Physiologie* 3 (1826): 327–76, where he published his speculative views on the homologies of the operculum of fish to the ribs. Owen cites this in his 1844 Hunterian lectures. (*Lectures on the Comparative Anatomy and Physiology of the Vertebrate Animals*, (London: Longman et al., 1846), Lecture 6, 138.)

stand in conflict. All relate to a larger project of natural inquiry in which the rigorous and even reductive analysis of life-phenomena is organized in terms of a more speculative science dealing with historical development and the purposes of nature.

A.4 Green's Three Views of Nature

The primary evidence that Green knew of and endorsed the Kantian philosophical positions outlined in the preceding discussion is the simple fact that he explicitly adopted the clear discrimination of descriptive, historical and physiological approaches in his Hunterian lectures.[92] This immediately serves to explain the otherwise curious organization of Green's lectures as they were delivered between 1824 and 1828.

Green's lectures, like Owen's later, were restricted by certain important material and institutional conditions. Their content was in many respects fixed by the statutory charge—the requirement that they demonstrate the Hunterian Collection itself. This Green did by discussing in detail the main functional systems of the animal groups represented in the collection. The 1824 series commenced with the lower forms and moved successively during the next four years through the main animal groups, terminating with the birds and mammals. The approach utilized in the exposition of these forms was primarily descriptive and analytical.

Previously, we have spoken of Green's practice of commencing these lectures each year by the display of two diagrams, one clearly "cuvierian" in inspiration, displaying the four main *embranchements*, and the other "lamarckian," organizing the organisms in an ascending and branching series. These "lamarckian" aspects were then relegated primarily to the status of the framework of exposition—i.e. organisms were treated over the years as parts of a complex ascending series that moved from the simplest to the most complex. This itself was an innovation in British comparative anatomy in 1824 when the series commenced. The typical pattern of exposition in the main public lecture series in comparative anatomy in Great Britain and on the Continent, by contrast, proceeded from the more complex groups to a termination in the lower orders.[93] Only Jean Baptiste Lamarck, in his

---

92 To my knowledge, no other British scientist made a similar set of distinctions. Green seems to be unique in importing this concept of three different, but related, inquiries into nature into the British scene. The distinction is also made by Coleridge in the first part of his "Opus Magnum," dated March 1827, Coleridge Papers, Victoria College SMS 18, fol. 14, and may be either the source of Green's distinctions or reflections by Coleridge after attendance at Green's 1827 lectures. See further comments below, p. 307.

93 This descending order was adopted by John Barclay and Robert Jameson in their lectures on comparative anatomy in Edinburgh. This order was also expounded in Cuvier's writings and in the organization of his *Cabinet d'anatomie comparée* at the Paris Museum. In London, Robert Edmond Grant had initially adopted a similar ordering in his lectures on comparative anatomy at University College in 1828. See his *Essay on the Study of the Animal Kingdom* (London: Taylor, 1828), 34–5, although this was reversed in his lectures of the 1830s.

invertebrate lectures at the Paris *Muséum,* seems to have publicly expounded the animal kingdom from the simplest forms upward. This fact had not escaped Green, who acknowledged at the outset his debt to Lamarck's exposition.[94]

The explicit debt of Green's lectures to the three Kantian distinctions was, apparently, made public only in 1827 and 1828 when he began his lectures on the vertebrate groups of the birds and mammals. Green used the opening lecture of 27 March 1827 to clarify his larger organizational intent. The three-fold distinction between a "description," "physiology" and "history" of nature were now given a clear definition:

> The three great divisions into one or other of which all Natural Science resolves itself are, Physiography or description of Nature, Physiology or Theory of Nature. Lastly Physiogony or the History of Nature—The office of the first or *Physiography* is to enumerate and delineate the effects & products of nature as they appear. Its Sphere is that of sensible experience of appearances in contradistinction from truths drawn from immediate facts by inference, the subject matter is not unhappily entitled by elder Naturalists Natura Naturata or Nature considered passively and the result may be compared to an immense family piece the figures of which are all portraits—the office of the Second or *Physiology* is to deduce by inference 1st the rules or principles by which the innumerable facts of Physiography may be reduced into manageable order, either in reference to the convenience of our faculties, which is the principle of all artificial classification or in relation to the objects themselves which should it ever be realized will be the ground of a Natural Classification. 2nd it is the office of physiology likewise to ascertain the powers which must be inferred from the phenomena and the *laws* under which they act: in other words to ascertain the idea of Life and its constituent forces—as far as it is common to all living bodies—That is in *Kind* without consideration of degree or other difference of the particular Subject—the 3rd or *Physiogony* regards the facts and appearances of the natural world as a series of actions & nature itself as an agent acting under the analogy of a will and in pursuit of a purpose, in what sense and whether by a necessary fiction of Science or with some more substantial ground we leave undetermined.[95]

The point of these distinctions was to provide a framework upon which Green could unify a purely a-historical and descriptive exposition of materials with an historical theory of organic development. The linkage between these was to be provided by a dynamic physiology.

In keeping with the Kantian definition of Physiology and Physiogony as seeking in the objects of sense both immanent law and subordination to a larger teleological plan, Green proposed that the various types and kinds of organisms could now be envisioned to compose a rational, historical system of ramifications, within which an underlying intelligibility could be perceived.

(contd.)
  Green's mode of presentation is, to my knowledge, unique in the mid 20s.
94 See note 62.
95 Green 1827: fol. [1]. See complete transcription of this lecture in the Appendix.

The Hieroglyphics [of the description of Nature] we can only expect to be mere
fragments while in nature we possess the whole before us as a book not indeed
without hiatus and interspaces to be filled up by future discoveries, yet no hiatus
of such magnitude or of such importance as to destroy or even obscure the manifest principles of arrangement that pervades the whole[;] if it be possible in one
sentence to convey the sort and the degree of interest which the object of physiogony or the history of nature is calculated to inspire, I might say that its object is
by means of evident principles, principles of reason supported in each step by the
facts corresponding[,] to exhibit nature, as labouring in birth with *Man*, to exhibit
every order of living beings, from the Polypi to the Mammalia as so many embryonic states of an organism, to which nature from the beginning attended but
which Nature alone would not realize[.] [96]

The linear developmental models envisioned by Green's contemporaries,
operating under the influence of early expressions of transformism by Lamarck and
the so-called Meckel-Serres law of development, were to be replaced by another
model. Nature, for Green, undergoes a "general development," but not in a simple
linear progression. This view of its development is best represented by the
metaphor of a branching tree rather than an arrow, with major groups representing
different branches. As this was expressed in the 1828 "Recapitulatory Lecture"
opening his lectures on the mammals:

The resulting forms of animal life present not a plan which we can consider as the
effect of any arbitrary combination, or of a regularity imposed upon nature by the
human fancy or understanding;—it is neither a scale, nor a ladder, nor a network; it
is neither like the combination of a kaleidoscope, nor the pattern of a patchwork; it
is not process by increase or superaddition:—but it is, as in all nature's acts, a
growth, and the symmetry, proportion, and plan, arise out of an internal organizing principle. This gradation and evolution of animated nature is not simple and
uniform; nature is ever rich, fertile, and varied in act and product:—and we might
perhaps venture to symbolize the system of the animal creation as some monarch
of the forest, whose roots, firmly planted in a vivifying soil, spread beyond our
ken; whose trunk, proudly erected, points its summit to a region of purer light,
and whose wide-spreading branches, twigs, sprays, and leaflets, infinitely diversified, manifest the energy of the life within. [97]

The tree metaphor, already utilized in his 1824 lectures, was now extended to
the relations of the mammalian groups. Owen's transcription of this lecture summarizes these points sketchily. "Nature like a tree," he writes, "& if she seems at
any time to recede tis only to gain strength for greater efforts."[98] In the second
lecture Green introduced a new diagram to illustrate the relationships of the

96   Ibid., fol. 1 verso. This closely parallels the discussion in the 1828 inaugural lecture, which
      does not survive in manuscript but was subsequently printed as an appendix to Green's *Vital
      Dynamics* (London: Pickering, 1840), 102–3.
97   Ibid., 109.
98   Owen's notes on Green's 1829 course are found in the Green Papers, RCS, MSS 67. 11, item
      3, fol. 3, hereafter cited as "Green 1828." Owen's notes on the inaugural lecture are brief,
      probably because he had made extensive notes on the similar lecture opening the 1827 series.
      (Owen Papers, RCS 275.b.21. See specimen below, p. 309.)

mammals, which Owen copied into his notes .[99]

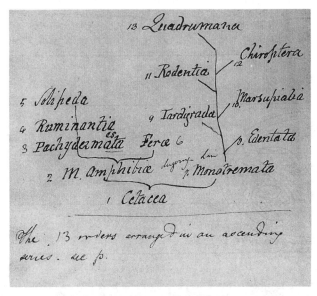

Fig. 2. Joseph Henry Green's Diagram Of The Thirteen Orders Of The Mammals, Arranged In An Ascending Scale (1828). This diagram is from notes taken by Richard Owen. *(Courtesy President and Council, Royal College of Surgeons of England)*

The nature of Green's impact on Owen's early thought is difficult to document. Owen took notes on these lectures, and he even has one expanded careful transcription of Green's discussion of the distinction of the three approaches to natural science.[100] Years later, Owen recalled that Green's great contribution was his presentation of a "vast array of facts . . . linked by reference to the underlying Unity, as it had been advocated and illustrated by Oken and Carus. The Comparative Anatomy of the latter was the textbook of the course."[101] But in the extant Green manuscripts themselves, we find neither a highly speculative transcendental anatomy nor a specifically Germanic content. Nor would there need to be such reference even if Green had been willing at the same time to argue that Carus was presenting the framework for a synthetic resolution of these questions. His express purpose in the lectures themselves was to reconcile descriptive and historical representations of organic affinity through physiology. The specifically German aspect of this lay more in the methodological approach

---

99 "Green 1828," fol. 10 verso. Owen comments on this diagram: "The 13 orders arranged in an ascending series."
100 This is at Owen Papers, RCS 275. 6. 21.
101 Richard Owen to John Simon, ca. 1865, in Green, *Spiritual Philosophy*, 1: xiv. See also Owen's comments on Green in the deleted section below, p. 126.

than in the specific content. His College lectures do not represent an ambitious imposition of transcendental *Naturphilosophie* upon unsuspecting surgery students, not because he was unaware of, or uninterested in, such speculative positions, but because he was precise about the distinctions and levels of philosophical justification one could claim for them. Consequently, his lectures are generally free of identifiable speculation in the tradition of German transcendental anatomy or *Naturphilosophie*. On his own report, they are primarily *physiographic* in their intent—i.e. they are concerned principally with the description of natural objects and their manifest laws.[102] For one following Kant, only this approach could claim the title of genuine science.

Furthermore, the adoption of this strategy by Green cannot be divorced from the material conditions imposed upon the lecturer by the Hunterian Collection. As we have seen in Table One, the arrangement of the Hunterian Collection was a complex one, obscure in its purposes, organized only in accord with some very general physiological principles. Within the various categories, the Hunterian displays typically presented separated organs from several different forms, illustrating the varieties of function found between the different groups. Expounding this complex ordering in a simple linear method of presentation would have been extremely difficult for reasons directly related to the organization of the museum. The display would also have presented severe difficulties for a presentation in terms of the arrangements suggested by Cuvier. Evidently Green found this problem most readily soluble by means of the metaphor of a branching arrangement of groups, similar to, but going considerably beyond, that suggested by Lamarck in his last works, ordering forms in an ascending and diverging order.

Green's contribution to Owen's thought can be seen, in light of the issues outlined above, as primarily giving him a conceptual project for reconciling historical, descriptive and physiological analyses with the demands of a given empirical collection, rather than in terms of speculative transcendental anatomy. The latter may, to be sure, be part of Green's impact on Owen's mature views, particularly if the views of Carl Gustav Carus were being utilized or discussed in the background.[103] However, Green's strictures on transcendental speculation would not lead us to expect much evidence of a highly speculative philosophical biology. Instead, the connection would lie more in the blending of medical and

---

102 See his introductory lecture, Appendix, fol. 2.
103 This might be particularly important in view of Owen's probable derivation of the concept of the vertebrate archetype from Carus' treatise *Von den Ur-Theilen des Knochen-und Schalengerustes* (Leipzig, 1828). For my understanding of the relations of Owen and Carus, I am particularly indebted to Nicolaas Rupke's forthcoming "Richard Owen's Vertebrate Archetype," which he kindly shared with me in manuscript form. Owen's recollection of Green's use of Carus in his letter to John Simon (see note 101 above) suggests that a stronger transcendentalism was being advocated by Green than the surviving lecture notes on the Green lectures taken independently by Clift, Owen and Thomas Egerton Bryant (Green Papers, RCS 42. 2. 19) would suggest. There is some possibility that by 1865 Owen was transporting back into his memory of the 1820s some of the strong transcendentalism he had come to embrace in the 1840s. None of the surviving transcriptions of the Green lectures have any reference to the theory of the archetype, nor is Carus even frequently mentioned.

functional physiology and empirical description, organized in some plan representing the speculative history of life. This kind of framework utilized in Owen's early lectures is the only evident connection with German transcendental biology .

Owen attended Green's lectures at least for the 1827 and 1828 terms. Shortly afterward he would encounter the urgent problem that would occupy him for much of his mature scientific life, the problem raised for comparative anatomists of the 1830s by the great debates in Paris between Georges Cuvier and his colleague at the Paris *Muséum*, Etienne Geoffroy St Hilaire, over the unity-of-type. Encounter with this issue supplied another layer to the intellectual formation of the young anatomist.

A.5 Reconciling Cuvier and Geoffroy

By the late 1820s, during the period when Owen appears to have been indirectly acquiring theoretical roots in the German tradition from Green, he was also engaged in the tedious and pressing task of assisting Clift in the completion of a usable catalogue for the Hunterian Collection. This proceeded in two stages, the first being a preparation of a brief list of the entire series with nothing more than the names of the materials on display. This was to be followed by a massive catalogue of the so-called "physiological" series, sufficient to satisfy the interests of the Museum Trustees, and was to include plates, some new research and extensive utilization of all the Hunterian manuscripts William Clift had been able to recover from various sources by this time.[104]

While the practical issues associated with ordering and arranging thousands of organisms and specimens occupied Owen's attention, he was made aware of the debates in Paris between two towering figures at the Paris *Muséum national d'histoire naturelle*, Georges Cuvier and Etienne Geoffroy St Hilaire, by personal contact with the participants themselves.[105]

On 6 August 1830, Georges Cuvier visited the Hunterian Museum in the company of five professors from the University of Edinburgh.[106] Apparently consid-

---

104 The manuscript problems created by Home's destruction in 1823 had been repaired to some extent by 1830 by recovery of notes and transcriptions from Hunter's associates and former students. Clift had also made private copies for his own use before Home's destructions, which, unaccountably, are not mentioned in the Parliamentary and Council inquiries into the destruction of the manuscripts. These were published only posthumously in 1861. See above, note 34.

105 On the full details see Toby Appel, *The Cuvier-Geoffroy Debate: French Biology in the Decades before Darwin* (Oxford: Oxford U P, 1987); and Pietro Corsi, *The Age of Lamarck* (Berkeley: U. California Press, 1989), chp. 8; see also Desmond, *Politics*, 41–59.

106 Visitor's Book, Hunterian Museum, RCS Archives 275. g. 42. Cuvier was in the company of two "Thomsons," presumably either Allen, William or their father John, all associated with the University of Edinburgh; W. B. Alison, Owen's former physiology teacher at Edinburgh; [?] Somerville; and a "Monroe"[sic] (probably meaning Alexander Monro, *tertius*), professor of anatomy at Edinburgh. They were authorized for entry by William Blizard.

ering it wise to be absent from Paris as the July Revolution ran its course, Cuvier was in London for approximately three weeks in August, and is reported to have become well acquainted with Owen.[107] As Cuvier's guide, and evidently the only French-speaking member of the Museum staff, Owen was able to display his anatomical knowledge and abilities to the famous Frenchman. Cuvier subsequently invited him to Paris in the following year to attend a meeting of foreign naturalists at the *Muséum* concurrent with the first anniversary celebration of the establishment of the July Monarchy. Owen's letters and diaries for this period testify to his delight and also his education during these travels. Meeting all the leading lights of French biological science, including both of the Cuviers, the expatriate Irishman John Pentland, Laurillard, Dutrochet, Humboldt, Etienne Geoffroy St Hilaire, Henri de Blainville, Jussieu, Latreille, Chaptal and Milne-Edwards, Owen had first-hand exposure to the theoretical debates taking place at the *Muséum*.[108]

The year following the initial meeting of Owen and Cuvier had been a year of political change in France, and it also was the year the Cuvier-Geoffroy debates of the early 1830s received greatest attention in the public and professional press.[109] Interested parties, such as the great Goethe himself, had wished to contribute to this debate. Owen, fresh from his extensive work with the Hunterian Collection and armed with Green's philosophical program, was in a unique position to assess these debates that pitted Cuvier's Aristotelian functionalism against St Hilaire's transcendental anatomy. Only three days before Owen's arrival in Paris, Cuvier had been planning another public attack on St Hilaire and Lamarck in his caustic *Éloge de Lamarck*, which was originally scheduled for delivery to the *Académie* on the twenty-sixth of July.[110] Travelling much of the time in the company of his former Edinburgh associate and now fellow London anatomist, Robert Edmond Grant, with whom he seems to have discussed several theoretical issues in biology of the day, Owen displayed an immediate interest in the issues raised by the transformist

---

107 The reasons for Cuvier's presence in Britain are not fully clear. At the meeting of 3 August, the Professors at the Paris *Muséum* decided to terminate all courses in progress in response to the political turmoil of July 1880, and it was observed that the students had already dispersed. Cuvier had apparently already left by this date, being listed in attendance at the meetings only through 20 July, and returned to these meetings on 24 August (*Procès verbaux des assemblées des professeurs*, Paris Archives nationales AJ15129). I have been unable to document any visits of Cuvier to the Hunterian Museum except that of 6 August, but he may have been able to return without signature, having been made an "honorary member" by William Blizard on his initial visit. The report in the *Life* (1: 49) that Cuvier had fled France is based upon Owen's annotation to a copy of a memoir on Cuvier.
108 Owen to Clift, 2 August 1831, Francis Hirtzel Owen Correspondence, Temple University, vol. I, fols. 12–13, also reprinted in *Life* 1: 51–8. Owen's contacts are also listed in his Paris Notebook, Owen Papers, BMNH L. O. C. O. 25, #5. He met several of these at a session of the Institut. There is some possibility Owen returned to Paris in 1832. See Lecture Seven, note 29–5, p. 301 below.
109 Appel, *Cuvier-Geoffroy*, esp. chp. 6.
110 This was not delivered on its scheduled date, and was delayed until November. See ibid., 278 n. 92.

and unity-of-type controversy.[111]

Owen was specifically concerned to learn more about the arrangements of specimens in the Paris *Muséum*, and in particular those of Cuvier's *Cabinet d'anatomie comparée*, to compare with those of the Hunterian Collection. However, the arrangements of displays encountered in Cuvier's *Cabinet* reflected the strong functionalism of Cuvier's comparative anatomy in which each species was treated essentially as an autonomous entity. The exhibits were ordered "so every Animal exhibits in a greater or less number of Preparations its Muscular, Digestive, Circulatory, Respiratory, Nervous and Generative systems."[112] In practice this consisted of an ordering of 16,665 displays in which the collection moved through the four main *embranchements* in a descending order from the vertebrates to the molluscs, articulates and radiates. Within these groups, taxonomic subdivisions were made down to the level of individual species. Each species, for example a specific vertebrate, would be displayed first in its skeletal system, and then its muscular, internal, nervous and reproductive systems would be dissected out in the same display.[113]

This Cuvierian ordering by individual organisms was in sharp contrast to the Hunterian arrangement we have seen previously (Table One). As Owen reported this, the Cuvierian ordering illuminated the "natural history" of the animals, whereas the Hunterian would illuminate their physiology.[114] By this difference Owen meant to underline the fact that the Hunterian displays presented to the viewer various systems rather than organisms.

This material difference of the two museums explains a great deal about Owen's position on the important theoretical questions at stake in the Cuvier-Geoffroy debate. Cuvier's arrangement directly resisted any tendency of the viewer to draw immediate connections between forms. His polemic against Lamarck and Geoffroy was precisely directed at the "superficial connections" one could make by treating organ systems in abstraction apart from their intimate correlation with one another.[115] The Hunterian ordering, on the other hand, immediately forced one to make these comparisons—but only to a limited extent. The Hunterian series suggested an affinity between animals deeper than mere functional similarity. In this respect its conclusions were non-Cuvierian, particularly in an 1830s context

---

111 Owen,"Paris Notebook." For example, on 17 August he enters "Bought *Philos. Zoologique* read till 11" (fol. 136) and otherwise seems interested in Lamarck's ideas, possibly growing from his conversations with Grant recorded on 12 and 14 August. The close and congenial relations with Grant at this time included several conversations on philosophical biology.

112 Owen, "Observations Respecting the Collection in the Jardin du Roi at Paris and Suggestions for Obtaining Room in the Museum of the College," Owen Papers, RCS 275.h.7, fol. 9, dated September 1831. This is reported by Owen and Clift on Owen's return.

113 See A. Valenciennes and J. Pentland's *Catalogue des préparations anatomiques* (see above n. 31). The RCS copy was presented to the College in 1832 by Madame Cuvier personally, and contains additional pencil notes and corrections by Pentland.

114 Owen, "Observation Respecting . . . ," fol. 9.

115 G. Cuvier and A. Valenciennes, "Résumé général de l'organisation des poissons," *Histoire naturelle des Poissons* (Paris: Levrault, 1828) I: 545.

Fig. 3. Interior View Of The Amphitheater Of The Galerie d'Anatomie Comparée, Muséum National d'Histoire Naturelle, Paris. Source: M. Boitard, *Le Jardin des Plantes* (Paris, 1844). (*Courtesy Library, Natural History Museum, London*)

where the issues had been so sharply focused by the Cuvier-Geoffroy debates. But the Hunterian display did not suggest an obvious solution in terms of a unity-of-type in the sense this was being advocated by Geoffroy St Hilaire. The collection presented an array of disconnected functional series, not a single order. Within these series the arrangements then generally displayed the ordering of an organ system from its simplest manifestation to its most complex. Any unification of these series with one another could only be accomplished by some kind of divergent model. We have seen that Owen had been exposed to exactly this solution in the Joseph Green lectures in 1827 and 1828, in which Green had displayed diagrams of a branching tree arrangement. After September of 1831 when Owen returned to England, he appears to have been convinced that some kind of theoretical reconciliation of the claims of Cuvier and Geoffroy in terms of his own developing theoretical vision was possible. Only gradually does that solution seem to have emerged. By October 1837, shortly after his first series of Hunterian lectures, Owen was writing to others that solution of this problem was his theoretical goal.[116]

As Owen entered the period in which his first Hunterian lectures were to be composed, the complexity of his scientific and intellectual background, along with his familiarity with current theoretical problems, placed him in a unique position to pull together several otherwise separated currents of inquiry. The lectures provide us with a window on this early synthesis.

B. Hunterian Professor

The accidental death, in 1832, of William Clift's son and heir apparent, William Home Clift, and the departure of Samuel Stutchbury opened up a permanent position for Owen at the College as Clift's Assistant Conservator. From this date, Owen was increasingly drawn away from hospital and surgical work to full-time occupation as a comparative anatomist. The increasing time demands occasioned by his concurrent appointment at the Museum, and his commitments at St Bartholomew's Hospital, were exacerbated by his growing interest in publication of his researches. The Council of the College, under insistent pressure from disgruntled visitors and attacks by the radical medical press, was emphatic that the main task before the curators was the completion of a finished extended catalogue of the collection that would make the collection accessible as a teaching resource. Owen's tendencies to follow more theoretically interesting pursuits were consequently resisted and provided continual sources of tension in his relations with the Council and the Museum Curators in these years.[117] In 1832 he drew the at-

---

116 In a letter to William Whewell, Owen explained that the aim of his researches in comparative anatomy was to formulate "a harmonious theory combining the transcendental and teleological views." Owen to Whewell, 31 October 1837, Whewell Papers, Trinity College, Cambridge, Add. MS. a. 210. 54. Quoted with permission of Trustees of Trinity College Library, Cambridge.

117 Owen was required to ask the permission of the curatorial board to publish on topics not

tention of the international scientific world with the publication of his *Memoir on the Pearly Nautilus,* describing this curious pelagic cephalopod whose internal anatomy and taxonomic relationships were almost unknown. This work, more than any single other contribution, made Owen a figure to contend with in comparative anatomical circles and drew the attention of foreign workers, especially those in Germany, to the young anatomist.[118]

The Hunterian lectures were, in some respects, a burden for the College. Minutes of the Council and the Board of Curators in the years before the 1834–37 suspension suggest that they were more often seen as a distraction, an annoying requirement of the Charter that diverted the curators from the urgent task of completing the catalogue, than as an offensive wing for College politics. More often than not, they created opportunities for vituperative savaging of the lecturers by the medical radicals.[119] They had even been forced into suspension in 1831 through a public disruption by Thomas Wakley and his friends, requiring even more awkward restrictions on admission to be enacted.[120]

Delays in completion of the catalogue also resulted from the incessant interruptions caused by visitors, whom either Clift or Owen had to conduct around the display. Also, inadequate space increased the difficulty of curating and describing the collection. As early as September 1831, following his trip to the Paris *Muséum,* Owen had recommended that the Hunterian quarters be expanded. Renewed appeals for more space from Owen and Clift and the need for uninterrupted time to work on the catalogue led to the approval by the Council in January 1834 of a closure of the Museum with extensive remodeling of the College premises, including expansion of the Museum by two-thirds, the rebuilding of the Library and the construction of a new lecture theatre.[121] These were carried out under the architectural direction of Sir Charles Barry, the architect for the rebuilding of the Houses

---

(contd.)
    directly related to the catalogue, with permission typically followed by a resolution in the minutes that he return promptly to the cataloguing task. By the early 1830s, the Curatorial Board was demanding that Owen and Clift produce proof-sheets, drawings and progress reports at each meeting of the Council. These incessant pressures may in part be the reason that Owen resigned for a short period from the College in early 1830 and moved to Birmingham to set up a surgical practice. See letters from RO to Caroline and William Clift, Hirtzel Owen Correspondence, TU, 1, letters for 9, 11, 12, 14 January 1830.

118 Owen's treatise was translated in Oken's *Isis* in 1832. From approximately this date, direct references and reprint exchanges begin with prominent German scientists, including Von Baer, Weber, Müller, Lichtenstein, and Tiedemann. See Owen Notebook covering the 1834–36 period, Owen Papers, BMNH, L. O. C. O. O. 25 # 11. Owen's reprint mailing list is found on the rear folia. Owen demonstrates that he had mastered enough German to compose instructions for a German publisher (ibid., fols. 84, 91).

119 See for example the review of Thomas Chevalier's unsuccessful surgical lectures in *Lancet* 3 (12 June 1824): 338.

120 On Thomas Wakley's disruption of the lecture of 8 March 1831, see Minutes of the Council, RCS Archives, 5, entries for 9, 14, 21 March. The lectures were not resumed until the spring of 1832 as a consequence.

121 Minutes of the Museum Curators 4, 357–8, entry for 13 January 1834.

of Parliament after the disastrous fire of 1834.[122]

Owen's interest in utilizing his talents more creatively than by drawing and describing preserved specimens also came to the surface in this period. He wished to make a public name for himself as a comparative anatomist, possibly inspired by the example of his long-time acquaintance Robert Edmond Grant.[123] In 1834 Owen approached the Council with a proposal to offer a series of lectures in comparative anatomy when the College reopened.[124] In the interim, he offered in April and May 1835 at St Bartholomew's Hospital his first public course of lectures on the animal kingdom.[125] In this series of twelve lectures, Owen displayed his concern to distinguish his presentation of the material from others currently available. Following the plan utilized by Joseph Henry Green, and now more recently by Robert Grant, Owen organized the course on a sequence moving from the lowest forms to the mammals. But Owen then organized the sequence of analysis in major plans according to the five-part classificatory system of William Sharp MacLeay, whose systematics were based on a circular, rather than linear or branching, arrangement of five main groups.[126] Somewhat as Green had done before, Owen combined an ascending order with an arrangement in major groups, beginning with MacLeay's newly defined group of simple invertebrates, the Acrita, composed primarily of the infusoria.[127]

Owen's participation in the 1835 lecture course at St Bartholomew's was predictably a matter of concern to his superiors at the College of Surgeons, frustrated by the slow pace of the appearance of the catalogue of the physiological series.[128] Their anxiety concerning the possibility that the College would still lack a catalogue of the collection at the projected grand re-opening of the new Museum two years hence is not difficult to understand. The College had suffered a distinct loss of image with the publication of Thomas Warburton's Parliamentary Commission Report on medical education in 1834, with a large section devoted to testimony

---

122 Zachary Cope, *The History of the Royal College of Surgeons of England* (London: Blond, 1959), chps. 28, 30 and 31. Although considerations of prestige were surely part of the issue in this redesign, documents suggest that the needs of space and a genuine concern to render the Museum more usable were uppermost in these decisions. See Minutes of the Museum Curators 4, 20 August 1833 and 13 January 1834.

123 Grant's lectures on comparative anatomy had been serialized in the *Lancet* in 1833–34. These presented Owen with a model both to emulate and, if possible, to surpass.

124 Owen was not applying for the Hunterian chair. He was later anxious about the prematurity of his election to this. See comments in letter of 6 May 1837 reprinted below, p. 65.

125 This course is outlined in a printed prospectus, Owen Papers, BMNH. O. C. 38.

126 Owen at the same time revised MacLeay's arrangements. He elevated three of the five groups of MacLeay's system (Mollusca, Radiata and Annulosa) to the Subkingdoms Heterogangliata, Nematoneura and Homogangliata, respectively.

127 See Owen, "Acrita," in Robert B. Todd, ed., *The Cyclopedia of Anatomy and Physiology* (London: Sherwood et al., 1836), 1: 47–9. The *Cyclopedia* was issued serially in sections that comprised the later bound volumes. The section with Owen's article first appeared in 1835.

128 Anthony Carlisle to Richard Owen, 10 September 1834, *Life* I: 86–8. Carlisle warns him that delivering these lectures at the hospital "will endanger your powerful position in the college." The first volume of *DIC* appeared in print on 5 March 1834.

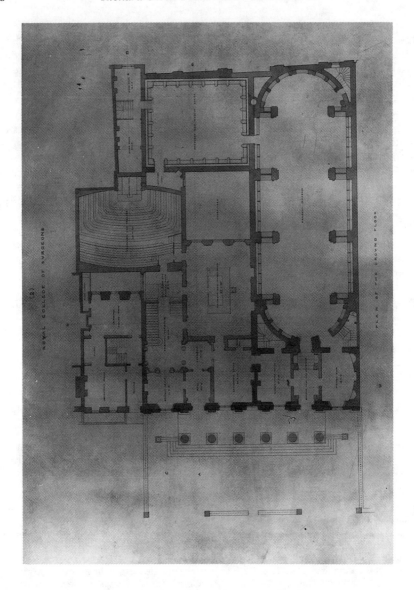

Fig. 4. Final Plan For The Proposed Remodelling Of The College Of Surgeons By Sir Charles Barry (1833). This displays the location of the enlarged main gallery to the right, the smaller new side gallery to the top, and the new lecture theater on the middle left. The College entrance is at the bottom. See also roof-top view, Fig. 14, p. 264. *(Courtesy President and Council, Royal College of Surgeons of England)*

from leading members of the College of Surgeons. This report detailed Everard Home's burning of the manuscripts, much to the delight of the College's enemies at the *Lancet*.[129] The catalogue, not Owen's lectures, was the presumed vehicle by which this loss of prestige was to be repaired.

The pressure of these tasks and the time required by the St Bartholomew's lectures forced Owen to resign from his hospital post in May after this single lectureship, turning the position over to his student Arthur Farre, a Cambridge-trained physician.[130] However, when Charles Bell resigned his position as Hunterian lecturer in comparative anatomy on 14 April 1836 to accept a position as surgical lecturer at the University of Edinburgh, a spot was opened for Owen, and on 28 April the Council of the College elected Edward Stanley to the surgical position and Richard Owen to the comparative anatomy lectureship.[131] Owen responded affirmatively to this invitation on 30 April.[132]

Reorganization of the lectures accompanied this appointment. Previously, the comparative anatomy lectures had been limited to twelve to fifteen presentations, with the remainder of the statutory twenty-four lectures to be surgical lectures "founded on Arris and Gale."[133] Owen's new series was to be expanded to twenty-four lectures, to be commenced in May 1837 after the completion of Stanley's series of six lectures.

Owen, as a generally unknown anatomist only thirty-two years old, was understandably anxious about delivering this opening series in the newly expanded College facilities. In the audience would be members of Parliament, the Council and Trustees of the College, major leaders of British medicine and several hundred students and invited guests, with a total crowd of possibly four hundred attendees, the capacity of the new theater.

The lectures became the subject of careful preparation and historical research immediately following his appointment to this position. By July 1836, Owen was reported to be "so much engaged in his Lectures" that he was unavailable for social

---

129 Thomas Wakley displayed obvious pleasure in reprinting this material in *Lancet* 1834–5, 471–2.

130 See "Minutes of the Medical College, St Bartholomew's Hospital" (Sept. 1834–August 1843), entry for 13 June 1835. See also Owen to J. S. Harrison, 30 May 1835, Owen Papers, Lancaster Municipal Library MS 7317, as quoted in D. L. Ross, "A Survey of Some Aspects of the Life and Work of Sir Richard Owen, K. C. B., together with a Working Handlist of the Owen Papers at the Royal College of Surgeons of England" (Unpublished Ph.D. Thesis, University of London, 1972), 25. On Arthur Farre, see below, p. 57.

131 Council Minutes RCS 6 (1833–36). V. G. Plarr's list of lecturers (note 40 above) fails to note Stanley's appointment. Stanley's lectures commenced on 18 April 1837 preceding Owen's own. They were serialized in *London Medical Gazette* 20 (1836–37): 342–5; 379–86; 421–4; 497–501; 577–80; 641–7.

132 Draft letter of Richard Owen to College of Surgeons, 30 April 1836, Hirtzel Owen Correspondence, TU 1: 37. This is partially transcribed in *Life* I: 95–6.

133 Council Minutes 6. The surgical series, predating the establishment of the College, had been endowed in 1645 by Alderman Edward Arris and a separate series had been established in 1698 by a Dr. Gale. See Plarr, *List of Lecturers and Lectures*, 19.

engagements.[134] The first drafting of the lectures reprinted in this volume began during this period. Written first in pen on sheets of foolscap in heavily reworked drafts, the lectures were then recopied by William Clift in his neat "copper-plate" hand onto large sheets of high-quality copy paper for public delivery, with Owen inserting new sections and revisions up to the time of delivery.[135] The stages in the development of these lectures can thereby be followed both in early draft form and in the final manuscripts through their revisions.[136]

No detailed account of the entire series of the twenty-four lectures survives. Secondary accounts have also created some uncertainties on their subject matter. According to the biography by Owen's grandson, the series was on the teeth, and this issue definitely formed at least a part of the lecture material.[137] Owen also confirms that in the context of discussing the teeth, these lectures dealt near the end with the dentition of Darwin's fossil of the unusual mammal *Toxodon platensis* collected by the *Beagle*.[138]

It is, however, difficult to connect these statements with the content of the surviving manuscripts printed here. The unity of the series emerges more clearly when Owen's lectures over the whole period from 1837 to 1856 are read in light of the Statutes governing the lectures.They were to expound the Hunterian Collection.

---

134 Anthony Carlisle to William Clift, 22 July 1836, located in William Clift's 1836 diary, Clift Papers, RCS 276. g. 1–33.

135 Clift had begun his career as John Hunter's copy-boy, and retained his skills as a penman through his later life. During the recopy Clift took several opportunities to comment upon or correct Owen's claims, as revealed in the emendations to the lectures.

136 Revisions of the drafts and final recopies can be followed in part through changes of the paper types. For the early drafts of the lectures, Owen utilized primarily a "J. Green and Son 1836" writing paper for the main body of the seven surviving lectures, with a few pages on an older "J. Tassell 1834" paper. The "Green 1836" paper was also being used for correspondence at this time in letters dated between 4 July 1836 (Richard Owen to William Buckland, Owen Correspondence, TU, I: 38) and 23 January 1837 (Owen to Charles Lyell, 23 January 1837, Lyell Papers, American Philosophical Library B D 25.L). New insertions were then made on occasion on a similar "J. Whatman 1836" paper, and invariably these insertions were then followed in the subsequent recopy by Clift. For further details on these matters see analyses of the individual manuscripts heading the lectures below.

137 *Life* 1: 109n2. The curious absence of the manuscripts in either form beyond the lecture for 16 May suggests that these lectures were incorporated directly into the *Odontography* text for printing. This long text was accepted for printing in March 1839, suggesting that it must have been prepared in a rather short space of time (see letter of Baillière to Owen, 20 March 1839, Owen Papers. BMNH, 2: 18). In the absence of manuscripts, it is impossible to determine the degree of revision that had taken place after the 1837 course.

138 R. Owen, "Preface," *Odontography, or a Treatise on the Comparative Anatomy of the Teeth*, 2 vols. (London: Baillière, 1840–5), ix: "In the first of these courses the teeth were considered in their relation to the Osseous System, and the intimate structure of their component tissues was more especially treated of." Owen reports that in the concluding lectures he had discussed, and apparently exhibited, the unusual characteristics of the fossil teeth of *Megatherium, Toxodon, Megalonyx* and *Mylodon* from the fossils brought back by Darwin from the *Beagle* voyage, and he had resolved at the conclusion of the 1837 course "to undertake a microscopic study of the anatomy of the teeth of these forms to determine more about their structure and affinities" (p. xiv).

Table Three:
The Seven Surviving 1837 Hunterian Lectures

Lecture #	Date of Delivery	Date of Probable Composition	Date of Completion of Final Recopy	Contents of Lecture
1	2 May 1837	Summer-Autumn 1836	22 Nov. 1836	Introductory remarks; the history of comparative anatomy from the Pre-socratics to Aristotle
2	4 May 1837	Summer-Autumn, 1836	22 Nov. 1836	Aristotle's importance in comparative anatomy; his functional classifications of organisms; the similar principles of Cuvier and Hunter
3	6 May 1837	Winter 1836–7	Before 11 April 1837	Comparative anatomy and physiology from Vesalius to Hunter; studies of generation; laws of animal organization
4	9 May 1837	Winter 1836–7	Main portion before 11 April, final portion after 11 April 1837	The vital principle; Harvey's and Hunter's views on the vital role of blood; the functions of digestion; Hunter on the nervous system; Hunter on the development of the chick; teratology; critique of transcendental anatomy of E. G. St. Hilaire and Tiedemann; recapitulation theory
5	11 May 1837	Late Autumn 1837	After 11 April 1837	Matter and organized matter; unity-of-type; role of vital force and organization; vital force and embryological development; law of the correlation of parts; relations of organization and species life and death; the germ theory
6	13 May 1837	Winter 1836-7	April 1837	Hunter's and Cuvier's views on relations of animal and plant kingdoms; chemical and structural differences of animals and plants; experiments on the "sensitive" plants; status of zoophytes; Ehrenberg's work on Infusoria
7	16 May 1837	Winter-Spring 1837	April 1837	Examination of the productions of the vital force; commencement of detailed examination of animal structure; chemical and microscopic studies on the blood by Berzelius, Brande, Müller, Prout; functions and properties of the blood; transfusion experiments

Owen's express task was to expound upon a subsection of this total collection, namely the "Physiological Series", which formed the focus of the *Descriptive and Illustrated Catalogue*.[139] Inspection of the arrangement laid out in this published catalogue would suggest an exposition beginning with the comparative anatomy of the "Organs of Motion and Digestion" with some extension into the circulatory, absorbent and respiratory systems. Hence, discussion of the teeth would be expected, forming a subdivision of the organs of digestion.

This structural ordering, in conformity with the Hunterian series, is in agreement with Owen's otherwise puzzling subsequent remarks on the series. Referring to his prior course at the commencement of his 1838 Hunterian lecture series, Owen reported that he had been concerned in the previous year with "the osseous and muscular structures, and the locomotive organs, all of which indeed may be held subservient to the process of digestion, as an animal was denominated by Hunter a moving digestive bag."[140] This would relate only remotely to the content of the surviving lectures. However, commencing the 1839 series, he elaborated again on the contents of the 1837 series, reporting that he had dealt with the history of comparative anatomy from Aristotle to the present

> with an especial reference to the advances made from time to time in those depart-
> ments of Physiological Research or discovery in which the Founder of our
> Collection more immediately interested himself. . . : the general character of
> Animal & Vegetable Organizations were laid down. The Comparative Anatomy of
> the blood: The Principles of the Classification of the animal Kingdom & the mod-
> ifications of the active and passive organs of locomotion in the different classes of
> Animals.[141]

This account directly connects with the material presented in the following transcription, and it suggests the linkage between discussions of teeth and digestion, and those of vitality, form and function and the other topics treated in the opening lectures. Unfortunately, manuscripts in either draft or final form exist only for the first seven lectures, and the exact content of the lectures beyond these can only be reconstructed with hypothetical principles.[142] With evidence that twenty-

---

139 See Lecture One below, fols. 13–14.

140 Owen, "Lectures on Comparative Anatomy," transcribed from notes taken by Wm. W. Cooper
     and revised by Owen, Owen Papers, RCS 42. g. 26 (8 May 1838).

141 Owen, "Hunterian Lectures 1839: Excretory & Tegumentary Systems," Owen Papers, BMNH
     O.C. 38, 1: 1. The introductory lecture to the 1842 series gives further clarification of the
     1837 series. There Owen remarks having treated in 1837 the "Series of preparations of the
     Elementary Tissues, [and] those of the preliminary organs of digestion including the Dental
     System. . ." Owen 1842 lectures, Lecture One, Owen Papers, BMNH O.C. 38: vol. 2.

142 The delivery dates of all the twenty-four lectures are carefully recorded in William Clift's
     daily diary for 1837 (Clift Papers, RCS 276. g. 1–33). Unfortunately, Clift almost never
     mentions the content. Some comments are of interest, such as the entry for the lecture of
     Tuesday, 23 May, "A Frog and Toad for Lecture." On 27 May he records the purchase of
     "Oysters and periwinkles for Lecture," suggesting that demonstrations of fresh material from
     invertebrates were being made. I have been unsuccessful in locating notes taken on these

four lectures were delivered, this requires some explanation. Some probably were utilized subsequently as manuscript for the published *Odontography* of 1840.[143] Others may have been taken directly from the published or manuscript text of the first volume of the *Descriptive and Illustrated Catalogue,* since this directly parallels the main material under consideration in this first series. Still others seem to have formed portions of Owen's introductory material to his edited volume of *Hunter's Works* that appeared in October of 1837.[144] At least some might have been destroyed or otherwise lost with the materials discarded or dispersed in the 1890s.[145] Whatever the explanation, the lectures reprinted here are the only ones to survive. They also supply the theoretical introduction, not only to the 1837 series, but also to those of the succeeding decade. The content of the seven lectures from the 1837 series can be determined from Table Two. The more general parallel of the contents of the Hunterian lectures as a series with that of the *Descriptive and Illustrated Catalogue* volumes in the 1837–42 period is displayed in Table Three.

The opening 1837 lectures were the prime opportunity for Owen to set forth his basic theoretical positions on fundamental issues of vitality, the connection of morphology and physiology, and the relations of organic groups. They were also his opportunity to give his opening statements on the transformist question, under debate in London circles at the moment, as he began his systematic exposition of the Hunterian materials. His understanding of his charge from the College was well described later:

> When I was honoured by the Council in 1837 with this arduous and responsible office, it seemed to me that the first obligation upon the professor was, to combine with the information to be imparted on the science of comparative anatomy an adequate demonstration of the nature and extent of the Hunterian Physiological collection. . . . The system adopted by Hunter for the arrangement of his preparations of Comparative anatomy was therefore made that of the lectures which were to be illustrated by them; and this plan was closely adhered to until the whole of the physiological department of the Collection had been successfully brought under

---

(contd.)
    lectures by individuals documented to be in attendance. There are no significant reports from a wide selection of the London medical and periodical press. Comments on the final lecture of 24 June are contained in the *London Medical and Surgical Journal* 1 N.S. (1837): 378–9. This describes the final lecture as on the "Comparative Anatomy of the Teeth" and complains about the compression of so much material into the allotted space of one hour, noting that Owen ran thirty minutes overtime. I am indebted to Adrian Desmond (personal communication) for this important reference.

143 This is particularly true of the discussions of the fossil teeth. However, the text of *Odontography* does not read as a lecture series, and incorporates new materials and references, suggesting extensive revision of any lecture text incorporated into it.

144 See notes to lectures below.

145 C. Davis Sherborn reported after Owen's death that "Owen's own MSS were on foolscap twelve feet thick" and after the writing of the *Life* were "distributed to those interested all over the world . . ." (BL Add Ms 42581 ff 247/50). I am indebted to Jacob Gruber for this reference.

your notice, and its demonstration completed, in the course of lectures which I had the honour to deliver last year. It is, I believe, generally known that Hunter has arranged his beautifully prepared specimens of animal and vegetable structures according to the organs; commencing with the simplest form, and proceeding through successive gradations to the highest or most complicated condition of each organ. These series of organs from different species are arranged according to their relations to the great functions of organic and animal life; and the general scheme is closely analogous to that adopted by Baron Cuvier in his *Leçons d'anatomie comparée,* and in the best modern works on Physiology. [146]

Owen was deeply immersed in the composition of these lectures, revising them, developing his own theoretical synthesis, and seeking to relate his theoretical views to the wide array of forms at his disposal in the Museum collection when, in October 1836, he came in contact with young Charles Darwin, just returned from five years of scientific work aboard HMS *Beagle.*

C. Converging Careers

The initial encounter of Owen and Darwin in London, in October 1836, brought together for the first time two professional trajectories that began with a common biomedical training under many of the same professors in Edinburgh. Shortly after the return of the *Beagle,* Darwin had apparently contacted Owen, probably by an initial visit to the College to discuss the fossil materials sent ahead to the Hunterian Museum through the intermediary of J. S. Henslow. Subsequently the two men met at a dinner party at Lyell's on 29 October.[147] Another meeting took place at a dinner at Roderick Murchison's residence on 5 November.[148] Additional meetings between Darwin and Owen cannot be documented until 6–8 December when Darwin again visited the College and supervised the unpacking of his fossil bones.[149]

These meetings were also bringing together two naturalists with similar scientific interests in a wide range of theoretical questions. On his departure on the circum-global voyage of the *Beagle,* Darwin was conversant with the contemporary scientific debates over such theoretical questions as plant reproduction and plant and animal microstructure, and he was concerned with the possibility of unifying plants and animals through reproductive processes. His work on insect taxonomy had also made him deeply aware of issues in classification and the analysis of variation. During the years aboard ship, Darwin continued to explore these ques-

---

146 Owen, *Lectures on the Comparative Anatomy and Physiology of the Invertebrate Animals* (London: Longman et al., 1843), 3–4.
147 Charles Lyell to Richard Owen, 26 October 1836, in *Life* 2: 102. Lyell notes that Owen will meet Darwin at the party, "whom I believe you have seen."
148 Entry from Caroline Owen's Diary, *Life* 1: 103.
149 Letter of CD to Caroline Darwin, ca. 7 Dec. 1836 in *Darwin Correspondence* 1: 524.

Table Four:
Owen's Hunterian Lectures, 1837-44[150]

Year	Lecture Topics	Corresponding Contents of the Published Hunterian Catalogues
1837	History of comparative anatomy; nature of vitality; plants and animals; classification; blood and solid component tissues; muscles, joints, bones and teeth	Vol. I: Organs of motion and digestion
1838	Digestion, absorption, circulation, respiration	Vol. II: Absorbent, circulatory, respiratory and urinary systems
1839	Excretory systems, renals, modifications of the skin and appendages in invertebrates and vertebrates	Vol. III, Pt. I: Nervous system and organs of sense; Pt. II: Connective and tegumentary systems
1840	Theories of generation and development; comparative structure of the ovum	Vol. V: Organs of generation Products of generation
1841	Organs of support and comparative osteology of invertebrates and vertebrates	Osteological catalogue
1842	Comparative anatomy and physiology of the nervous system	No correspondence
1843	Comparative anatomy and physiology of the invertebrates	Published in 1843
1844	Comparative anatomy and physiology of the vertebrates; defining characteristics of the classes of vertebrates	Published in 1846

150 This schema has been constructed primarily from a study of the contents of his lectures as revealed by the surviving manuscripts in the Owen Papers at the BMNH. I have also been assisted by Nicolaas Rupke's valuable study, "Richard Owen's Hunterian Lectures on Comparative Anatomy and Physiology," *Medical History* **29** (1985): 237–58. Owen's own description of the contents of his lectures through 1842 is found at the beginning of the manuscripts of his 1842 lectures, Owen Papers BMNH O. C. 38, vol. 2. This text attests to Owen's aim in the lectures to construct these on a parallel to the Hunterian Collection, and states that the 1842 course will complete the survey of the physiological material. No manuscripts can be located for the 1843 lectures, the first to be published in completeness from notes made by W. W. Cooper and revised by Owen. The lectures on generation for the 1840 course were published in summary form in *Lancet* **39–40** (9 May 1840–20 March 1841).

tions in depth.[151]

Owen had been drawn independently to many of these same questions in the early 1830s. In this period, he made several studies of live invertebrates during trips to the seacoast, very often the same groups that interested Darwin on the *Beagle* during the same years.[152] Owen was also seeking to clarify, by his own observations, details in John Hunter's views on reproduction in plants and animals for the preparation of the fourth volume of the catalogue. The two men consequently had mutual interests in an unusual set of functional questions in this period as well as immediate concerns with the fossils themselves.

Contacts between Owen and Darwin remained sporadic in the early months of 1837. Darwin apparently visited the College on his trip to London on 3 January, when he seems to have supervised the opening of materials sent from Cambridge to the College on 30 December.[153] Discussion between Owen and Darwin on the taxonomic position of his fossils probably occurred on this date. On 23 January, Owen communicated to Lyell his tentative placement of the fossils into Orders.[154]

---

151 See M. J. S. Hodge, "Darwin as a Life-long Generation Theorist," in Kohn, *Darwinian Heritage*, 207–43; Desmond, *Politics*, chp. 9. I have also discussed important aspects of Darwin's formation in my "Darwin's Invertebrate Program," in Kohn, *Heritage*, and idem, "Darwin, Vital Matter, and the Transformism of Species."

152 Aspects of this work can be seen in Owen's letters covering the period between 1830 and 1836 (Letters 11, 22, 26 and 38 in the Hirtzel Owen Correspondence, TU, I). These report such items as Owen's observations on microscopic studies on the circulatory patterns in the colonial zoophyte *Sertularia*, exactly the issues that were of interest to Darwin in these same forms. Owen was also a close friend of Joseph Jackson Lister and Arthur Farre, both of whom were working on these same invertebrate groups in these years. See J. J. Lister, "Some Observations on the Structure and Functions of Tubular and Cellular Polypi, and of Ascidiae," *Philosophical Transactions of the Royal Society* **124** (1834): 365–88; and A. Farre, "Observations on the Minute Structure of Some of the Higher Forms of Polypi, with Views of a More Natural Arrangement of the Class," *Philosophical Transactions of the Royal Society of London* (1837), 387–426. Owen communicated the paper to the Society on 11 May.

153 Darwin had offered his fossil materials to the Museum on 19 December 1836, and the Curators recommended acceptance of this offer to the Council of the College at their next meeting on the twenty-first (Minutes of the Curatorial Board, 4: 457; entry for 21 December). See also CD to Anthony Carlisle, 19 Dec.1836, *Darwin Correspondence*. 1: 525, with Carlisle's reply of the twenty-third (ibid. 526). William Clift mentions the receipt of the Darwin materials on 3 January (Clift Papers, RCS 276.g.1–33). The range of the donation had been expanded by Darwin to include twenty-eight mammalian skeletons, "all new to the collection," and twenty specimens of pickled birds "highly interesting for dissection accompanied with notes of the locality and habits of each of the species" (Minutes of the Curators 5: 3, entry for 4 January 1837). Darwin had come to London on 2–4 January for a dinner party with Lyell and the meeting of the Geological Society. On 4 January the Board of Curators also approved Darwin's proposal to have plaster casts made by the College as a condition of the receipt of the donations. This casting commenced on 18 January.

154 Owen to Lyell, 23 January 1837 (American Philosophical Society Lyell Papers B D25. L). This is reprinted in Leonard Wilson, *Charles Lyell* (New Haven: Yale University Press, 1974). Owen refers in this letter to his tentative placement of the fossil camelid into the genus *Auchenia*, but has no generic name for the *Toxodon* fossil by this date. On the complexities of these identifications, see Stan P. Rachootin, "Owen and Darwin Reading a Fossil: *Macrauchenia* in a Boney Light," in: Kohn, *Darwinian Heritage*, 155–83. I have not been

More sustained personal contacts between the two men waited upon Darwin's moving his permanent residence to London on 6 March of 1837. After this date, opportunities for meetings and conversations at the gatherings of the Zoological and Geological societies supplemented contacts at the College.[155]

In the early months of 1837, the College of Surgeons finally entered the upper echelons of public scientific, and not simply medical, life in London. The new Hunterian Museum, the renovated Library, and the greatly expanded lecture theatre of the College at Lincoln's Inn Fields were opened officially, with much fanfare, on 14 February, the date of the traditional Hunterian Oration. The orator for that august occasion, Benjamin Collins Brodie, looked out on an audience that included Prime Minister Robert Peel, the Duke of Wellington, the Bishop of London, leading luminaries of London science, including, in the front row, Peter Mark Roget, author of one of the Bridgewater treatises and Secretary of the Royal Society, Charles Babbage, all members of Council and the Trustees, and a remaining audience totalling over four hundred persons.[156] Brodie predicted that the College would establish through its new Museum and lectures "a school of what may be called 'the science of life,' such as has never existed in this metropolis before . . . contemplating the phaenomena and laws of life, generally; not as they are exhibited in our own species only, but as they exist in the whole animal creation." Brodie also announced the forthcoming lecture series in comparative anatomy, to be "delivered by the junior conservator, Mr. Owen, whose zeal and

---

(contd.)
  able to document a subsequent meeting between Darwin and Owen during Darwin's mid-February trip to London to attend Lyell's Presidential Address to the Geological Society on 17 February. Owen was not yet a Fellow of the Geological Society and does not seem to have been present at the 17 February meeting. This was also the week of the grand opening of the new Hunterian Museum, 14 February.

155 Unfortunately visitor books for the Hunterian Museum and Library during these crucial months had not been kept due to the long closing of the facility and were not resumed when the facility reopened in February. Clift was already a member of the Geological Society by 1837. Owen was introduced as a visitor to the meeting of the Geological Society by Clift on 2 November 1836. By this date Darwin had already been proposed for membership, and his admission to Fellowship was balloted on the sixteenth, with Fellowship approved on 4 January. Owen was proposed for Fellowship on 3 May 1837 and admitted on 14 June. They can be documented to have attended the same meetings of the Geological Society on 19 April and 31 May when they both read papers (Minutes of Meetings of the Geological Society, 19 Feb. 1836–17 Jan. 1838, Geological Society of London Archives). Owen and Darwin can also be documented to have jointly attended the Zoological Society meetings of 14 March, and probably those of 11 and 25 July, since the minutes for these two meetings are in Owen's hand (Minutes of the Open Meetings of the Zoological Society, Oct. 1835–Aug. 1840, Zoological Society of London Archives). Other joint attendances at these society meetings are probable but cannot be documented.

156 Exact details on the seating arrangements of the notaries and a precise calculation of the capacity of the new theatre are contained in Clift's 1837 Diary (Clift Papers, RCS 276. q. 1-33) entries for 14 February. Clift computed the theater to hold 413 persons, but the facility was unable to accommodate all who wished to attend that evening. See *Lancet* (1836–7), 822.

talents are known to you all."[157] Such public announcements to the assembled leaders of London civic and scientific life that evening could not have eased the pressure on the young anatomist.

Owen's activities in the months of February, March and April were divided between his official duties involved with completion of the ever-present catalogue, arrangements of the displays in nearly complete form for the formal opening of the Museum to the public on a regular basis in April and his duties as a member of the governing board of the Zoological Society. What free time he could spare— typically in the late evening and early morning—from these daily duties was re- quired for the final preparation of the forthcoming lectures. He was also at work on his paper describing Darwin's *Beagle* fossil *Toxodon platensis,* scheduled for presentation to the Geological Society on 19 April, which would offer the first public results of his study of the *Beagle* materials.

Some record of the contacts between Owen and Darwin in these months can be compiled from careful examination of the Owen lectures, the pocket notebooks of both Owen and Darwin, and documents of professional meetings and corre- spondence from this period.

On his trip to Paris in 1831, Owen had commenced the practice of keeping a series of small pocket diaries, similar in form to the small clasp-bound notebooks used by Darwin for his own reflections.[158] The twelfth notebook, dated from February 1836 to December 1837, covers the relevant period of the initial Owen- Darwin meetings. In the Darwin materials, a similar period is covered by three notebooks. The earliest, designated "RN," is acknowledged by Darwin scholars to bridge the period of the latter months of the *Beagle* voyage until approximately June 1837. The second, labelled "A," dealing primarily with geology, is difficult to date precisely, but was probably begun around June 1837.[159] The third, the well- known "B" notebook, is stated by Darwin to have been begun "about July" of 1837, and it runs until February of 1838.[160] By working between these texts from the Owen and Darwin materials, several interesting issues begin to emerge.

Owen's twelfth notebook opens with the date of February 1836 and is kept from both ends by inverting the book, a common practice at this time, also followed by Darwin in some of his notebooks. This enabled a single notebook to be used sys- tematically for serial entries on at least two different topics. The two series of en- tries would in this case meet someplace in the middle, at which point another note-

---

157 Benjamin C. Brodie, *The Hunterian Oration. . .14th February, 1837* (London: Longman et al., 1837), 35–6.

158 These fifteen notebooks, covering the period from October 1830 to November 1839, have been rebound in small buckram bound individual volumes in the Owen Papers, BMNH, L. O. C. O. 25.

159 See introduction by Sandra Herbert to her edition of this text in *Charles Darwin's Notebooks, 1836–1844: Geology, Transmutation of Species, Metaphysical Enquiries,* ed. by P. Barrett et al. (London: BMNH Press, and Ithaca: Cornell University Press, 1987), 83. Hereafter cited as *Darwin's Notebooks.*

160 David Kohn (ibid., 167) argues that this date should not exclude a commencement as early as mid-June.

book would be started. In Owen's case, this meeting of entries in the February 1836 diary took place about December 1837.

Read from the front, the Owen text begins with a discussion of the anatomy of various cephalopods, apparently representing notes taken on the dissection of fresh specimens obtained from the public market. Detailed accounts of the internal anatomy and reproductive structures of the common squid, *Loligo sagillata,* constitute the most extensive discussion. Occasional dated entries through March 1836 cover such topics as the microscopic anatomy of hydatid cysts taken from cows, an anatomy of the lanceolate *Amphioxus* and a discussion of the anatomy of a porpoise brain. Extensive entries in April and May describe shipments of specimens received from his friend George Bennett in Australia. During the summer and early autumn of 1836, the main entries focus on the pathological anatomy and descriptions of intestinal parasites obtained from animals in the Regents Park Zoo, and there is a long discussion of the internal anatomy of the Rock Kangaroo. Although these observations appear miscellaneous, Owen was concerned in that period with following up on some of John Hunter's claims in preparation of the fourth volume of the *Catalogue* dealing with generation.[161] He was particularly interested in determining if there were "granular bodies" within the ova of various invertebrates. For example, in describing a specimen of the cephalopod *Argonauta* sometime after 11 October, he speaks of the ova containing "only a tenuous granular matter."[162] In remarkable parallel to Darwin's own microscopic studies aboard the *Beagle,* Owen wanted to understand the importance of the microscopic organelles that the new achromatic microscopes were revealing to observers in the 1830s.[163]

At page fifty-four of the notebook, the writing reverses, indicating the point of meeting from the other end, presumably marking the closing date of December 1837. It is in reading the notebook from the rear that we encounter the first evidence of Owen's study of the fossil materials conveyed to the College in the autumn of 1836.[164] In an undated pencil entry seven pages from the rear, and thus presumably early in the dating of the notebook, we read:

Pampas—
Mammilary toothed anl
Common
Great Armadillo next
Megatherium third
300 miles south of

---

161 See below, note 178.
162 Owen, Notebook 12, MS p. 46.
163 See Sloan, "Darwin, Vital Matter," for discussion of the background of this question.
164 The fossils were apparently delivered to the College in packing cases when the *Beagle* was moved from Portsmouth to London (Darwin, letter to Caroline, 7 Dec. 1836, *Correspondence* 1: 524). See above, note 125. Catherine Darwin was also concerned about their safety after receiving Hensleigh Wedgwood's report that the *Toxodon* skull was simply lying about "in a room with workmen" (Catherine Darwin to CD, 27 December 1836, ibid., 1: 534).

On the next page there is a series of measurements headed "Toxoreodon," later written over as "Toxodon," that had been unmistakably taken of the unusual skull of Darwin's *Toxodon platensis*, the fossil Owen would describe at the meeting of the Geological Society on 19 April. Following seven pages of rough entries on this skull is a series of entries that are referenced in the notebook index page as "Farre Obs." I have described these entries elsewhere in detail.[165] This latter series of entries initially describes observations on specimens of invertebrate zoophytes and bryozoans. These are then followed by drawings taken through a microscope and comments on the muscle fibres in both insects and sea-cucumbers. Finally, there are drawings and comments on the germinal disc of an embryo, possibly that of a bird. We can thereby follow a chronologically ordered sequence of entries in which the rough notes on the *Toxodon* fossil precede the entries related to Arthur Farre.

The dating of this latter series of entries can be determined with some precision. Between 23 January and 22 March, Owen seems to have settled on the name of *Toxodon* for the large fossil mammal that had been recovered only as a skull.[166] The entries that follow the use of this name in the notebook thus narrow down to a period after late March. We then read in Caroline Owen's diary entry for 11 April: "Dr. A. Farre and Mr. Darwin here this afternoon. After tea muscular fibre and microscope in the drawing-room."[167] I have presented evidence in another context indicating that these notebook passages on muscular fibres, microscopy and embryology in Owen's text were probably written during this meeting with Farre and Darwin on 11 April.[168]

The events of the week following this meeting are of interest. On 19 April, Caroline Owen was to enter in her diary "R. wrote the latter part of his third lecture and read it to me."[169] The importance of Caroline's notation deserves comment. The drafting of the Hunterian lectures was, as we have described, taking place in two main stages, the first in the form of rough hand-written manuscript by Owen and the second in the recopy for final delivery by William Clift. As the final version was being completed by Clift, Owen had begun, in December of 1836, to read these lectures to others and circulate them for comments. As a consequence, the lectures often underwent revisions and subdivision before public delivery. By late November 1836, Clift had finished his copy of the first lecture, and in December, Owen decided to divide it, after reading

---

165 Sloan, "Darwin, Vital Matter."
166 The form is unnamed in the letter to Lyell of 23 January. The *Toxodon* name is first used by Caroline in her diary entry for 22 March (*Life* I: 108). The overwriting of the name "Toxoreodon" suggests the transitional period in which these entries were made. The name designated the unusual curved teeth, which for a considerable period of time convinced Owen that the form was transitional between the Rodentia and Pachydermata.
167 *Life* 1: 108.
168 Sloan, "Darwin, Vital Matter," 427–30.
169 *Life* 1: 108.

Fig. 5. Oil Portrait Of Arthur Farre By Saverio Altamura, Ca. 1862, Located At The Royal College Of Physicians, London. *(Courtesy President and Trustees, Royal College of Physicians of England)*

it in the new theatre, into two parts.[170] The stages in the manuscript preparation are summarized in Table Two.[171]
Analysis of the manuscripts also displays that Owen decided at some date to split the completed second lecture into the third and fourth lectures of 6 and 9 May, with the Clift recopy of the original third lecture now made into the fifth lecture of 11 May. Since Clift was only copying the manuscripts, Caroline Owen's remarks must refer to Owen's own drafts. Analyses of these reveal some interesting issues.

The nineteenth of April, the date on which Owen was reported to have rewritten the end of one of his lectures, was also the date on which Owen delivered his paper on Darwin's *Toxodon* fossil to the Geological Society. Owen's work schedule was such that he often had time for work on the lectures only very late in the evening and early morning, with Caroline assisting him until near midnight. It is probable that it was in the late evening of 19 April, following the Geological Society meeting at 7 pm, that this redrafting took place.

Owen's description of the *Toxodon* fossil at the meeting of the Geological Society that evening had concentrated on its unusual, large skull and peculiar dentition. This presented a puzzling problem concerning its similarities to the rodents and pachyderms. Owen's conclusion, still maintained in the eventual publication of his findings in his *Fossil Mammalia* for the *Zoology of H.M.S. Beagle* in 1840, was that the fossil "manifests an additional step in the gradation of mammiferous forms leading from the *Rodentia*, through the *Pachydermata* to the *Cetacea* . . . ."[172]

The *Toxodon* fossil inevitably raised questions concerning the life and death of species, the causes of their extinction, and affinities of these forms to recent and fossil creatures in the same regions of South America. These are exactly the issues we see discussed in the fifth Hunterian lecture of 11 May, the original third lecture.

Owen's autograph draft of this lecture shows evidence of having undergone significant reworking near its latter half, with a new section inserted between pages twenty and thirty, followed by a renumbering of the remaining manuscript, precisely the sections dealing explicitly with the life and death of species, and the

---

170 Ibid., 109.
171 See also details on the manuscripts heading each lecture below.
172 *Proceedings of the Geological Society of London* **2** (1833–38): 542. This is an unsigned summary, apparently prepared by Whewell and reviewed by Owen. It displays some important alterations from the original summary in the minutes of the meetings of the Geological Society, Geological Society Archives OM1/8: 357–60, probably reflecting the work Owen and Darwin had done on these materials in July. The April paper was originally intended for publication, and it was favorably refereed and approved for publication by the Council of the Society, but was subsequently withdrawn by Owen in November (ibid., entry for 1 November 1837). This was evidently to allow its direct incorporation as the initial portion of his *Zoology of H. M. S. Beagle; Part I: Fossil Mammalia*, ed. C. Darwin (London: Smith, Elder & Co., 1840), 16–29. Owen's footnotes to the *Toxodon* section of the *Beagle* report indicate that this discussion was written earlier and published without incorporating his later researches on a lower jaw of the *Toxodon* subsequently discovered among the Darwin fossils (ibid., 19n.).

relationship between the duration of a species and its degree of organization.[173] This section also demonstrates directly Owen's debt to Johannes Müller's theory of the relationship between the conservation of life-force and the duration of species.[174] The draft indicates that Owen struggled with these issues, some sections heavily rewritten and others deleted. Owen is concerned to tie together in this section the "germ" theory of embryological formation, the concept of an immanent, teleologically acting vital force, and the relationship between the level of organization attained and the duration of the individual. Owen's solution to this problem was to claim that the death of species occurred by the "wearing out" of a fixed quantity of life-force endowed in the original germ of the species, with the rate of this expenditure dependent on the degree of organization achieved. Hence higher mammals, undergoing a great deal of development from their primordial germ, will have a shorter species duration than lower infusorians.[175] Explicitly denied as an explanation in this section is the hypothesis of the transformation of species.

The Geological Society paper of 19 April had posed indirectly the problem of the extinction of species and discussed the relationships holding between contemporary and fossil forms. The new draft material added to the manuscript of lecture three fits this issue into a more general physiological theory.[176] Darwin would follow Owen's discussion at the next meeting of the Society, 3 May, with his own discussion of the relation of these same fossils to the elevation and subsidence of the land. Before turning to this, we must review another unusual alteration of the manuscripts in this period.

Sometime in the spring months, Owen had also decided to split the original second lecture into the lectures delivered on 6 and 9 May. This text displays an unusual alteration directly related to the transformist issue. The Clift recopy of the final portion of this lecture contains a smoothly integrated new section of nine full folio pages of discussion that replaces a single concluding page of the original draft manuscript.[177] This new section inserts a discussion of the Cuvier-Geoffroy debate, raising for the first time in Owen's public writings the distinction between homology and analogy, and expounds on the relationship envisioned between

---

173 This is a ten-page interleaf between pages 20 and 21, corresponding to the final manuscript folia 34–44 of Lecture Five as given below. The other possible reading of Caroline Owen's comment is that Owen drafted the final added section of Lecture Four on this occasion. However, there is no evidence that this lecture was ever numbered as "Lecture Three." See further comments heading Lecture Five below.

174 On this see Sloan, "Darwin, Vital Matter," 406–15.

175 See below, pp. 68ff.

176 I have been unable to locate rough minutes or other notes that would reveal details of the discussion that followed his paper on the evening of the nineteenth.

177 This new section, comprising 93–101 of the final manuscript below, exists only in the final recopy without a corresponding autograph. The paper type of Clift's copy at this point shifts from a "J. Bune 1832" used on the first and second lectures, and the subsequent third lecture, to a "J. Whatman 1837" utilized for the remainder of Clift's recopies. This implies a revision after the reworking of the ending of the third lecture on 19 April. On these changes from the autograph, see Notes to Lecture Four below.

developmental embryology and determination of affinity.

Since this section of Owen's fourth lecture has engendered some important comment in the literature, it is useful to see the larger context of these additions provided by the full lecture framework.[178] Originally, the autograph draft of this lecture had ended simply with a short paragraph praising "the learned and eloquent Professor of Comparative anatomy at the London University," i.e. Robert Grant. The recopy now terminated with a discussion of the excesses of transcendental anatomy and of Meckel-Serres recapitulationism, coupled with a plea for the reconciliation of classification with developmental embryology. Is this simply a strategy in a political struggle with the Lamarckians and Geoffrians, using the new embryology to disarm transformism? The possibility cannot be ruled out. But the context would suggest that something else was at stake.

Owen's lectures, it should be recalled, had been immediately preceded by the surgical lectures of Edward Stanley. In his first lecture dealing with human and comparative anatomy Stanley had proposed the possibility of an underlying unity of plan governing the animal skeleton, citing Newton as his authority.[179]

As Owen approached the same issue in his own lectures, he saw, however, that a simple unity of plan of the form envisioned by Geoffroy St Hilaire was very difficult to reconcile with the Hunterian Collection itself. Neither Geoffroy nor his opponents had the issue correct, at least if this was to be useful for expounding the

---

178 Owen may have added this new section directly as a result of reading Martin Barry's "On the Unity of Structure in the Animal Kingdom," Pt. 1 (*Edinburgh New Philosophical Journal* 22 [No. 43, January 1837]: 116–41; and Pt. 2 [No. 44, April 1837]). This has been commonly claimed in the literature, often with the allusion that Owen had no original ideas on the matter himself (see Desmond, *Politics*, 337–51, and earlier Dov Ospovat, "The Influence of Karl Ernst von Baer's Embryology, 1828–1859: A Reappraisal in Light of Richard Owen's and William B. Carpenter's 'Paleontological Application of von Baer's Law,'" *Journal of the History of Biology* 9 (1976): 1–28). The *ENP* volumes containing Barry's serialized publication were published in its first part on 14 January, and the second was issued on 7 April (see *London Times*, issues for 14 January and 7 April 1837). This dating would have allowed Owen to incorporate this material in his addition of the new material to the conclusion of Lecture Four, which must have been after the redrafting of the latter part of Lecture Five on 19 April. The second part of Barry's paper is most directly similar to Owen's new discussion. However, the issue is not immediately resolved by this. The *ENP* is not held in the RCS collections, and the similarities between Owen's discussion and Barry's are not compelling. Owen later would cite Barry's "masterly Papers" on generation in the *Philosophical Transactions* for 1838 and 1839, but makes no reference to the 1837 papers (*DIC* V(1841): 148n.). Owen was fully aware of the work of Von Baer and the other German anatomists in their original German and Latin prior to the appearance of Barry's papers. Generation was, as we have seen, a direct concern to Owen in the preparation of the fourth volume of the *DIC* in its early stages when the lectures were being drafted.

179 ". . .throughout the animal kingdom, whatever may be the function which an organ is called on to perform, an uniformity of plan in its construction is to be observed, being carried on to more and more complexity and near to perfection as the amount of function necessary to be performed increases." Summary of Stanley's opening lecture of 18 April in *London Medical Gazette* 20 (1836–7): 342. Owen would also appeal to a limited notion of a unity-of-type on the authority of Newton in his second lecture of 4 May. See below, Lecture Two, fol. 65.

specific collection at hand. Owen is not so much concerned with Geoffroy-bashing as he is with a more practical question. He does not discount restrained claims in support of the proposition that there is some principle of unity greater than Cuvier's purely functional similarity. Edward Stanley had just made such a point. Furthermore, Geoffroy St Hilaire himself is always treated with respect.[180] It is only the excesses of unnamed enthusiasts for a "transcendental anatomy," probably those of Johann Spix and the French embryologist Etienne Serres, that Owen is concerned to combat.[181] Against St. Hilaire, Owen agrees that the reduction of the skeleton to the repetitions of a single structural element could be no more plausible than the claim of a geometrician who "should gravely advocate their unity of composition, and put forth the observation as one of high philosophical importance," because all mathematical figures can be derived from the mathematical line.[182]

Cuvier, however, was considered by Owen to have underrated the value of resemblances and homologies. Some deeper principle of unity must be sought that reconciled these positions. Owen concluded in these revisions of the late spring that organic affinity was disclosed by the diverging embryological development of forms, the solution he saw advocated by "Tiedemann, Purkinge [sic], Baer, Rathke, Wagner, Valentin, Müller, and the disciples of their school."[183] Owen envisioned organic relationship best revealed by a branching model. We have seen that Green had surely introduced this notion in the lectures of the 1820s. Such models were now supported by the new embryology in the immediate process of introduction to the English-speaking community from Germany. To give this concrete expression, Owen had sketched out a schematic pattern of the relationships of the animal kingdom probably sometime in early 1837 (facing).[184]

---

180   See also Owen's remarks on Geoffroy's interpretations of the mollusc-fish relationship in Lecture Two, fol. 59 below.

181   Serres had explicitly sought to define the concept of an "anatomie transcendante" in his "Anatomie transcendantes.—quatrième mémoire. Loi de symétrie et de conjugaison du système sanguin," *Annales des science naturelles* 21 (Sept. 1830): 5–49. Owen's comments bear some similarity to the cautions of Peter Mark Roget, who had previously singled out Serres for criticism on this point. See *Animal and Vegetable Physiology Considered with Reference to Natural Theology*, 2nd ed. (London: Pickering, 1837), 2, chp. 4. The second edition of this had only momentarily appeared near the date of this new redrafting. See *London Times*, 24 April 1837. Roget was probably expected to be in the lecture audience. See above, p. 54.

182   Owen Lecture, 6 May 1837, Owen Papers, RCS 42. d. 4, fol. 96. In some respects Owen would appear to make an about-face on this claim by the mid-1840s in positing the composition of a transcendental vertebrate archetype from the repetitions of a single unit. The difference between this latter position and St Hilaire's is that Owen rejected the notion of the empirical vertebra as an elementary unit. It is an *ideal*, not a material, vertebra that composes the archetype. I am indebted to Nicolaas Rupke's forthcoming "Richard Owen's Vertebrate Archetype" for deepening my understanding of Owen's theory (personal communication).

183   See Lecture Four, fol. 97.

184   Owen Papers, RCS, File L Misc. This is sketched on the rear of a memorandum dated 20 January 1837 notifying Owen of the Zoological Society Publication Committee meeting on the twenty-fourth. Owen used similar Zoological Society memoranda in late 1836 in the

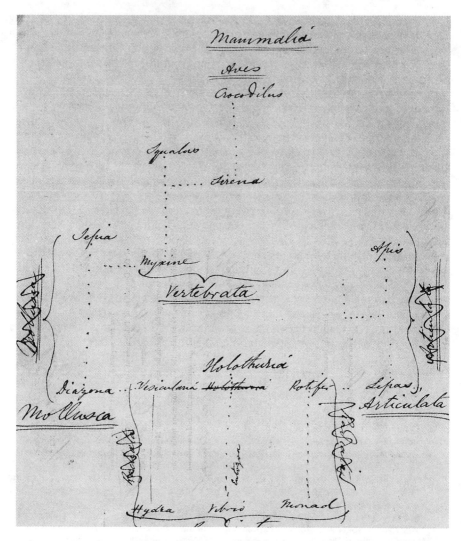

Fig. 6. Richard Owen's Diagram Of The Branching Relations Of Animal Groups, Ca. 1837. *(Courtesy President and Council, Royal College of Surgeons of England)*

It is important that such a sketch not be seen purely as a theoretical speculation. Owen was faced with a distinct practical issue. He had no liberty to alter the order of arrangement of materials in the collection. It functioned almost as a given of nature, at least as far as his own speculations were concerned. Owen's unique approach to the issues of morphology, which culminated in his theory of the archetype in the 1840s lectures, displays the interplay between theory, practice and an externally given body of phenomena that constituted a creative line of research inquiry.

Revising the lectures in more minor ways up to the moment of delivery, Owen was preparing to present these ideas to a waiting public.

D. A New Star of London Science

As the inaugural date of 2 May approached, Owen was preoccupied with the final stages of preparation, and he was understandably anxious about the event, given the advance build-up the series had received. Edward Stanley's opening series of six lectures on surgery had commenced on 18 April, and was due to end on Saturday the 29th. Owen was still concerned about the length of his lectures, and on the evening of 1 May, he read the first to Caroline and decided to shorten it even further.[185]

The new lecture theater in the College, now capable of holding over four hundred persons, was packed to overflowing at 4 pm on Tuesday, 2 May for Owen's opening presentation. Caroline's diary describes that day from the standpoint of a dutiful wife, excluded by gender from admission to the theatre:

> May 2. — So busy all the morning, had hardly time to be nervous, luckily for me. R. robed in the drawing-room and took some egg and wine before going into the theatre. He then went in and left me. At 5 'O' clock a great noise of clapping made me jump, for I timed the lecture to last a quarter of an hour longer, but R., it seems, cut it short rather than tire Sir Astley Cooper too much. All went off as well as even I could wish. The theatre crammed, and there were many who could not get places. R. was more collected than he or I ever supposed, and gave this awful first lecture almost to his own satisfaction! We sat down to a large party to dinner. Mr. Langshaw and R. afterwards played two of Corelli's sonatas. [186]

Owen's lectures of Thursday the fourth and Saturday the sixth completed the long historical survey of comparative anatomy up through the work of John Hunter. Presentation of this historical framework was a novel event in the Hunterian series, and Owen had been anxious about its reception. Therefore, its

---

(contd.)
    drafting of two pages of the sixth lecture of 13 May; see "Analysis of Manuscript: Lecture Six" below.
185 *Life* 1: 109. It is probably at this time that the rather hasty pencil deletions of several passages in this lecture were made. See emendations to Lectures One and Two.
186 Ibid.

apparent success increased his confidence in the series as is evident in a letter to his mother:[187]

Royal College of Surgeons
May 6th. 1837.

My dear Mother,
   Having this day delivered my third Lecture, and to an audience which has increased rather than diminished, (which I take to be a fair guarantee of my having so far afforded satisfaction to the College in my new capacity) it is with much pleasure, as well as my duty, to acquaint you of my success. The President, Sir A. Cooper, has done me the honour to attend each day, and has taken notes—but that I take to be an act of his good nature, and meant as an encouragement to the young beginner.
   It is a formal, and therefore somewhat awful affair, our Lectures—First the members & Students assemble in the Gallery and body of the Theatre—then as the Clock strikes 4 the honorary visitors who have previously congregated with the Council in the Council room are ushered down, the President, in his robes, being preceded by the Mace, which is reverently deposited on the Lecture-table by the Beadle—when lastly walks in the Professor and then—when the clock strikes 5, your obedient & affectionate Son makes his bow and exit, with a much lighter heart than <when> he entered— I am truly thankful for the health and strength which has thus far supported me through a severe trial—My Colleague, an old experienced Lecturer, found it so; and most have acknowledged the same; I trust to complete the Course, which lasts till the end of June, without greatly disappointing the expectations of those who have, (earlier than I could have myself wished it,) placed me in this sphere—Pearson Langshaw was I believe the only Townsman who witnessed my debut, and expressed afterwards his wish that you could have taken a birds-eye view of the affair.
   The subject of my opening Lecture was a Disquisition on the Knowledge of Comparative Anatomy possessed by the Ancients—and today I brought up the history of the Science to the time of John Hunter. The days of Lecture are *Tuesday, Thursday & Saturday.* We called yesterday on Mrs. Rawsthorne & saw both her & M^r. R. & family—all very well[.]
   We shall be very glad to hear of the safe arrival of dear Grace's silk—and tell her we deeply sympathize with the shock she must have sustained in witnessing what, with every preparation, is always a harrowing scene: indeed you must all have suffered from the additional anxiety and cares which a fatal illness in a large family involves. We hope, however, you are quite recovered, and are all well, and that the Summer which has, at last, visited us, has also enlivened and warmed the good old town.
   I leave domestic news to Caroline, who continues to enjoy the best of health,

---

187 R. Owen to Catherine Perrin Owen. Courtesy of the Francis Hirtzel Collection of Richard Owen Correspondence, Temple University, Vol. I, fol. 43. A partial transcription, incorrectly dated 16 May, is in *Life* 1: 111. Quoted with permission of Temple University Libraries. Punctuation as in original.

and with my kindest love to my dear Sisters and yourself. Believe me, Dear
Mother,

Your affectionate Son,

Richd *Owen*

On Tuesday the ninth, Owen began the transition to more theoretical issues,
delivering the fourth lecture, in which he developed his positions on the Cuvier-
Geoffroy debate and advocated the importance of von Baer's embryology for
determining animal relationships. This theoretical level of exposition he continued
in the crucial fifth lecture of Thursday, 11 May. In this lecture, Owen presented a
general theoretical synthesis of physiology and anatomy and outlined his intent to
explain the "law governing the longevity of species" in terms of the German
physiologist Johannes Müller's theory of the conservation of life-force, relating
species and individual longevity to the degree of organization.[188] On the thirteenth,
he then extended these issues to the examination of the differences between plant
and animal existence, and discussed the puzzling problem of the boundaries
between the two kingdoms in light of the properties of the infusoria, particularly
important following Christian Ehrenberg's landmark study with the achromatic
microscope on the internal anatomy of these minute forms.[189] Owen was forced to
conclude from his survey that a rigid boundary between the plants and animals
was indeed difficult to sustain.

Owen's success as a lecturer convinced him that he could rely less and less on
reading directly from a text,[190] and by 16 May, he was confident enough to deliver
the seventh lecture, on blood, "entirely without notes," illustrating points with the
assistance of diagrams.[191]

With completion of the discussion of blood, Owen concluded the theoretical
preamble to the lectures and was ready to begin a detailed exposition of the
Hunterian physiological series, commencing with the organs of motion and digestion, probably following the order of exposition that he and Clift had recently followed in the first volume of the *Descriptive and Illustrated Catalogue*, and moving
on eventually to an elaboration of the displays on the teeth similar to those he

---

188 For analysis of this see Sloan, "Darwin, Vital Matter."

189 C. Ehrenberg, *Die Infusorienthierchen* (Berlin, 1834).

190 Owen's success as a lecturer is even praised at this point by the normally hostile *London
Medical and Surgical Journal*. Reporting on the new College facilities, it comments that
"without a desire to make an invidious comparison, the lectures which Mr. Owen has already
delivered form a striking contrast with those morbid ossific compositions which immediately
preceded them" (ibid., N.S. 1 (1837), 175). Report is dated 13 May. The comparison is being
made with Edward Stanley's lectures that opened the 1837 series.

191 *Life* 1: 110. The dating of the lectures given in the diary excerpts in the *Life* is difficult to
reconcile with the dating I am proposing. Some of this may be due to careless editing by R.
S. Owen (see misdating of letter above, n. 187). Caroline's reported diary entry on the
eleventh noting this as the date the *second* lecture was delivered is surely in error, and is
inconsistent with the dating in Owen's hand on the cover sheet of the lecture.

would later publish in the *Odontography*.[192]

The lecture series proceeded without interruption on its Tuesday-Thursday-Saturday schedule throughout May and most of June, continuing to attract large audiences and enthusiastic response.[193] The strain of which Owen speaks in the letter to his mother of 6 May was, however, intensifying. Caroline writes in her diary on the thirtieth that Owen was "very queer on coming back from lecture; if he is not better by next lecture I shall try and get it postponed."[194] By 8 June, she reported alarmingly that "R. was scarcely well enough to lecture."[195]

In spite of these stresses, Owen maintained his strength sufficiently to impress important members of the audience with his abilities and his grasp of the material. On 19 June, Owen was informed of his election to the prestigious position of Fullerian Professor of Comparative Anatomy and Physiology at the Royal Institution, an honor he felt it necessary to decline on the twenty-ninth.[196] Owen had clearly transformed the Hunterian series into one of the major events of London scientific life.

But the physical toll on Owen was considerable, and he was near exhaustion as the series drew to a close. The final presentation was made on the twenty-seventh, and three days later Owen left for a short holiday by boat to the seaside village of Ramsgate, returning to London on 4 July. [197]

It was in the following week that Owen and Darwin had a long meeting at the College to work on the *Beagle* fossils.[198] This crucial meeting, near the date of the commencement of Darwin's first transmutation ("B") notebook, is of particular interest. Darwin's immediate concern was to clarify the taxonomic status of his South American fossils, an issue that Owen had discussed publicly at the Geological Society in April and to a limited extent in the just-concluded lecture series. Darwin had returned to this issue in his paper to the Geological Society on 3 May. Both Owen and Darwin were, therefore, seeking to find a secondary cause

---

192 *DIC* 1 (1833). See note 138 above.
193 The twenty-second lecture, scheduled for Tuesday, 20 June, was cancelled due to the national mourning for the death of King William IV (Clift Diary, Clift Papers RCS, 276.9.1–33).
194 *Life* 1: 112.
195 Ibid.
196 *Life* 1: 113–114. The ever-present catalogue was surely a major cause adding to these pressures. The Curatorial Board was concerned that the lectures had already taken too much time from this more urgent task (Curator's Minutes 5, entry for 21 June 1837). The arrangement of the displays in the new museum had also been left unfinished by the interruption required by the lectures and was seen in urgent need of attention by Owen and Clift.
197 Clift Diary, entry for 30 June 1837: "Mr. Owen left London bridge at 10 o'Clock A.M. by the 'City of London' Steamer for Ramsgate till Tuesday next."
198 Clift reports that Owen departed for Lancaster "at 6 A.M" on Friday, 14 July, for a vacation trip of over a month (ibid.). Darwin had returned to London from a trip to Shrewsbury on Thursday evening, 6 July (CD to W. D. Fox, 7 July, *Darwin Correspondence* 2: 29). Consequently, the July meeting between Owen and Darwin to work on the fossils, described in the letter to Charles Lyell of 30 July (*ibid.* 2: 32), could only have taken place in the period of 7–13 July.

of the extinction of species, especially those of the large South American mammals. For Owen, it was to be explained by his theory, borrowed from Johannes Müller, of the conservation of the *Lebenskraft*. Only two months previously Owen had expressed this idea in the lecture of 11 May:

> Each species of Organism is self-existent from the period of its Creation; the Individuals have a transient existence, and some perish: but the Species long remains; and its duration was once thought to be necessarily coeval with that of the existing Sphere of its actions:—but the Species must disappear with the extermination of the reproductive Individuals, for the Genus has no power to reproduce the Species, nor the Family the Genus.
>
> The history of the revolutions of the crust of the Earth teaches us that many species have become extinct;—some belonging to existing Genera, and others to Genera which are no longer represented by living beings. The study of Geology shows us, moreover, that the beings whose organized remains are discovered, have not all existed contemporaneously. In their succession they manifest a gradual approach to the forms of the beings at present in existence;—and in the Strata of the later formations, remains of animals referrible [sic] to living species occur:— but the different organized forms which have succeeded each other do not display regularly progressive stages of complication, or perfection of Structure. Plants and animals exhibiting different degrees of complication of Structure have co-existed at different periods; their existence, and fertility, and well-being seems, as now, to have been regulated by the conditions of the then external World; and a change in those conditions disturbing the harmony of the relations necessary for the well-being of those existing plants and Animals, has caused their extermination; while new species appear on the stage, endowed with powers and forms adapted to the new conditions of the external world, but not necessarily superior in their Organization to the extinct Species which they have replaced. Thus, the higher organized Sharks and Sauroid fishes which prevailed in the antient carboniferous and secondary formations disappear, and are replaced by other & lower forms in the tertiary Strata. And the functions performed by the Chambered Cephalopods in the antient seas in which the secondary strata were deposited, are now principally assigned to trachelipods, which display a much lower Type of Molluscous organization.[199]

Darwin's thoughts on this matter were still transitional. It is evident from his paper to the Geological Society on 3 May that he also held to a similar view, derived from Lyell.[200] It would appear that, through the month of July, Darwin still rejected an explanation of extinction solely in terms of the action of external conditions. The problem remained an "extraordiny [sic] mystery" to him. [201] A

---

199 Owen, Lecture Five, 11 May 1837, fols. 33–35. See full text below.
200 As Darwin put this point in his paper to the Society on 3 May: "This is analogous to what has been observed in Europe and particularly insisted on by Mr. Lyell, of the longevity of the species of warm blooded animals being of shorter duration than among Molluscs." Darwin, "Sketch of the Deposits Containing Extinct Mammalia in the Neighbourhood of the Plata," summarized in Minutes of the Meetings, Geological Society Archives OM/8, 364–7. Quoted with permission of the Geological Society of London. This summary was later published in *Proceedings of the Geological Society* 2 (1833–8): 542–4.
201 Darwin to Lyell, 30 July 1837 (*Darwin Correspondence* 2: 32).

sequence of passages from Darwin's *Journal of Researches,* known to have been written in the June–July period—i.e. subsequent to Owen's important lecture of 11 May— take on particular interest:

> One is tempted to believe in such simple relations, as variation of climate and food, or introduction of enemies, or the increased numbers of other species, as the cause of the succession of races. But it may be asked whether it is probable that any such cause should have been in action during the same epoch over the whole northern hemisphere, so as to destroy the *Elephas primigenus*, on the shores of Spain, on the plains of Siberia, and in Northern America; and in a like manner, the *Bos urus*, over a range of scarcely less extent? Did such changes put a period to the life of *Mastodon anguistidens*, and of the fossil horse, both in Europe and on the Eastern slope of the Cordillera in Southern America? If they did, they must have been changes common to the whole world; such as gradual refrigeration, whether from modifications of physical geography, or from central cooling. But on this assumption, we have to struggle with the difficulty that these supposed changes, although scarcely sufficient to affect molluscous animals either in Europe or South America, yet destroyed many quadrupeds in regions now characterized by *frigid, temperate,* and *warm* climates! These cases of extinction forcibly recal [sic] the idea (I do not wish to draw any close analogy) of certain fruit-trees, which, it has been asserted, though grafted on young stems, planted in varied situations, and fertilized by the richest manures, yet at one period, have all withered away and perished. A fixed and determined length of life has in such cases been given to thousands and thousands of buds (or individual germs), although produced in long succession. Among the greater number of animals, each individual appears nearly independent of its kind; yet all of one kind may be bound together by common laws, as well as a certain number of individual buds, in the tree, or polypi in the Zoophyte.
>
> I will add one other remark. We see that whole series of animals, which have been created with peculiar kinds of organization, are confined to certain areas; and we can hardly suppose these structures are only adaptations to peculiarities of climate or country; for otherwise, animals belonging to a distinct type, and introduced by man, would not succeed so admirably, even to the extermination of the aborigines. On such grounds it does not seem a necessary conclusion, that the extinction of species, more than their creation, should exclusively depend on the nature (altered by physical changes) of their country. All that at present can be said with certainty, is that, as with the individual, so with the species, the hour of life has run its course, and is spent. 202

We may compare this to Owen's similar point in the lecture of 11 May:

---

202 Darwin, *Journal of Researches,* vol. 3 of *Narrative of the Voyage of H. M. S. Adventure and Beagle,* ed. Robert Fitzroy (London, 1839; reprinted New York: AMS Press, 1964), 211–212, hereafter cited as *JR.* Darwin finished the draft of *JR* on approximately 28 June, after which he took an eight-day holiday to Shrewsbury (CD to W. D. Fox, 7 July 1837, *Darwin Correspondence* 2: 29). He continued to revise and work on this through July after his return (CD to J.S. Henslow, 13 July 1837, ibid., 31). It was at the July meeting that he and Owen decided there were five species of fossil Edentata in the Bahia Blanca materials (CD to Lyell, ibid., 32). Reference to this revised number of species and a direct reference to a meeting with Owen at the College appear in the published text of the *JR* (96; see also 149), indicating a revision of the manuscript of the *JR* following this meeting.

In the germ the organizing energy exists in its state of greatest concentration:—the developing power is at a maximum. But when the organism is carried beyond youth, we no longer perceive a simple condition of the whole with energy undivided, but a complex whole with energies variously distributed. This distribution or division of the organizing energy of the whole is accompanied with a corresponding impairment of incitability. Development therefore being at an end, and the general external stimuli of life ceasing at length to be responded to, Death ensues, and the continuance of Life in the Species depends on the previous generation of a germ in which the undivided energy developes a new Organism.[203]

These texts illuminate the ways in which Darwin and Owen were mutually struggling to find an explanation for extinction with remarkably similar theories. For Owen, species could perish by the wearing out of their contained vital energy in the germ, but never transform in response to changing environmental conditions. This same vital energy was expended as the organism developed embryologically from its primordial germ to its appropriate level of organization. Darwin embraced a similar thesis in the *Journal of Researches*, giving an explanation for Lyell's observation of the longer duration of molluscan in comparison to mammalian species.[204] Both saw branching and diverging relations as the key to the issue of affinity, and both were concerned at this time with the internal rather than external causes of species life and death.

From this common point in the spring of 1837, the two naturalists pursued divergent paths. Owen was not the source of Darwin's views in this period, but instead was the author of a coherent theoretical framework incorporating the latest Continental and domestic biomedical science against which Darwin could react. Relations between Darwin and Owen would remain close into the late 1840s. But they were passing like two ships moving in opposite directions, directions which would eventually lead them into nearly hostile relations after the publication of the *Origin* in 1859. The "B Notebook" would soon reject internal limitations on species life and death, and endorse a thesis of unlimited species change in response to environmental conditions. Owen would remain concerned with the demands of physiological theory and the interplay between life-force and organization. Owen's apparent concern to hold together relations of inquiries similar to those outlined in Green's program for a "description," "history" and

---

203 Owen, Lecture Five, fols. 51–2. See full text below.
204 M. J. S. Hodge (see note 1 above) has offered penetrating criticisms of arguments I have advanced earlier in Sloan, "Darwin, Vital Matter." Hodge has suggested that Darwin's views are consistent with Lyell's arguments in the *Principles* and show no impact of Owen's notions. I would suggest by reply that Darwin seems to hold to no explanatory theory of species extinction in his presentation to the Geological Society on 3 May, and only mentions this theory in his remarks in the *JR* in passages that seem to be inserted as revisions in the manuscript. Hence they would follow upon Owen's lecture presentation of the theory of the conserving life-force on 11 May, including his reference to the conservation of this force contained in "germs." The interplay of these issues exactly concurrent with the combined study by Owen and Darwin on the fossils in this period is too close for coincidence. See also passages in "B Notebook," *Darwin's Notebooks*, 175–7. This is not to claim a simple derivation of Darwin's views from Owen, but an interaction between them on these issues.

"physiology" of nature gives us a clue to his intellectual enterprise, which would otherwise appear fragmented and incoherent. With this Owen could retain his physiological and functional perspective, appealing to the explanatory role of vital forces in the development both of the individual and of the species. The archetype concept Owen would develop in detail after 1845 presented this concept both as a transcendental organizational plan and as an immanent law, manifested in the action of causal forces acting dialectically with matter to bring about an increase of organization.[205]

This difference in explanatory structure would also serve to separate Owen and Darwin on the complex question of natural teleology. Little in Owen's thought connects him with the common picture of British natural theology and Bridgewater Treatise design-argument biology. Owen's teleological view of nature was in many respects that of Kant and his interpreters, an immanent teleology that manifested itself in a complex historical process. Nature is a system of natural forces, but a system working toward the realization of an immanent goal, the goal of human consciousness and moral freedom. As Darwin moved increasingly to an adaptational, rather than an internal-force, view of species life and death after 1837, teleology of an external variety would become more and more problematic in his thought and the arguments of the tradition of British natural theology would become an easy target for his natural selectionist perspective.

Owen's positions on the major questions of biology were not fully to be engaged by Darwin in the *Origin*. The critical German tradition Owen had first encountered in the Green lectures was concerned to explore a different solution to the problems of organic form, function and relationship than the one pursued by Darwin. Whether Darwin was among the four-hundred plus auditors at these lectures in the months before he opened his first reflections on the transformist question remains impossible to answer. Whatever is to be the conclusion on the character and importance of the relationship between Darwin and Owen, deeper insight into Owen's contribution to nineteenth-century science is now possible.

Phillip Reid Sloan

---

205 R. Owen, "On the Archetype and Homologies of the Vertebrate Skeleton," *Reports of the British Association for the Advancement of Science* (1846), 169–340, esp. 339–40. See the discussions of Owen's archetypal concept in H. Haupt, "Das Homologieprinzip bei Richard Owen," *Sudhoffs Archiv* 28 (1935): 143–228; R. MacLeod, "Evolutionism and Richard Owen, 1830–1868," *Isis* 56 (1965): 259–80; and most recently E. Richards, "A Question of Property Rights: Richard Owen's Evolutionism Reassessed," *British Journal for the History of Science* 20 (1987): 129–71. This is all to be updated by Nicolaas Rupke's forthcoming "Richard Owen's Vertebrate Archetype" (MS in press).

PART TWO

The 1837 Hunterian Lectures

Editorial Methods

The text reproduced below has been based upon the final copies of Owen's first series of lectures. The copy presents Owen's final version as it was intended for delivery, and incorporates all of Owen's final revisions.

With this goal in mind, several decisions needed to be made in the editing of these lectures. In common with several recent text editors, I have found the methods of Fredson Bowers ("Transcription of Manuscripts: the Record of Variants," *Studies in Bibliography* **29** (1976), 212–64) and G. Thomas Tanselle ("The Editing of Historical Documents," ibid. **31** (1978), 1–56) most useful as basic guides to the complex problems presented by this set of manuscripts. These methods are designed to produce a "clear text" free of attempts to reproduce cross-outs, interlineations and other revisions, and maintain all spellings (and misspellings), capitalizations and punctuation exactly as found in the original manuscripts. Hence evident spelling errors, abnormal capitalizations, or unusual punctuation should be assumed to be as encountered in the original copy text.

In use I have also made some important revisions of the Bowers conventions to render these methods less cumbersome in practice. The awkward line numbering method of recording emendations has been replaced by the use of footnotes at the bottom of the page keyed to small superscript numbers. This inserts editorial intrusion into the transcription, but prevents the need for referring continuously to appendices to determine the emendations.

In use of this convention, one issue must be clarified. The small superscript numbers will appear at the site of the emendations described in the notes. This will produce non-conventional use of the superscript numbers. For example, the first note on the first page of Lecture One 'audience[1].' indicates that the word was interlined to replace the earlier 'class' *before* the final period. Conversely, the interlineation of 'as it now exists—[10]' on folio four of Lecture One (p. 80) is indicated by this use of superscripts to be immediately prior to the phrase 'so that'.

Page breaks of the manuscripts are indicated by the folio numbers in the left margins, and the page break, when it occurs within the line, is designated by the editorial insertion of an '[/]' symbol. Where this symbol does not appear, the new page can be assumed to commence at the beginning of the line indicated by the folio number. On a few occasions, where it was immediately necessary for grammatical sense, I made minor grammatical revisions. These have been indicated in the footnote emendations by *[ed.]*. In recording the emendations I have followed the Bowers conventions. The footnote gives an abbreviated line reading to left of the bracket "]" symbol, followed by a descriptive summary (*sigla*) of the emendation in italics to the right. The emended text is given in single quotes. Where the emendation contains a substantial deletion or replacement, the text in the note is reproduced in a form that will allow it to be read directly as it appears without additional footnotes. Internal emendations within these quoted passages in the sigla are given by editorial description. The "*" symbol initiating a phrase in these secondary emendations, followed by the editorial description in "{ }" brackets, will indicate an internal emendation. Tertiary or quaternary internal emendations might also be indicated by "**" or "***" symbols if necessary. For example, in recording the erasure on manuscript folio three of Lecture One (p. 80), the erased phrase 'by a member of the College' is indicated by these conventions to contain the further deletion of 'some'. Paragraph breaks in these quoted passages are indicated by a "{¶}" symbol. Page breaks are indicated in these notes by the [| *fol*. n] symbol in the text.

On occasion there are substantial differences between Owen's original autograph draft manuscripts and the final recopied delivery versions used as the copy text. A few of the more important differences are reproduced in the endnotes to the lectures.

I have made the following silent alterations: William Clift, the copyist, typically repeats the initial word, or on occasion the first two or three words beginning the next page at the bottom of the previous page. I have deleted these repetitions, recording the text as it begins each new page unless there are significant differences. I also have not indicated hyphenations of words on line breaks. In cases where there may be uncertainty concerning the intention of the hyphenation it has been recorded.

I have not given a record of pen rests. Underlined words have been uniformly rendered in italics. For quotations in the text, Clift typically places double quotation marks along the left margin of each line of the quoted passage without otherwise setting off the text. I have reduced and offset these passages.

Content notes and references are located for convenience as separate endnotes at the end of the transcription of each full manuscript. This will mean, for example, that the notes for Lectures One and Two and for Three and Four are found at the ends of Lectures Two and Four, respectively, since these lectures were formed by a late splitting of the original manuscript. Those for the remaining lectures appear at the ends of these manuscripts. These notes have been keyed to the folio and line numbers of the manuscript pages. For example, the note listed as 1–9 for Lecture One (p. 126) is a note to folio one, line nine of the first lecture.

Lectures One and Two
Analysis of the Manuscript

The copy text for the first two lectures is based upon the manuscript in the Owen Archives at the British Museum of Natural History, bound as the fifth item in the volume O. C. 38. These are copied in ink in William Clift's hand on large folded folio sheets of a high-quality paper measuring 34.2 x 40 cm. that has been folded to give individual pages. These large double sheets are uniformly on a "J Bune 1832" paper. Annotations have then been made in both pencil and ink. Substantial revisions have typically been added by means of a facing replacement section written by Owen in ink on the back of the blank facing verso of the succeeding page. Occasionally notes, queries or comments have been appended by William Clift or W. H. Bennett with replies by Owen.

This lecture was originally intended for a single delivery, with the recopy completed by William Clift on 22 November 1836, a date attested to by Clift's pencil note at the end. It was then read to Pearson Langshaw on 27 November, and after finding that it required two and one-half hours to read, Owen decided to cut this into two introductory lectures. This later division is confirmed by the addition in Owen's hand of an "& 2" to the original title "Lecture 1" and Owen has then added in a different pen the dates "Delvd May 2 & 4th" to the cover page. The division point between the two lectures is not clearly indicated, but would seem best located at the break on page 40 of the manuscript where the division has been made in this transcription.

The holograph draft for this lecture is located in the Owen Papers at the College of Surgeons (MS 67.b.12.A). This manuscript is written on small sheets of lightweight vellum measuring generally 21 x 26 cm. The first two pages of this manuscript are on a smaller "J Whatman 1836" paper. The body of the text is then written on "J Green & Son 1836." No other indications of dating of this draft are found on the manuscript. This draft is often heavily reworked by Owen and the revisions he has made are generally followed in Clift's recopy of the delivery text. The BMNH manuscript has deleted, subsequent to its recopy by Clift, pages 4–11 corresponding to folia 5–11 of the Royal College draft. This material summarizes previous lectures in the Hunterian series. The deleted material, as it appeared in the original RCS draft, is reproduced in the endnotes to folio 12–1 on page 126 below.

Fig. 7. John Flint South Delivering The Hunterian Oration In 1844 As Illustrated In *Pictorial Times*, London, 17 February 1844. This is the only complete illustration known of the interior of the new Lecture Theater as it would have existed at the time of Owen's lectures. The seating of the President and other dignitaries of the College, the placement of the Mace and other details alluded to by Owen in his lectures are clearly illustrated. Description of this event in 1844 indicated that Benjamin Collins Brodie is depicted in the Presidential chair, and in attendance were Owen, Joseph Henry Green, Peter Mark Roget and several other important dignitaries. (*Courtesy of Library, Royal College of Surgeons of England*)

Lecture One
2 May 1837

1   Mr. President

Sir

    In commencing the Course of Lectures allotted to me as Hunterian Professor to this College, my mind is deeply and painfully impressed with a sense of the disproportion of its faculties to the adequate discharge of the important duties of that office. While, at the same time, the honour of your choice, and my peculiar obligations to this College impel me to exert my utmost endeavours to satisfy your desires and expectations.

    Yet in attempting to address the illustrious circle which now surrounds me 2   I confess that many circumstances, of which [/] each is apt to disturb faculties of the human Mind, present themselves in a crowd before my eyes;—for whether I consider the dignity of the place, the learning of my audience, the multitude of my superiors in Science, rank, and age; or lastly, my little talents in the arts of demonstration; all these circumstances combine to throw me into no small confusion. For if the most eloquent and experienced Lecturers have trembled and been embarrassed in addressing an Audience constituted like the present, what must I feel who do not participate in their advantages of frequent practice and familiarity in addressing a numerous audience[1], nor possess either from art or nature a readiness or elegance of speech.

    A sense of duty,—the hope of increasing the sphere of my utility in the College—alone have prevented me from shrinking from the anxious honour. And when I remember your sanction and approval of the share which I have taken with my Senior officer and friend Mr. Clift, in the published exposition of the Collection—this gives me ground to hope for that favour and indulgence which you never refuse to those who have addressed you on these occasions.[2]

3   I have passed some years in the Museum whose treasures these Lectures

---

1   audience] *pencil interline above* 'class'*without deletion.*
2   A sense. . .occasions.] *RO insertion in dark ink from facing fol. 3ᵛ with vertical pencil deletion on fol. 2 of* 'Some considerations which lead me to hope I may be useful in this present sphere, have alone prevented me from [*top of fol. 3*] shrinking from the anxious honour.—'.

are destined to explain, and have constantly laboured to acquire such a knowledge of the preparations as would enable me to explain them with advantage to the Student of Comparative Anatomy; and those leisure hours which were available for the purposes of study, have been, I can conscientiously affirm, devoted chiefly to the advancement of that Science.

These circumstances I am induced to urge solely as affording me grounds to hope for that favour and indulgence which you never refuse to those who have addressed you on these occasions;—and thus, I doubt not, but that however deficient I may be from want of talent or want of exercise, I shall not wholly fail of the ends I have in view.

3[1]   The duties, Sir, of the Hunterian [/] Professor of Anatomy in the College are defined in the Parliamentary Trust Deed which placed the Hunterian Collection under your control, to be: to illustrate that collection in a course of 24 lectures to be delivered annually in the theatre.[2]

In approaching the consideration of those duties, and in endeavouring to form a just idea of what I have undertaken, I have first carefully examined the share which my Predecessors have already accomplished in the exposition

4   of[3] the scientific treasures accumulated by the [/] illustrious founder of the Collection.[4]

I find that there is, scarcely a single Series of the preparations[5] which has not been brought before you in the Theatre of the College, where they have been described and commented on, with more or less detail; and these demonstrations have been associated with enunciation of some of[6] the most important discoveries which have graced the history of the progress of Physiological Science during the present Century.

In[7] going over the same ground which has been trodden by my able and accomplished predecessors—since I cannot bring before you the fruits of the strides of previous years tending to the improvement of any particular department in Physiology—my object will be to convey a[8] precise and adequate[9] knowledge of the scope & nature of the contents of the museum as it now exists—[10]so that these Lectures may serve I hope[11] as a *Catalogue*

---

1   3] *fol. number is repeated on this page with erased pencil* 'The duties, Sir, of the Hunterian' *followed by three erased words at top of page.*
2   to illustrate . . .theatre.] 'annually' *transposed from after* 'theatre' *followed by erasure of* 'by *some [del]* a member of the College' *and 1/3 page blank* .
3   'the exposition of'] *inserted over erasure.*
4   Collection;] *followed by erasure of pencil* 'I find that there is' *with* 'I find that there is,' *written in same ink as facing insertion. Period inserted to replace comma [ed.].*
5   preparations] *before deletion* 'of the Hunterian Collection'.
6   some of] *interlined with caret.*
7   In] *paragraph break inserted [ed.].*
8   a] *over deletion* of 'as'.
9   adequate] *followed by deletion of* 'a'.
10   as. . .exists—] *interlined in ink with caret.*
11   I hope] *interlined with caret.*

raisonnée of the Collection.

Yet[1] when I remember how well, and by what Masters in Physiological Science the Hunterian Labours have been explained from this chair, I can hardly hope to interest you by the more humble duties which I feel myself limited to endeavour to fulfil.[2]

12[3]   I[4] am willing however[5] to believe that a series of Lectures, the aim of which is strictly and consecutively[6] to explain the Hunterian Collection may be acceptable to many[7] :—

I need scarcely observe to you, Sir, that[8] The Hunterian Collection consists of a combination of several distinct Series of Preparations[9].

—There is 1st[10] a Department of Natural History;—adapted to convey a knowledge of the principal external and internal characters of the different Zoological Divisions of the Animal Kingdom.

—2d. A series of Monstrosities and Malformations, which may hereafter be

13   made subservient to the illustration of a course of [/] Lectures on the highly interesting[11] subject of Teratology.

—3rd. An extensive Collection of Morbid Anatomy, both Human and Comparative, arranged by M[r]. Hunter himself on a plan which agrees with, and serves to explain, some of his peculiar opinions and views on that important branch of Medical Science.

---

1   Yet] *above deleted* 'And' *and double space to indicate paragraph.*

2   In going. . .endeavour to fulfil [*sic*]] *All inserted from fol. 12ᵛ facing replacing deleted section on fol. 4* 'Of that grand department of the Collection, termed, from its object, Physiological, the preparations relating to the Mechanical Functions of the Animal Frame have been illustrated by the Lectures of Sir Everard Home, and Sir Charles Bell:—the various organs of the Digestive System have received ample demonstrations from Sir Everard Home and Sir Astley Cooper;— and following the [l *fol.* 4] the *scope and nature of its contents —{deletion on fol. 12ᵛ}.*

3   12] *manuscript pagination skips from four to twelve at this point. See end notes.*

4   I] *inserted in margin. Paragraph break inserted [ed.].*

5   however] *interlined with caret.*

6   strictly and consecutively] *interlined over erasure with* 'and' *interlined with caret.*

7   many] *before colon and deletion of* 'and compensate by their utility for what may be deficient in novelty, either of facts or views. I propose, therefore, to make these Lectures strictly subservient, in the first instance, to an exposition of the Hunterian Collection, as it was arranged by its* illustrious {*above in pencil* 'Immortal'} Founder, and as it now exists.' *then on facing fol. 13ᵛ in Owen's hand, deleted* 'and that should such demonstrations be available to the visitor of the Museum in acquiring a more precise and adequate knowledge of the scope and nature of its contents,—they might compensate by their utility for *whatever might dev {crossed out} any want of novelty either of facts or views.'.

8   I. . .that] *interlined.*

9   Series. . .Preparations] *after deletion of* 'Departments of' *with* 'S' *over lower case before deletion* of 'illustrative of those different branches of Science, which are either essential to, or dignify and adorn, the Character of the British Surgeon.'.

10   'There is 1st'] *transposed with lead line from after* '1st'.

11   interesting] *before deletion of* 'and important'.

—4thly. A series of Fossil remains, remarkable for its extent, and value[1] considering that it was formed at a period when the importance of these subjects had hardly begun to be appreciated by Men of Science.

—5thly. An extensive collection of the Osteology of Recent or Existing species of Animals:—

—And lastly, the far celebrated Physiological Series of Comparative Anatomy in the Gallery of the Museum, which may be regarded as the essential part and feature of the Hunterian Collection[2].

It is with the Physiological Series that I propose to commence, in these Lectures, the descriptions of the Hunterian Museum.

14   Those beautiful preparations, which we admire in the gallery of our Museum, form the book which contains the record of John Hunter's labours in the philosophy of Animal Life, and Organization[3]. Would that I could bring the requisite powers to the task of perusing this book; so as to comprehend the full and deep meaning of the Author, which every line might, and perhaps ultimately will, unfold!

Until the scope and full[4] extent of Hunter's[5] labours shall be made manifest to the World, our great Physiologist will[6] continue to hold a false position in the History of Comparative Anatomy.

I know not, however, whether his Character has suffered more in the estimation of Continental Anatomists from ignorance of his Achievements in Science, or from the too-zealous ascription, by Friends & admirers, to him, of Opinions and facts[7] to which he had no Title as the Original Discoverer.

Both causes[8], have contributed, and still tend to prevent his assuming his

15   true position in the history of Science. Yet, independently of our [/] interest in the Character of the Man who has contributed so much to raise this College to the rank which it now holds among[9] the truly[10] Scientific Communities of Europe:

---

1 and value] *interlined with caret.*
2 Collection.] *line across 1/3 page and note on 14ᵛ facing by WC in pencil:* 'Great number of laborious and valuable drawings and pictures—&c—&c—of evanescent appearances,—Casts—nearly a thousand.'.
3 Organization.] *pencil deletion of insertion on 15ᵛ facing page:* 'The student of anatomy cannot fail to acquire many valuable ideas even from a cursory examination of these preparations,'.
4 full] *interlined with caret.*
5 Hunter's] *interlined above deleted* 'his'.
6 our . . .will] *interlined to replace deletion* 'Mʳ. Hunter must'.
7 facts] *interlined above deleted* 'Discoveries'.
8 causes] *followed by deletion in pencil of* 'but the latter I am inclined to believe, more especially'.
9 has. . . among] *insertion from facing 14ᵛ with deletion of* 'mainly' *after* 'has' *to replace deleted* 'this [interlined in pencil] made our College what it is,—one among' .
10 truly] *interlined in pencil above deleted* 'truly'.

—Independently,[1] of our personal sympathy in all that relates to Hunter, it must be interesting to us in an abstract scientific point of view, to ascertain the true relation which this Physiologist bears to those other gifted Mortals for whom Meditation was alike a want; and who, directing their energies into similar channels of inquiry, sought the true Laws of Nature in the contemplation of her Works.

I have thought, therefore, that before entering upon an exposition of the characteristic labours and achievements of Hunter[2] a brief survey[3] of the progress of Comparative Anatomy and Physiology considered with especial reference to the nature and amount of the knowledge already acquired on those points which are regarded as peculiarly Hunterian Discoveries and Doctrines, would be a subject both suitable to the present occasion, and to the Office I have the honour to[4] be engaged in by your will and suffrages[5].

16   Comparative Anatomy is that branch of Natural Science which treats of the Organization of Animals, and which considers it with reference to the Peculiarities of the Human Structure, or as it illustrates[6] the functions, the laws of development, the mutual dependencies and coexistences of the different organs:—or as it indicates[7] the natural[8] affinities of Animals, one to another:—or, lastly, as it elucidates[9] the doctrine of final causes.—

If we penetrate into remote antiquity for the Origin of Comparative Anatomy, we shall find this, like other branches of Human Knowledge at first indicated[10] by vague, obscure, and scattered allusions.

From the earliest periods in the History of Man we find him waging unavoidable[11] war with[12] the wild animals with which he was surrounded;—and their slaughter, whether for food or in self-defence, must have forced upon his notice some rude conceptions of their internal structure;—especially of the viscera of the larger[13] cavities.

17   The accidents of the chace, the wounds, and slaughter[14] of the Battle-Field must have exposed to Man's observation numerous proofs of analogous structures in his own frame:— And the casualties of Starvation, and disease, and violence, causing the death of both beasts and men in the fields, would

---

1   Independently,] *before deleted* ' I say'.
2   before. . .Hunter] *interlined with caret with* 'characteristic' *interlined above deletion of* 'principal'.
3   survey] *interlined above deleted* 'sketch'.
4   I. . .honour to] *interlined with caret*.
5   suffrages] *followed by line across 1/3 page* .
6   illustrates] *interlined above deleted* 'relates to'.
7   indicates] *interlined above deletion of* 'throws light on'.
8   natural] *interlined with caret*.
9   elucidates] *over erasure and before deleted* 'to'.
10 indicated] *before deleted* 'adverted to'.
11 unavoidable] *interlined above deletion of* 'a necessary'.
12 with] *interlined above deletion of* 'on'.
13 larger] *below* 'cavities'.
14 slaughter] *after deletion of* 'mutual'.

sometimes expose to view whole Skeletons.

The very early origin of the Sacrificial mode of Worship, and the practice of deducing Auguries from the position, and healthy or morbid appearance of the Viscera of the Victims, presupposes in the Pagan Priests, a considerable practical acquaintance with Comparative Splanchnology:—While those of Egypt, who were the depositories of the[1] Mysteries of Isis, had additional and extensive opportunities of observing the Anatomical Structure of Animals in the practice of embalming the various species which were the objects of their worship, or reverential regard.

But let me not be misunderstood;—I would not found the Antiquity of our
18   Science on the vague and feeble notions forced upon Men in the ordinary occupations or casualties of Life:—nor regard that knowledge as Science which was acquired during a superstitious epoch of the history of the Human mind:—which was shut up in Temples and cultivated solely by the Priests who made it[2] a Mystery to their votaries, and presented it to them in an Emblematic form.

When Philosophy, whose germ had been founded in Egypt, began, after a long interval, to be developed in Greece;—when the Sciences were intirely separated from Superstitious rites and no longer cultivated by Magi and Haruspices, but by Sages who communicated the results of their researches without reserve or disguise;—at that epoch only may we begin to trace with satisfaction the progress of Man's knowledge of the organization of his own frame, and of[3] the lower[4] animals.

To investigate this part of the early history of Zootomy with advantage, as well as pleasure, it must be connected with a consideration of the abstract principles which guided the Greek Philosophers in the pursuit of knowledge and we must trace the progress of the science with reference to the influence which these principles, or philosophical creeds, exercised over it.[5]
19   The absence of the principle of the[6] division of Labour in the acquisition of knowledge[7] was the prevailing characteristic of the Grecian Epoch of Philosophy;—and but little progress can be looked for in a Science of pure Observation at a period when every Philosopher was at the same time a

---

1   of the] *deleting repeated* 'of the' *[ed.]*.
2   it] *interlined with caret.*
3   of] *interlined with caret before deleted* 'that of'.
4   the lower] *insertion over erasure.*
5   To. . . over it.] *insertion from facing fol. 19ᵛ to replace* 'This progress, we shall see, was much influenced by the abstract [| *fol.* 18] principles which guided the Greek Philosophers in the pursuit of knowledge—' *with* 'this part of', 'be' 'we must be' *and* 'reference to' *interlined with carets and deletion of* 'and trace the connexion of the various success with which the Sciences of observation were cultivated with the influence exercised over' *after* 'knowledge'.
6   The absence. . .of the] *marginal insertion with* 'the principle of the' *interlined with caret over deleted* 'a'. *Paragraph break inserted [ed.].*
7   the acquisition of knowledge] *interlined below deletion of* 'that pursuit'.

Metaphysician, a Moralist,[1] a Naturalist and[2] a Geometrician[3].

One glorious exception[4], we find in Aristotle, whose universal and commanding Intellect raised every branch of Science which it surveyed, to a height of perfection unknown before.

He first defined the objects of the different Sciences, and assigned to each part its natural limits.

But unfortunately[5] he left no successors capable of appreciating, and profiting by, the System of Division which he had marked out;—and it is not the fault of the Stagyrite that the foundation of the Sciences, on well regulated and divided labours should date but three Centuries back.[6]

20   The astonishing extent to which this Philosopher had carried his researches into the history and organization of animals lends an interest which the subject would otherwise hardly possess[7] to the consideration of the amount of knowledge of Comparative Anatomy acquired by the Greeks before the time of Aristotle. And here we find[8] the different Sects contributing their proportion of facts and inferences to the general stock of knowledge exactly[9] according as their peculiar Philosophy[10] approximated to the present recieved[11] acknowledged doctrines of the power and nature of the Human Understanding.

To a Metaphysician[12] who had succeeded in persuading himself that all he saw and felt around him was ideal, and that all sensible Nature was an illusion; the external World could present few attractions. He who with Parmenides[13] believed that general ideas in man are not formed by means of Abstraction, but that they are a remembrance of those the mind possessed when it was united to the divine intelligence would carefully[14] avoid a debasing intimacy with matter, and[15] would seek in the secluded recesses of his study to regain by Meditation what the Soul had lost in the process of its union with gross body.

Whatever progress might be made in Metaphysics, it is not therefore from
21   the followers of the Elean or Academic [/] sects that we derive the Germ of

---

1   Moralist,] *comma inserted [ed.]*.
2   a Naturalist and] *interlined with caret.*
3   Geometrician] *before deleted* 'a Naturalist, and a Physician.'.
4   exception,] *before deleted* 'indeed'.
5   But unfortunately] 'But' *added before line and* 'Unfortunately'*altered to lower case [ed.]*.
6   back.] *before deletion of* 'Returning' *at bottom of page.*
7   The... possess] *interlined at top of page with caret above erased* 'Returning'.
8   And...find] *interlined with caret above deleted* 'we find'.
9   exactly] *interlined with caret.*
10   Philosophy] *interlined above deleted* 'dogmas'.
11   present recieved [*sic*]] *interlined with caret.*
12   Metaphysican] *interlined below deletion of* 'Philosopher'.
13   with Parmenides] *interlined with caret.*
14   carefully] *interlined above deletion of* 'naturally'.
15   and] *interlined above deleted* 'but'.

our Physiological Doctrines.[1]

We detect the first appearances of a scientific knowledge of Comparative Anatomy in the disciples of Pythagoras. This Philosopher was born at Samos about six centuries before the Christian Æra[2], and he appears to have deduced the principle that it was possible to estimate all the powers and magnitudes of Nature in numbers, and thus render them comparable, and capable of being submitted to calculation.

As this Idea is the same which at the present day serves as a basis to all Mathematical Physics, it was natural to expect that some of the followers of a Man who had given them such an example of the healthy exercise of the human Intellect,—could not fail to elicit some truths in Natural Science.

Alcmaeon appears to have been the first of those who engaged in Anatomical[3] researches on Animals. He made observations on the formation of the Embryo;—he said that the head was first formed, probably from observing that at the early [/] period of the fœtal life it is proportionally of very great size. He thought that the foetus was nourished by the Skin, and with a just perception of physiological analogies, compared the period of puberty in Man with the Flowering of Plants.

Empedocles, another Pythagorean, extended his observations on Embryology, and on the Analogies of the two great divisions of Organic Nature. He discovered the Amnios, and showed the analogy between the Egg of Animals and the Seed of Plants. He is also supposed to have known the cochlea of the ear—While some have attributed to[4] Alcmaeon[5] the discovery of the Eustachian Tube, in consequence of his well known[6] statement that certain animals respire by their ears.[7]—in which some Anatomists *imagine* they have seen a proof of his having discovered the Eustachian tube, by which the Air in fact penetrates from the back of the mouth into the internal Ear.

Such are the results which have been [/] handed down[8] of the labours of the Pythagorean Philosophers, in Physiological Science:—and when we reflect on the numerous writings of other Philosophers of the same School,

(margin numbers: 22, 23)

---

1 Doctrines.] *inserted after deletion first of* 'Sciences' *and then* 'Doctrines'. *Period inserted [ed.].*
2 about. . .Æra] *interlined above deletion of* 'in the Year 58A before Christ'.
3 Anatomical] *after deletion of* 'the'.
4 some. . .to] *interlined with caret.*
5 Alcmaeon] *followed by deletion of* 'having observed that'.
6 his. . .known] *interlined with caret.*
7 He. . .ears] *entire passage inserted from facing 23ᵛ to replace pencil deletion on fol. 22 of* 'It has also been supposed, from a verse of his which has been preserved, that he knew the Cochlea of the Ear:—but in attributing to him this discovery, the same reserve ought to be maintained as in considering the value of the well known statement by Alcmæon that Goats respire by their Ears;' *with pencil query by WC in margin* 'reservation?'.
8 down] *followed by deletion of* 'to us'.

embracing the subject of Medicine with Natural and Moral philosophy, of which only the record now remains, but some of which, as the Works of Epimarchus were held in high esteem; it is not an unreasonable conjecture that Aristotle might have derived some of the materials for his astonishing System of Comparative Anatomy and Natural History, from these Sources.

The Sect of Philosophers from which a further progress in a science of Observation might be expected, is that which was founded by Leucippus;—who, disgusted with the Idealism of Zeno, and the abuse that had been made of it, fell into the opposite extreme, and became intirely a Materialist. He recognised nothing in the Universe beyond Vacuity and Atoms:—to the latter[1] he allowed only[2] figure and motion; the eternal Circle of destruction and reproduction of beings resulted from their Motion;—and the Soul itself was by an aggregation of Atoms in a particular mode of Combination.

25[3]　　Democritus of Abdera: first comparative anatomist was the continuator of the Atomistic School. He[4] was contemporary with Socrates.

As motion was to him the great and ruling phænomenon of the Universe, what was more natural than that his attention should be more especially attracted by the Members[5] of the Animal Kingdom, in which the phænomena of Motion were most conspicuous and most diversified?

With whatever view or object Democritus was actuated in his inquiries, this is certain, that they were directed to the discovery of the differences of Organization in different Species of Animals; and that he tried to deduce from them the differences in their manners and habits:—that he was, in fact, the first who instituted Comparative Anatomy.

He knew the Biliary passages, and seems to have been acquainted with the Sympathies subsisting between the organ of the mind and the chylopoietic viscera; since he attributes Mania to an alteration of the Viscera of the lower Venter, or Abdomen.

26　　You[6] will have seen, Sir, that a Principle was still[7] wanting to guide the

---

1　latter] *before deletion of* 'only,'.
2　only] *interlined with caret.*
3　25] *fol. misnumbered in original.*
4　He] *insertion to replace deletion of* 'It is known that he lived in the Year 399 before Christ, and' *with pencil* 'X' *in margin.*
5　Members] *insertion over erasure.*
6　You] *Paragraph break inserted [ed.] and follows deletion of paragraph beginning fol. 25 and continued to fol. 26* '{¶}*It is familiar to all, that {deleted} Hippocrates visited Democritus in his retirement; and thus [l fol. 25] describes his Occupations:—{¶} "He found the Philosopher sitting on a Stone, under the spreading shade of a Plane-Tree:—a number of books arranged on each side; one on his knee; a pencil in his hand; and a number of Animals which he had dissected, lying around him. His complexion was pale, his person thin, his countenance thoughtful;—at times, he laughed, at times he shook his head, mused for a while, and then wrote; —then rose up, and inspected the Animals,—sat down, and wrote again."{¶} Hippocrates acknowledged the great importance of his inquiries, and regretted much that his own professional employments, his domestic concerns, and other avocations, did not permit him to indulge in similar

investigator of the Animal Mechanism; without which, his researches must
ever have remained barren of any great Physiological results:—I mean, the
consideration[1] of the conditions of 'existence' or of the designed adaptation
27   of structures to a special end[2]—For the discovery of[3] [/] this fruitful principle[4], however, juster notions of the Nature and cause of the Universe were
absolutely necessary.

The followers of the Sects to which we have already alluded, must of necessity have remained blind to the harmonious relations and mutual dependencies of the objects of the material world. These could only be appreciated
by the Philosopher, who, admitting the reality of *matter*, at the same time
raised himself to the Conception of *Mind* [5] as distinct from Matter; and of an
Intelligence by which it was governed and arranged.

Such a Philosopher was Anaxagoras, the master of Socrates. His writings
unfortunately, are only known to us by the allusions to them in those of
others:—some of his Apothegms have been preserved;[6] we may infer that he
was practically acquainted with Comparative Anatomy from the following circumstance; It is related that the People of Athens[7] having looked upon a Ram
which had only one horn, as a fearful prodigy;—Anaxagoras dissected the
Animal, and explained the physical cause of the Monstrosity. Hence it is obvious that he sought the reason of things in Observation.[8]
28   The influence which the discourses of Socrates[9] exercised over[10] the
progress of Science is too well known to need further allusion; But the profession which owes so much to the discoveries of Harvey, ought ever to regard with Sentiments of admiration and gratitude the memory of[11] the Man
who first introduced into Philosophy the principle of final Causes.

Socrates informs us that the Idea was suggested to him by reading a work
of Anaxagoras' entitled[12] *On the intelligence which has arranged the*

(contd.)
   pursuits. The events recorded in this visit, and the object of Democritus's Dissections
   are almost all the results of his labours with which we are acquainted.'.
7   still] *interlined with caret.*
1   consideration] *over erasure.*
2   of the. . .end] *interlined at bottom of fol. with lead to insert point after deleted* 'final
   causes'.
3   discovery of] *before deletion of* 'which' *at bottom of page.*
4   this . . . principle] *inserted in margin with* 'fruitful' *interlined with caret.*
5   Mind] *over erasure.*
6   preserved;] *followed by deletion of* 'and'.
7   of Athens] *interlined with caret.*
8   Intelligence. . .Observation] *pencil line along margin of entire passage. Pencil cross-out
   of final line of fol. 27 continued to top of fol. 28* 'He died at the age of 72, [I *fol.* 27]
   428 Years before the Christian Æra—'. *Period inserted [ed.].*
9   Socrates] *followed by deletion of* '(the most celebrated of the Pupils of Anaxagoras),'.
10 exercised over] *interlined above deleted* 'had upon'.
11 the memory of] *insertion in margin over erasure.*
12 entitled] *interlined with caret.*

Universe.[1]

His contemporaries who had never experienced the delight occasioned by studying the harmony subsisting between the conformation of an animal, and its powers & exigencies[2], naturally ridiculed the Philosopher who could condescend to measure the (extent of) the leaps of a flea[3].

There is yet another Distinguished Ancient[4] whose writings must have been useful to Aristotle, I mean[5] Hippocrates.[6]

29   Hippocrates was the principal contributor, but not the exclusive author of the Records of the practice of the Asclepiadean Priests and Physicians which have been handed down to us under his name. The merits, as a Physician, of him who is justly called[7] the Father of Physic are universally known and acknowledged; but as an Anatomist he has little claim to our Admiration.

His description of the Veins is almost intirely imaginary; and yet the different bleedings which he described were in accordance with the alleged distributions of the bloodvessels. Thus, He speaks of a vein which goes from the forehead to the Anterior surface of the Arm; and another which passes from the side of the head to the back part of the Arm, &c.

The brain is considered by him in one place, as a spongy organ destined to absorb the moisture of the body. In another he mentions two nerves arising
30   from the Brain, which he makes the seat [/] of the Soul; and calls it the Organ by which we see, hear, feel, and reason.

He had no knowledge, however, of the Nerves, as we now know them; ΣΕΟΣΟΡ in his writings signifies the Tendons, ligaments, and, in general, the different white tissues:—But it must be remembered that in the time of Hippocrates it was considered as a profanation even to touch the dead body with any other intent than that of paying it the last rites; and, therefore, it was next to impossible to obtain in Greece at that time any knowledge of the internal organization of the human body.

---

1   Universe.] *quote marks surrounding title deleted [ed.].*
2   & exigencies] *interlined above deletion* 'of action,'.
3   flea] *in margin* 'X'.
4   Distinguished Ancient] *interlined above deleted* 'Philosopher'.
5   I mean] *interlined above deletion of* 'viz.'
6   Hippocrates] *followed by pencil deletion at bottom of fol. 28 and presuming continued deletion at top of fol. 29 of:* 'This great Physician belonged to the family of Asclepiadae, a limited sect of Philosophers, who may be said to have formed an Utilitarian [I *fol.* 28] School, since they cultivated the Sciences solely with a view to their application to Medicine and Surgery. Most of the Priests of the Temples of Æsculapius were of this Family' *and deletion of originally intended insertion from facing fol. 29*ᵛ 'But to recount the Anatomical and Physiological doctrines which are contained in the writings of the Asclepiadean Priests & Physicians which have been handed down to us in his name, would be superfluous before this learned audience. And my apology for tracing the gradual progress of Anatomy to the time of Aristotle must be the desire to appreciate more justly the advantages which this great man might have derived from his predecessors.'.
7   called] *interlined in pencil above deletion of* 'termed'.

Hippocrates has least deviated from truth in the department of Osteology:—He has given names to some Muscles, and mentions the pulsations of the Heart, which he calls a strong Muscle.

My apology for alluding to these facts which must be both trite and familiar to many of my hearers, is the desire of thus tracing the gradual progress of Anatomy to the time of Aristotle, in order more truly[1] to appreciate the advantages which that great man might have derived from his [/] predecessors while[2] devoting his master-mind to the study of Natural History. And these advantages were many and great; for even allowing the details respecting the habits and organization of Animals already[3] determined by the Philosophers who preceded the Stagyrite[4] to be twice as numerous as those recorded in the antient writings which we have been considering; yet this previous knowledge was of far less consequence to the success of Aristotle than the gradual ascent of the human mind to the right[5] modes of Studying the phænomena of Nature[6].

And we find[7], that the intellect of Aristotle, while[8] elevating and improving every subjective branch of Science, did also successfully[9] grapple with that highest department[10] which teaches the right application of the reasoning faculties. He saw that the doctrine of innate Ideas as taught by Plato, his contemporary, as a necessary consequence, condemned[11] the Senses to inaction in the pursuit of Truth:—He saw, that, however flattering to the pride of Man,[12] such a doctrine must ever keep him in ignorance, by discouraging the exercise [/] of those obvious and natural powers of investigation which had been[13] assigned to him for the purpose[14] of cultivating the study of Nature.

Aristotle therefore taught that Man can acquire general Ideas only by abstraction;—that nothing occurs in the mind which has not first passed through the Senses; and hence that all knowledge necessarily takes its source in Observation and Experiment.

31 (margin)
32 (margin)

---

1   in. . .truly] *interlined with caret, with* 'more' *over* 'to'.
2   while] *interlined above deletion of* 'before'.
3   already] *interlined with caret.*
4   the Philosophers. . . Stagyrite] *interlined with caret above deletion of* 'his predecessors'.
5   right] *interlined with caret.*
6   Nature] *period inserted [ed.] before deletion of* 'which are alone adapted to the Nature of man himself' *inserted over erasure and continued deletion of* 'real nature and faculties of the Human *man [double deletion]* man's Understanding.'.
7   And. . .find] *inserted over erasure.*
8   while] *interlined above deletion of* 'in'.
9   successfully. . .with] *inserted over erasure.*
10 department] *inserted over erasure.*
11 condemned] *inserted with lead line from after* 'contemporary'.
12 Man,] *followed by deletion of dash mark.*
13 which. . .been] *interlined with caret.*
14 for the purpose of] *inserted over erasure.*

The sanatory influence of this principle[1] is manifested in every one of Aristotle's Treatises:—His whole Philosophy assumes in consequence of it, a peculiar character, and Zoological[2] Science sprang from his labours, we may almost[3] say,[4] like Minerva from the Head of Jove, in a state of noble and splendid maturity.

In Aristotle's Analytics—

> Particular facts are first appreciated by sense—, for nothing is in the understanding which was not before in the sense. And although the ratiocination by Syllogism is naturally first and more known, yet that which is made by induction is most conspicuously[5] clear to us: and therefore we define singulars[6] with more ease than universals: for there lies more equivocation in universals; Wherefore we must pass from particulars to generals. [7]

On perusing[8] "Aristotle's History of Animals" says Cuvier,

> it[9] is difficult to understand how the Author could have obtained from personal observations of many [/] generalizations, so many Aphorisms whose accuracy is perfect, but of which his predecessors seem not to have formed the slightest Idea.

33

This Work is not, properly speaking, a Treatise on Zoology:—it is a summary which bears the same relation to this branch of Science, as the *Philosophia Botanica* of Linnaeus holds in another department.

The first Book treats of the parts which compose the Body of Animals, and Aristotle arranges the *partes similares* [10] or proximate components[11] as follows:

Blood, Lymph, Fibre, Flesh, or Gland, Bone, Cartilage, Skin, Membrane, Nerve, or Tendon, Hair, Nail, Fat, [12] Excrements.

In beginning the Study of the Physiological Department[13] of the Hunterian

---

1   principle] *before deletion of* '(which, obscured by the Schoolmen in the dark ages, required the Genius of a Bacon to *re-{interlined with caret }* establish,)'.
2   Zoological] *interlined above deleted* 'the'.
3   we. . .almost] *interlined with caret.*
4   say,] *comma inserted [ed.].*
5   conspicuously] *interlined.*
6   singulars] *interlined above* 'particulars'.
7   In . . .generals.] *final period inserted [ed.] and remainder inserted from facing fol. 33ᵛ to replace deleted* 'It is in the application of his method to the study of Living objects that he is most happy.' *And on fol. 32 in margin WC query in pencil* 'X Qu it here'.
8   On perusing] *inserted in margin.*
9   it] *following deletion of* 'is truly a Masterpiece. on reading this Treatise'.
10   partes similares] *interlined with caret above deleted* 'similar'.
11   proximate components] 'proximate' *interlined with caret over pencilled interline, and* 'components' *over deleted* 'component parts'.
12   Fat] *before deleted* 'Suet,'.
13   Physiological Department] *interlined with caret.*

Collection, with a previous knowledge of the writings of Aristotle[1], we
might almost suppose it to have been expressly formed in illustration of the
Historia Animalium.

The Series which would first strike the eye of the learned visitor[2] would be
a similar Category of the component parts of the Animal Body,—Blood,
Fibrous flesh, Tendon, Ligament, Cartilage, Bone, Gland, Brain, Nerve,
Cellular substance, Membrane, Skin, Earth, Oil[3].

34   Considered[4] as a Series[5] of the component proximate elements of an
animal body to which we arrive without having recourse to Chemical analysis, nearly the same defects, both in kind and in degree occur in the enumera-
tion of these constituent parts by Aristotle, as in the corresponding Series
with which the Hunterian Collection commences.

In both we find substances which differ in their degrees of simplicity or
elementary character arranged in the same Category[6].

In both we find modifications of the same Component Tissue set forth as
distinct elementary parts.

35   The present improved knowledge of the characters of the proximate
constituents of the animal body,[7] the subject of[8] General Anatomy or
Histology[9], it is termed, are chiefly due to the improvements[10] in Organic
Chemistry, & to the genius &[11] Research[12] of Bichat[13].

We shall find other striking similarities in the views and observations of
these two great and original minds in the progress of an analysis of the
Historia Animalium. These coincidences reflect honour alike on both; for it
will be readily granted, I apprehend, that John Hunter derived none of his
Physiological Knowledge from a perusal of the Works of Aristotle.

Both drank deep draughts[14] of the pure stream of knowledge from its
original Source[15].

Proceeding with the Text of Aristotle we find the whole of the

---

1   Aristotle] *in margin inserted pencil* 'X'.
2   the eye . . .visitor] *inserted with* 'the' *written over* 'his'; 'eye of' *inserted in right
    margin, and* 'the. . .visitor' *over deletion of* 'eye'.
3   Oil.] *before deletion of* 'We' *at page break.*
4   Considered] *after deleted paragraph inserted on fol. 35 of MS. (q.v.) with marginal
    brackets and* 'next p.' *in margin.*
5   Series] *interlined above deleted* 'Category'.
6   Category] *deletion line through* 'Category' *with stet marks below.*
7   body,] *comma inserted [ed.].*
8   constituents. . .subject of] *interlined above deletion of* 'Elements; and component
    Tissues, or'.
9   or Histology,] *interlined with caret.*
10 improvements] *after deletion of* 'recent'.
11 &. . .genius &] *interlined below deletion of* 'and also to the'.
12 Research] *deletion of terminal* 'es'.
13 Bichat.] *in margin* 'p. 34}' *to indicate insert of material from fol. 34.*
14 draughts] *interlined.*
15 We. . .Source] *inserted from fol. 34. See note 14 supra.*

Commencement of His first Book in some degree detached from the rest, to which it seems intended to serve as an Introduction.

It is principally composed of Aphorisms or general propositions which indicate the observation of an immense number of particular facts; as may be judged from the following:

> All Animals, without exception are furnished with a Mouth, and possess the sense of Touch; but these Characters are the only two which are indispensable, and we cannot find a third which is not absent in some Species.

Here we have[1] a combination of the definitions of Hunter and Linnaeus'— The first characterized[2] an animal by the possession of a stomach; the second by sensation—But the digestive cavities in some animals, as the Taenia—can hardly be called a stomach—although these present distinct mouths,[3] & the Phenomena of the sensitive plants afford an apparent objection to the definition of Linnaeus.[4] Hence the combination of the digestive & sensitive characters which characterizes the definition by Aristotle renders it perhaps the best that has yet been given.[5]

36   Permit[6] me to notice a few more of his general propositions.

> Amongst Terrestrial Animals, says the Stagyrite, there is not any which are fixed to the Earth: Amongst Aquatic animals, on the other hand, there may be many that are fixed.
> Every Animal which has wings has also feet.[7] Many Insects have stings: those which possess this organ in the anterior part of the body have never more than two wings: those that have it posteriorly, possess four.

37   Such propositions, it is well known, cannot be formed *a priori*, they are

---

1   have] *interlined above deleted* 'find'.
2   characterized] *interlined with caret above deleted* 'defined'.
3   these. . .mouths,] *interlined.*
4   Linnaeus.] *period inserted [ed.].*
5   Here we have. . .given.] *insertion from facing fol. 36ᵛ to replace deletion of passage marked with* 'X' *in margin at bottom of 35 and top of fol. 36* 'Such is Aristotle's definition of an Animal. Linnæus made Sensation alone the grand Characteristic of the Animal Kingdom. "Mineralia crescunt; Vegetabilia crescunt et vivunt; Animalia [l fol. 35] crescunt, vivunt, et sentiunt [sic]".—Mʳ. Hunter's favourite definition of an Animal was its possession of a Stomach. But a digestive Bag cannot be predicated of Animals which like the Tape-Worms are nourished from simple canals continued uniformly through the body:—But the Teniæ [sic] have a mouth; and Linnæus's definition is weakened by the well known phænomena which follows the irritation of the Dionæa, Mimosa, and other Sensitive Plants;—but these do not combine a Mouth with their sensitive property. Hence Aristotle's is yet, *perhaps {inserted with lead line from after 'best'} the best definition of an Animal that has been given.'.
6   Permit. . .his] *interlined with caret to replace deletion of* 'Let us proceed with his other'. *Paragraph break added [ed.].*
7   Many] *followed by pencil line and in margin* 'Gnats may be'.

necessarily based on a profound observation of facts, and indicate an almost universal examination.

In the same Introduction, Aristotle establishes the foundation of his Classification of Animals[1].

We have already seen that he distinguished Blood from Lymph.[2] What Hunter here[3] shows us as the Blood of the Lobster[4], was with Aristotle Lymph:—red colour in short, entered essentially into his idea[5] of Blood; and he therefore primarily divides Animals according as they have Blood, or have it not:—in other words, he separates the red-blooded from the white-blooded Animals. He was, however, ignorant that[6] certain Worms had red blood.

Modern Naturalists have shown that the true essential character of these two great divisions of the Animal kingdom reposes[7] on a condition of the Nervous System instead of that of the circulating fluid:—and deriving their 38   Terms from the protecting Case of the Cerebro-spinal axis of the higher division, have called the Enaima of Aristotle, Vertebrata;—the Anaima, Invertebrata. The defect of this binary division was first pointed out & remedied by Cuvier[8].

It is rare that we find a[9] general proposition defective from a limited observation: it is most commonly from using a word in a different sense to that in which we now understand it. The Animals with red blood he divides into Viviparous Quadrupeds, Oviparous Quadrupeds, Serpents, Birds, Fishes, Cetacea.

Although both the two latter Classes live in water, and closely resemble each other in external form, Aristotle in connecting them is far from confounding them. He knew the essential nature of the Cetacea as well as Hunter did: He knew that these animals have warm blood; that they bring into the World a living Offspring; and that they nourish their young with Milk secreted by mammary Organs:—in short, His Anatomical researches on this Class of Animals have perhaps never been equalled by those of any other individual until the time of Hunter:—and the celebrated paper in the Philosophi-39   cal Transactions [/] adds to our knowledge of the Anatomical details of the Cetacea, but nothing as[10] to their essential nature and affinities, which Aris-

---

1   Animals] *With note over light erasure on facing fol. 38[v] with line for insertion.* 'Here demonstrate the Tabular View of Aristotle, Linnaeus & Cuvier's Systemata Animalium—'.
2   Lymph.] *period inserted [ed.].*
3   here] *interlined.*
4   Lobster] *with pencil insertion of* 'Na—' *alongside in margin.*
5   idea] *after deletion of* 'complex'.
6   that] *with WC insert in margin of* 'X 'and' 'Qu.{'.
7   reposes] *restoring deletion of* 'reposes' *for grammatical sense [ed.].*
8   The. . .Cuvier.] *interlined over erased pencil with* 'Cuvier' *over pencil* 'Aristotle'.
9   a] *lightly deleted in pencil and restored for grammatical sense [ed.].*
10 as] *interlined with caret.*

totle first wrought[1] out.

The Anaima of Aristotle, or invertebrate white-blooded Animals, are divided into Malakia, Ostracoderma, Malacostraca, and Entoma, corresponding respectively to the Cephalopoda, Testacea, Crustacea, and Insecta of Modern Naturalists.

It is wonderful, considering that the Nervous System, the true key to[2] the primary divisions of the Animal Kingdom was wholly unknown to Aristotle, that he should have approximated so nearly in propounding these Classes, to our modern Systems. For his arrangement, although not irreproachable, is more accordant with Nature than that of Linnæus.

The sketches which Mr. Hunter has left of his views of the arrangement of the[3] Animal Kingdom more nearly correspond with that of Aristotle and Nature, than the more artificial system of Linnæus;—but the attempts in Zoological Classification which the Hunterian Manuscript contains cannot of course be compared to the bold and clear enunciations on this subject[4] which the Stagyrite gave to the World; and the Merits of [/] which it required the profound researches of a Cuvier to enable us fully to appreciate[5].

40

---

1 wrought] *interlined over deleted* 'worked'.
2 key to] *interlined above deleted* 'character of'.
3 arrangement of the] *interlined with caret.*
4 subject] *interlined above deleted* 'head'.
5 appreciate] *at this point MS has pencil deletion of* 'I shall not dwell on the Subdivisions which Aristotle established in his primary Classes, but have embodied them in the Tabular sketch behind me. {¶} [Refer to Table. Aristotle compared with Linnaeus & Cuvier.] '. *MS then marked with horizontal line to indicate lecture division.*

Fig. 8. The New Facade of the Royal College of Surgeons, Depicted in *The Mirror of Literature, Amusement, and Instruction*, London 26 March 1836. *(Courtesy British Library, London)*

Lecture Two
4 May 1837

I now proceed to consider the Physiological and Anatomical parts of the Historia Animalium.[1]

The Observations of Aristotle on the locomotion of Animals are general. He was ignorant of the properties of the muscular and nervous fibres on which those motions depend. With respect to the hard parts of Animals, he has many Osteological Observations, but here, there are some errors;—as when he speaks of the Lion having but one bone in the neck, probably from having[2] observed the cervical vertebræ anchylosed in an old Individual[3] as in the present specimen[4].

The Cephalopoda, like the red-blooded tribes, Aristotle says, are externally fleshy and internally solid;—but the Malacostraca and the Ostracoderma are Animals which have their solid part without, and the fleshy, within. He further points out a difference in the powers of Mechanical[5] [/] resistance[6], or mode of fracture of the *Crusts*[7] of the one Class, (Crustacea); and the shells of the other, (Testacea).

My Learned Colleague has already observed that Aristotle has pointed out corresponding differences in the density of the osseous system, though necessarily ignorant of the chemical conditions on which those differences depended[8]—

The Entoma, in Latin, Insecta, are Animals of which the body[9] is distinguished by incisures; and which has no solid or fleshy substance separate, but something intermediate;—being equally hard within and without.

41

---

1 Animalium.] *period inserted [ed.].*
2 from having] *pencil deletion of second repeated* 'from having'.
3 Individual] *deleting period [ed.].*
4 as. . .specimen] *interlined in ink after* 'Individual' *with light pencil line across 1/3 page.*
5 Mechanical] *inserted by WC in pencil with underline* 'Greek quotation somewhere here'.
6 resistance] *with* '?' *inserted in margin.*
7 *Crusts*] *underlined in pencil with pencil* '?' *in margin.*
8 My. . .depended] *insertion on facing fol. 42ᵛmarked for insertion with symbol* 'Ø'. *Paragraph break inserted [ed.].*
9 body] *period deleted [ed.].*

97

—These generalizations on the Modifications of the organs of support are illustrated by the Preparations in the commencing Series of the Hunterian Collection.

With respect to the Digestive Organs, Aristotle dwells[1] largely upon the Food proper to each Animal in connection with their modes of life, and the influence of external circumstances of Climate, of Seasons, and of the medium in which the different Species live. He notices the most remarkable peculiarities in the modes of taking food;[2] and observing how the structure of the Elephant renders it difficult for this animal to[3] [/] gather nourishment from the Earth; and he shows that the Trunk supplies this inconvenience, and becomes an organ of prehension essential to such a form of body.

In the Hunterian Series, the preparations of the Teeth are placed according to their physiological relations to the Digestive function, and precede the preparations of the[4] Stomachs:—

Aristotle treats of these Organs after having given the History of Hair and[5] Horns.

It is only of late years that the correspondence of Teeth in their mode of Growth and vital properties to the other extra-vascular parts has been again prominently insisted upon by the Philosophic Anatomists of Germany. I am told however by M^r. Clift that the situation actually assigned to the preparations of the Teeth, at one time, by Hunter himself, was in close proximity

42

---

1 dwells] *interlined above deletion of* 'enters'.
2 ;] *inserted in pencil followed by deletion of* 'describing the means whereby the Angler-Fish, (Lophius) allures the little Fishes to devour them:—and, *in describing the Elephant {*interlined with caret*}, observes, in Language as precise as a writer of a Bridgewater Treatise would use in the present day,—*in describing the Elephant, {*deleted*} how the length of the fore legs and the nature of the joints, render it difficult for this Animal to drink and to gather'.
3 and. . .animal to] *interlined in pencil at top of fol. 42.*
4 preparations of the] *interlined with caret.*
5 Hair and] 'X'*inserted by WC in margin and note on facing 43* 'It ought to be recollected, and perhaps to be noticed, that the Teeth do not stand in the same relative Situation that M^r. Hunter *placed* and *left* them—Sir Everard *removed them* from being successors to Peculiarites (& preceding the Generative organs, or last in the first great Subdivision) to *follow* the Stomachs, and precede the intestines. This was afterwards overruled (when Sir Everard fell off) by M^r. Abernethy[,] M^r. Cline, Sir W^m. Blizard &c—as being an *erroneus {*interlined above* 'outrageous'} situation, and consequently they were subsequently placed where they now are, as "Parts preparatory to Digestion" as the Drawings of the Teeth *were so described. i.e. Parts preparatory to Digestion.*' *RO has then replied in pencil in the margin :* 'Sir Everard's reason that some animals had no teeth and consequently the teeth must follow after such *stomachs.*' *WC then continues* '{¶} No one at that time could understand why the Teeth followed "Peculiarities" and thought that M^r. Hunter *had placed them there because he did not know where to find a place for them.* This was the general opinion—and they were in consequence removed to where they now are; but it was not *Mr. Hunter's* arrangement as the old Catalogues will testify. W. C.' *This is then followed by pencil deletion of* 'I had written so far before I had read the succeeding paragraph W.C.'.

with the Cuticular Organs;—which is a very interesting fact.

Aristotle describes the different forms of Teeth in different Species of Animals, according to the nature of their food; being, as he observes, sharp and pointed in the Carnivora (Karkarodonta); flat and grinding in the Herbi-

43   vora. In some [/] Animals, certain teeth protrude and form tusks, but no animal he observes[1] is armed at the same time with Tusks and Horns:—This is a general proposition of which the experience of 2000 years has only served to establish the truth.

In the Elephant, he observes, the Tusks of the Female are small, and directed to the ground. Here, says Cuvier, is one of the propositions wherein we should have thought that Aristotle was wrong. The Indian Elephants, indeed, do not exhibit any difference, in this respect, indicative of Sex; but the African Elephant, which is that described by our Philosopher, has really the peculiarity mentioned.

Aristotle does not omit to describe the manner in which the teeth are removed in Man and Animals;—and the accurate remarks which occur on this subject, which once[2] astonished me in perusing the Sylva Sylvarum[3] of Lord Bacon, I find are derived from this Source.

With respect to the abdominal viscera Aristotle distinguishes the Stomach, the Omentum, the Liver, the Gall-bladder,[4] the Spleen, the Bladder,

44   the Kidneys, [/] and their appendages. He says, The right kidney is placed higher than the left, which shows that he had derived this part of his Anatomical knowledge from the dissection of the lower animals[5].

Aristotle describes the Cæcal appendages of the intestine in Birds and Fishes:—in the former, he says, they are placed near the end of the gut; in the latter at its commencement. He also notices the singular[6] exception in the position of the gall-bladder which occurs in Serpents.[7]

45   Aristotle knew nothing of the Absorbents?

In treating of the vascular System, he first quotes the opinions of his predecessors on the course of the Veins, which he deems erroneous:—he

---

1   he observes] *interlined at top of page with caret at insertion point.*
2   once] *interlined with caret.*
3   Sylva Sylvarum] *interlined above* 'Natural History'.
4   Gall-bladder,] *comma inserted [ed.].*
5   animals] *followed by deletion of* '; as the relative position of the kidneys in Man forms an exception to the rule *as {interlined} observed in Brutes.' and followed by faint pencil interline of* 'as my words have ____{illegible word}'.
6   singular] *interlined with caret.*
7   Serpents] *followed by deletion of* 'His notion of Digestion was nearly the same as that of Hippocrates: he imagined the stomach to be a passive organ in which the food is boiled or concocted; but that it was assisted by the heat of the liver, the spleen, and the viscera lying in the neighbourhood:—a Theory, however, which is more rational than that which attributes the function of a warming pan to the Spleen alone; for the liver in all animals bears such a relation to the Stomach as to be able to communicate to it a certain heat if required, while the spleen is sometimes remote [l *fol.* 44] from the Stomach'.

adds, however "but the cause of their ignorance arises from the difficulty of surveying these parts."—

Let us imitate his forbearance, and view the imperfections of his work with a similar Spirit. Considering the difficulty of investigating this intricate part of Anatomy without the helps of Injections, Aristotle in his description of the Vascular System comes wonderfully near to truth; and far surpasses all his predecessors.

In investigating the Veins, he recommends that they be observed in Animals which have first wasted away to leanness, and are afterwards strangled. He first speaks of the veins whose principal trunks have their origin in the heart:—he distinguishes well the *Venæ cava* from the pulmonary vein: he describes also the Aorta from its origin to its division at the inferior part of the trunk: But he had no idea of the Circulation as we [/] now know it. He did not know that the Arteries contained blood; and even seems to have thought that the Air penetrates to the Heart;—an organ which he describes as having only three Cavities.

Of these he distinguishes the left Ventricle, which he regarded as[1] the seat of the Soul; from whence it diffuses its influence through the Arteries;—and the Heart is accordingly the Organ of motion, sensation and nutrition;—of the different passions;—and of the vital flame!

It is not certain what Idea Aristotle had formed of the use of the Brain;—we may infer, perhaps, that he considered[2] it of more importance than Hippocrates did:—since he commences his anatomical description with this Organ. He states that the Brain[3] is found in all red-blooded Animals without exception; but that among White-blooded animals it is found only in the Malakia, which corresponds to the[4] Cephalopods, where it is protected by a cartilaginous Cranium.

This observation is very remarkable, as it is only since the publication of Cuvier's Introduction to Natural History that this fact has been known to Modern Anatomists. However, some years before Cuvier's time, Hunter had determined the existence of the Brain of the Cuttle-fish, and exhibited it in this preparation [N^O. 1629.] of which a figure is published in the third volume of the Physiological Catalogue.

Man, Aristotle observes, possesses, of all animals, the largest and the moistest Brain. Where the cranium is largest, the face is small and round. Animals, on the contrary, which have small crania[5], have long jaws. We

---

1 regarded as] *interlined above deletion of* 'makes;' *with* '??'*in margin.*
2 considered] *interlined above deletion of* 'regarded'.
3 Brain] *with pencil indicator arrow alongside in left margin.*
4 the] *interlined above deletion of* 'our' *with second pencil arrow indicator mark in left margin.*
5 crania] *interlined above deletion of* 'heads'*and on facing fol. 48^v pencil note by RO without indicated insertion point:* 'This, whether taken in the absolute or relative sense is an error—The Elephant has a larger brain than Man, the Sparrow has a larger brain in proportion to its body.—'.

know how interesting this idea of the receding proportions of the Cranium to the face, as we descend from Man, was to Hunter; by his having had the Sketches (which illustrate Aristotle's proposition) introduced into his Portrait by Sir Joshua Reynolds.

The Brain, says Aristotle, in all Animals is without blood. This error, relative to a part to which more blood is transmitted than to any other in the whole body, is obviously owing to the extreme minuteness to which the[1] blood vessels are subdivided before they [/] penetrate the cerebral substance.

48

Aristotle describes the Dura Mater and Pia Mater which envelop the Brain;—[2]its division into cerebrum and Cerebellum; and the two halves or hemispheres of the former, and a[3] cavity in its middle.—He also describes the different nerves which leave it, to be distributed to the Eye:—but to these points all his neurological knowledge was confined.

Under the Term νεϋρον, Aristotle, like his predecessors, comprehended small arteries, ligaments, tendons, as well as the true nerves; whence he philosophically observes, the νεϋρα are derived, not from one principle, but are divulsed about the limbs and joints.

Aristotle in that part of his work where he speaks of the Sensations, is particular in mentioning those animals which are naturally[4] deficient in[5] any organ of Sense, and those in which these organs present certain peculiarities:—Thus, on Vision, he speaks of the eye of the Mole, which, he says, is hidden under the Skin, but is similar in its configuration to that of other animals, and is furnished with a nerve.

49

An English Naturalist might here suppose the Stagyrite in error, in describing the relation of the eye to the integument;—but the mole of [/] Greece[6] is a different species from ours, and is really distinguished by the permanently closed state of the eye-lids.

Aristotle describes the third eye-lid, the *membrana nictitans* in Birds, but, being ignorant of the properties of the muscular fibre, he could not appreciate[7] the beautiful mechanism concerned in its motions.

With respect to the organ of Hearing, it is certain that Aristotle knew the Eustachian tube; for he expressly mentions it where he refutes Alcmæon who contended, as we before stated, that Goats respire by their ears:—and if any should be disposed to smile at such a statement, let it be remembered that some of our ablest Naturalists have gravely advanced, and many persons still believe, that Deer and Antilopes can and do respire by the ant-orbital sinuses

---

1 the] *inserted in margin.*
2 which. . .Brain;—] *interlined in ink by WC with caret.*
3 a] *interlined above deletion of* 'the'.
4 naturally] *interlined with caret.*
5 deficient in] 'icient in' *over erasure.*
6 of Greece] *on page break with* 'Greece' *repeated on next page.*
7 being . . .appreciate] *insertion from facing fol. 50ᵛ with caret and* 'x' *at insertion point to replace deletion of* 'he was not aware of'.

while their nostrils are immersed in water:—a notion for which there are even fewer anatomical grounds than for that Idea above mentioned of the Antient Philosophers.

   In an excellent description of the Crocodile Aristotle describes the disposition of the Organ of Hearing. He also treats of the Audition of Fishes, and determines that water must serve them as a medium for the transmission

50   of Sound. [/] The time is past when the zealous admirer of Hunter could claim for him the merit of having first established the accuracy of Aristotle's opinion by the discovery of the organ of hearing in the Class of Fishes:—for although the observations of Hunter on this subject were most numerous and exact, and bore date, according to his own statement, prior to those of Monsr. Geoffroy; yet we shall hereafter see that both Anatomists were anticipated in this discovery many years, by Casserius.

   Aristotle shows that Insects also enjoy the faculty of Hearing, and even that they have the Sense of Smell; since they are driven away by certain odours, and attracted by others. M^r. Hunter, we know, came to the same conclusion respecting Bees.[1]

   The sense of Taste is considered in the same philosophic spirit, and the modifications of its organ is traced[2] through the whole Animal Kingdom. Aris-

51   totle mentions or describes almost all the peculiar forms of [/] the tongue[3] which are noted in the present day.

   He speaks of its shortness in[4] the Elephant:— of its great length, peculiar extensibility and prehensile powers in the Chameleon:—He describes the disposition of the tongue of the Frog, which instead of being, as in most animals, free anteriorly, and fixed; behind; has its root attached to the anterior part of the lower[5] jaw, and its free extremity directed towards the palate.

   He describes the bifid tongue of the Seal; and mentions the fleshy structure of the tongue[6] in the Parrot, and its relation to the power of imitating Articulate Sounds:—and states that in the Crocodile this part is wanting, as in Fishes; overlooking its rudimental condition in both Cases.

   Mr. Hunter has preserved examples of all the modifications above mentioned, in his Series of the Lingual Organs.

   If we compare what Aristotle says of the Adipose and Tegumentary Organs with the preparations of the same parts by which Hunter records his investigations therein, we shall find an extraordinary parallelism in the Researches of these two Philosophers, both as to manner and extent.

---

1   Bees.] *followed by deletion of* '{¶}With reference to the sense of Smell, we may also observe that Aristotle knew that the proboscis of the Elephant was part of that Organ.'.
2   traced] *followed by deletion of* 'almost'.
3   the tongue] *inserted at page break on fol. 50, and interlined above deletion of* 'that organ' *on fol. 51*.
4   shortness in] *before deletion of* 'of the Tongue of' *with* 'in' *inserted in margin*.
5   lower] *interlined with caret*.
6   of the tongue] *interlined with caret*.

52   Both arrange the Adipose substances [/] according to their degrees of density at a given temperature. With respect to Hair, Aristotle distinguishes that which grows from the time of Birth from the hair peculiar to the period of Puberty. He considers Animals under the relation of the distribution of Hair.

    Among those which carry a Mane he cites the Bonassus or Aurochs (Bos Urus) and then three animals of India, some of which have only recently been re-discovered:[1]—and all happen to be living at the present moment in the Gardens of the Zoological Society:—I allude to the *Hippelaphus* or Horse-Stag; the *Hippardium* or Hunting Leopard;—and the Buffalo.

    After having terminated all that relates to hair, Aristotle next speaks of the Horns; and on this subject he lays down general principles whose accuracy has been confirmed by all succeeding observations.—Let us instance the following:

> No Animal, he says, has Horns which has the Hoof undivided: but the in-
> verse rule does not hold; and thus the Camel which has a divided hoof,
53   bears no horns. Those Animals which have divided hoofs, [/] Horns, and
> no Teeth in the upper jaw, all ruminate and reciprocally there is not a
> single ruminant which has not all those three characters.
>
>   Horns are hollow, or solid; the former are persistent; the latter caducous,
> and renewed every year.

    With respect to the other modifications of Cuticle we find observations which show the same astonishing extent of observation, and powers of accurately generalizing:—thus Aristotle remarks that no Bird is armed at the same time with Spurs and with Claws. This, in the Opinion of Cuvier, is a proposition which one is astonished to find in the Science almost at its Birth.

    Nearly all the most interesting Preparations which Hunter has assembled in his Series of Peculiarities, illustrate powers and habits of Animals which were known to, and described by, Aristotle:—Thus he speaks of different poisonous Animals;—of the Shock which the Torpedo gives when taken in the hand;—of the manner in which the Cuttle fish hides itself from the pursuit of its enemies, by discolouring the water with its Ink:—of the strong odours defensively emitted by certain animals:—the powers which Snakes possess of 54 casting [/] and renewing their Cuticle;—and the singular reproduction of the tail in Lizards.

    He also notices the presence of hair instead of Teeth, in the mouth of the Whale, apparently from having observed the fringed extremities of the whale bone, which would be alone visible in looking into the mouth of a stranded Whale.

    In speaking of the Voice, he distinguishes properly the true voice produced by the transmission of Air from the lungs, from the different noises which some Animals make. He describes on this occasion the peculiar musi-

---

1   discovered:] *followed by erasure of* 'the Hippelaphus, Horse-Stag,—the Hippardium or Hunting Leopard;'.

cal apparatus of Grasshoppers and Cicadæ.

All the knowledge possessed by Aristotle on the Organization of Animals which we have just been considering, is, however, trifling in comparison with the astonishing extent of his Researches on Generation. It is here that we find some of the beautiful Observations recorded by Hunter in the Animal Œconomy anticipated. Aristotle was the first who noticed the periodical variation of size in the Testes of Birds. He knew that there were certain quadrupeds in which the Testes did not descend into the Scrotum; and notes
55   the Hedge-hog, the [/] Elephant, and the Dolphin as examples of these Testicouda.[1]

56   We have seen that the development of the Embryo and Fœtus (that most interesting of all the branches of Comparative Anatomy) [/] had attracted the attention of Alcmæon, and, the facilities which the Ova[2] of Birds afford for pursuing this inquiry did not escape the penetration of Aristotle; and he accordingly studied with remarkable perseverance and success the development of the Chick during Incubation.

He describes it day by day, and speaks of the heart as the first point which appears:—of the vitelline veins which then extend from the superior to the inferior parts of the body;—and of the allantoid vesicle which soon incloses the whole Egg.

It must be remarked that all these Observations were made with the eye alone, and that the[3] errors and omissions[4] which occur, arise intirely from Aristotle not having had the assistance of Magnifying Glasses. The first emendation of these errors, the first additions to this branch of Physiology appears in the time of Harvey, nearly two thousand years after the existence of the extraordinary Work we are considering.

Yet a few more facts on Generation.

57   Aristotle studied the Ova of Reptiles and Fishes:—he divides the former into eggs with a hard shell, as those of Crocodiles [/] and of Tortoises; and eggs with a soft envelope, as those of Serpents. He knew that certain species

---

1   Testicouda] *followed by deletion by vertical pencil scoring of:* "{¶} In this part of his Work, however, where most is attempted, we find errors as well as omissions:—but the former will frequently be found to depend on a peculiarity of thinking, rather than defective observation:—Thus, being influenced in his Idea of a Testicle by **form* and *substance {pencil underline }* —He does not admit their existence in animals where they are flattened, elongated, and pulpy;—as when he says that "No Fishes have testicles, or any Animal that has Gills:—nor the whole Genus of Serpents;—nor, in short, any Animal without feet; judiciously excepting the Cetacea, because they generate an animal within themselves.["] {¶} Although Aristotle asserts that the Testes prepare the Seminal fluid by a *vis insita* he afterwards*expresses a {interlined above deletion of 'seems'} *doubt {modified from 'doubtful'} concerning their use and importance; and seems to think that in many animals they are rather organs of *mere {interlined with caret } convenience, designed by Nature to prevent a retraction of the Spermatic Chord.'.

2   Ova] *modified from* 'Ovum'.

3   the] *deleting* 'slight'.

4   and omissions] *interlined.*

of this Order of Reptiles brought their young into the World alive; but he takes care to distinguish this mode of production from that of his true viviparous quadrupeds; and states, in language which might be used by the best informed Physiologist of the present day, that the Viper generates by Eggs, but that these eggs are hatched in their insides.

With respect to the Ova of Fishes he appears to have studied them as extensively as the eggs of Birds, for he advances to the Enunciation of this general proposition:—That the allantoid membrane does not exist in the Eggs of Fishes, nor in those of any Animal which respires by Branchiæ.

Here also, as usual, he notes the exceptions to the ordinary mode of Reproduction in the Class, with reference to the Fishes which are viviparous; but being apparently unacquainted with the Blenny, his generalization is not exact where he says that all viviparous fishes are Cartilaginous.

With respect to the Oviparous Fishes, Aristotle well knew their[1] characteristic indifference to the fate of their Ova; but he also mentions a remarkable
58 exception with[2] [/] reference to a Fish which he calls Phycis.

This species, he says, makes a nest *like a bird's*.[3] For a long time the thing was treated as a fable; recently, however, an Italian Naturalist has discovered that the Gobius niger has this very habit. The male, in the season of Love, makes a hole in the sand; surrounds it with Seaweed, making a true nest, near which his mate waits, and which he never leaves until the eggs which have been deposited are hatched.

Aristotle's knowledge of the Generative Œconomy of Cephalopods still[4] surpasses that of the moderns, and much observation of their living habits is necessary to verify and understand the descriptions of the Stagyrite with reference to this singular and interesting class of Animals.

He enjoyed an advantage that no Modern Naturalist can hope to possess:—the labours of a thousand Fishermen were placed by his truly great Sovereign at this command.

But I shall not dwell on this wonderful part of the *Historia Animalium*.[5] It is literally true what Cuvier states, of the superiority of Aristotle's Anatomical
59 knowledge of the Cephalopods, [/] as compared with that of any of the Moderns before Cuvier.

But I have been surprized to find, in reading the elaborate refutation by Cuvier of the proposition maintained by Geoffroy Saint Hilaire, relative to the unity of composition in the Vertebrate and Molluscous Animals that Cuvier no

---

1 their] *inserted over erasure of* 'the'.
2 With. . .exception with] *faint vertical pencil score as if deleted and then restored.*
3 *bird's*.] *period inserted [ed.].*
4 still] *interlined with caret.*
5 *Historia animalium.*] *double pencil underline followed by vertical pencil deletion of* 'as *men having [interlined in pencil to replace* 'I have'} elsewhere contrasted Aristotle's Anatomical, Physiological, and Zoological observations with our present state of Knowledge regarding the Cephalopodous class.'. *Period inserted after* '*animalium*'[ed.].

where alludes to the fact that the same comparison of a Cuttle-fish[1] with a vertebrate Animal bent double, by which Geoffroy believed that the interval between the two was destroyed.—This very comparison I say, was made by Aristotle in his Philosophical review of the different plans of Organization which the Sanguineous, Molluscous, and Crustaceous Animals present.

In this part of his Work, Aristotle refers to a Diagram or illustration (the first Anatomical figure on record) by which the text was rendered more intelligible. After remarking that some animals have their locomotive members placed at the two extremes[2] of the body, he says "But the Cuttle-fish has all its members proceeding from the Head, as if the posterior part had been bent to the anterior part of the body, and the two extremes blended together."

He then proceeds to explain his idea by referring to the letters of the Diagram:[3]

60   With respect to Insects, Aristotle gives a multitude of most interesting details, whose accuracy we have the Authority of Cuvier for considering[4] as perfect.

He speaks of the œconomy of Bees, and alludes to a belief that what some persons consider the King, was assuredly a Female. He describes the kind of cell constructed for this privileged individual, which shows that he had observed the interior of hives.

He treats then of the domestic œconomy of Wasps, of Hornets, of Mason-Bees, and of Drones. He describes the singular covering which envelopes the larva of *Phryganea* or Caddis-fly[5] and speaks of the Spiders

61   which carry under the [/] abdomen the capsule which contains their eggs.

He exposes the Metamorphoses of Insects which before acquiring their last form, pass through the states of Larva and of Chrysalis. He knew also those incomplete Metamorphoses in which the larva differs from the perfect insect merely by the absence of Wings, and undergoes only a single transformation.

Aristotle admits however spontaneous Generation in those Animals, and worms, which he calls Apods;—and thinks that when the constituent

---

1   Cuttle-fish] *note on facing fol. 60*[v] '(The *mollia* & *Turbinated Testacea*)'.
2   extremes] *with* 'emes' *interlined in pencil above deletion of* 'ities'.
3   Diagram:] *Followed by* 'De part. Anim.' *at page break with pencil deletion beginning at top of fol. 60 of* 'Anim: lib:IV.C:9. {¶} With respect to the Generation of the Cephalopods, Aristotle's knowledge surpassed that of Hunter. From a MSS. description of a dissection of the Cuttle-fish which is preserved, it appears that Hunter believed them to be Hermaphrodite:—Aristotle knew that they were of distinct Sexes. {¶} He describes the characteristic coverings of the Egg of the different Genera of Cephalopods, only one of the forms of which appears to have been known to Hunter. Aristotle also contrasts the peculiar mode in which the yolk is connected to the Embryo in this class and in Birds.'.
4   for considering as] *inserted, with* 'considering as' *interlined above deletion of* 'stating to be'.
5   or Caddis-fly] *interlined with caret.*

Elements are found in certain proportions, and in favourable circumstances, in the hands of Ehrenberg has only the other day dissipated the same error which had attached itself to the production of the Animalcules of Infusions.

I regard the contemplation of the extensive Series of Preparations of the Generative Organs, accumulated and arranged by Hunter, as greatly aiding the[1] study and comprehension[2] of those parts of the writings of Aristotle which relate[3] to the same function:—perhaps there is no anatomical collection in Europe which more fully and more closely illustrates the Text of the Stagy-
62 rite, than that of Hunter.

But[4] it must always be remembered that the facts which are connected together by a full perception of their physiological relations, in the Museum of Hunter, are scattered in a certain confusion through the pages of Aristotle; and I have brought them together in parallel series to facilitate a comparison of the two observers in the light of *simple collections of facts,* by no means intending to extend the comparison to the higher department[5] of deducing the physiological consequences from the observed Phenomena—Here, however, we must at the same time[6] allow that the labours & discoveries of his predecessors placed Hunter on an eminence from which Aristotle could never have surveyed the varieties of animal organization: He was ignorant of the Circulation[7], of the property of the muscular fibre and of the great[8] functions of the nervous system.—

But the state of Physiological knowledge in the time of Aristotle only serves to increase our admiration of the amount of positive[9] facts which, without[10] the guide of physiology[11], he succeeded in determining. Besides,[12] Natural History[13] formed a small proportion of that vast expanse of knowledge which the all-commanding intellect of Aristotle could grasp & elevate[14], & it is only in this branch that we can venture to compare the labours of Hunter with those of the Stagyrite—Here, however[15], we find the same spirit

---

1   greatly . . .the] *interlined above deletion of* 'highly necessary in' *with* 'studying' *modified to* 'study'.
2   comprehension of] *interlined* 'sion of' *above deletion* 'ing'.
3   relate] *deletion of terminal* 'd'.
4   But] *inserting long note from facing 63ᵛ.*
5   department] *interlined above deleted* 'justification'.
6   at the same time] *interlined with caret.*
7   of the Circulation] *interlined with caret.*
8   great] *interlined with caret.*
9   positive] *interlined with caret.*
10 without] *after deleted* 'the energy of his *character {in pencil above deleted* 'intellect'}'.
11 physiology,] *before deletion of* 'enabled him to accumulate'.
12 Besides,] *interlined with caret. Period inserted for grammar [ed.].*
13 History] *before deleted* ', too'.
14 grasp & elevate] *interlined in pencil to replace* 'comprehended {illeg. word}'; *with deletion of* 'and if we acknowledge him inferior to Hunter in Zootomy—in all else'.
15 But. . .however] *inserted from facing fol. 63ᵛ with insertion point marked with* 'Ø' *and deletion of* '{¶} And here as elsewhere'.

of universal inquiry; the same ardent love[1] of truth guiding both Philosophers in their investigations.

In seeking a knowledge of Nature[2] they applied to no inferior or secondary source of Information.—Her alone they interrogated, and she[3] responded freely to them as to Friends.

To some, perhaps, I may appear to have dwelt at unnecessary length on the writings of an Antient Author; which, allowing them to be what I have represented, may still[4] be supposed to have but a remote relation to[5] the Science of Comparative Anatomy, as it is now known. I have been desirous to be as brief as possible; and have limited myself to the notice of such observations in the 'Historia Animalium'[6] as relate to their Organs, and to their uses. These, however, form but a small proportion of the Work:—Besides the Classification and External descriptions of Animals, Aristotle treats of their Intelligence and Instincts; their Animosities and Friendships; their Stratagems;—the Hiding and hybernation of Bears and Reptiles:—the concealments of Insects;—the Nidification and Migrations of Birds;—the Migration of Fishes;—the Geographical distribution of Animals;—the influence of Place, of Sex, and of Gestation on their habits. He treats also of their Maladies, and the appropriate remedies.

63   The diseases of the Elephant are no-where more fully described. Among[7] the Entozoa or Internal Parasites of animal bodies[8] Aristotle distinguishes[9] the *Elminthes platesii* or Tape Worms, the *Elminthes strongulus* or Nematoid worms, of which he mentions the Parasites of the small intestine, the Ascaris lumbrusudi of the fundus; & the smaller species which infests the rectum—the Oxyurus vermicularis—In alluding to[10] the diseases of the Hog he mentions the[11] parasitic[12] hydatids which occasion what is called[13] measly Pork:—And gives a graphic account of Hydrophobia. In describing the diseases of Fishes he makes the curious observation that they have no contagious Maladies.

As these Chapters belong to Zoology more exclusively, a cursory allu-

---

1   love] *deleting repeated* 'love' *[ed.]*.
2   Nature] *after deletion of* 'The works of'.
3   she] *interlined above deletion of* 'Nature'.
4   may still] *inserted in margin after deletion of* 'must' *with* 'still' *interlined with caret.*
5   have. . .to] *interlined above deletion of* 'afford but a very inadequate' *with* 'relation to' *interlined below deleted* 'notion of'.
6   Historia Animalium] *emended from* 'History of Animals'.
7   Among] *inserting at this point long note from facing fol.* 64[v].
8   of. . .bodies] *interlined with caret.*
9   distinguishes] *over pencil erasure.*
10 In. . .to] *interlined above deletion of* 'He describes how'
11 Among. . .mentions the] *inserted.from facing fol.* 64[v] *with insertion point marked with* 'Ø'.
12 parasitic] *deletion of repeated* 'the' *[ed.]*.
13 what. . .called] *interlined with caret.*

sion to them is all that our present purpose demands;—but I should be doing injustice both to the extraordinary work under consideration, and to the proper subject of these Lectures if I were not lastly to notice the Principles laid down by Aristotle at the Commencement of his Treatise[1] as our Guides in the Study of the Organization of Animals.

The parts of Animals, he says, either agree with, or differ from one another, in four principal ways:—and accordingly, they may be considered or arranged

1st. Κατα το γενοσ, or Κατ. ειδοσ; i.e. According to the Genus, or Natural Group of Animals.

Now this is the plan[2] on which the parts of Animals were described by Professor Green, in the noble[3] Course of Comparative Anatomy which he delivered[4] in the Theatre of this College in the years 1824, 5, 6, & 7.[5] It is the principle on which the Specimens in the Cuvierian Museum of Comparative Anatomy are arranged;—It is the mode in which the organs of Animals are studied in reference to the[6] Zoological affinities of the[7] Natural System.[8]

64   *2ndly*. Aristotle says, Organs may be arranged according to Excess, or Defect. (καθ' υπερχην και ελλειψιν.) As Examples of this rule, he instances the Classification of the Skeleton of Fishes according as they are more or less hard, into Bony or Cartilaginous:—That of the Beaks of Birds into long and short; &c.—Stomachs are arranged according to this plan; where those consisting of one cavity are grouped together;—while those having many cavities form another Class. The differences of Multitude or paucity of parts of an organ, of its relative magnitude, shape & colour, are also considered under this head—[9]

In short, it is precisely according to this Second Principle laid down by Aristotle, that Hunter[10] has arranged all the Organs of the Animal Kingdom in the Gallery of the Museum. It is the principle upon which Comparative Anatomy is studied with reference to Physiology.

3rdly.[11] Organs may be arranged or considered[12] according to their Situation (κατα τψν θεδιν). In his illustrations of this rule Aristotle however exempli-

---

1   at. . .Treatise] *interlined with caret.*
2   plan] *over erasure.*
3   noble] *interlined above deleted* 'only complete'.
4   which. . .delivered] *over erasure.*
5   in the years. . .7.] *interlined.*
6   the] *altered from* 'their'.
7   of the] *interlined in pencil to replace* 'and the'.
8   Natural System] *over erasure. Period inserted [ed.].*
9   The difference. . .head—] *insertion from facing fol. 65ᵛ with insert point marked with caret.*
10 Hunter] *with deletion of interlined* 'studied and'.
11 *3rdly*.] *period added for consistency [ed.] Also inserted after* θεδιν.
12 or considered] *interlined in pencil with caret.*

fies classifications of animals rather than that of their[1] organs, as where he instances animals[2] having Pectoral Mammae which in others are abdominal[3]. But This mode of studying the Variation of the position of similar parts has been justly stated by one of our own most eminent Naturalists, M[r]. Macleay[4], to be one of the most important considerations in Zoology.

4thly: Organs may be arranged (κατ: αναλογιαν[5]) according to their Analogies:, as for Example, when we compare the Claw of a Lion[6] with the[7] Hoof of a horse[8];—or the feather of a Bird with the Scale of a Fish:—Or, as

65   Aristotle does, in another [/] part of his Work, compare the Wing of a Bird with the fore-foot of a Quadruped.

Now this comparison of Organs, with reference to their analogies or homologies[9], has been one of the objects[10] of the Comparative Anatomist ever since Aristotle established the Science. Belon, one of the earliest Naturalists after the revival of Literature, extends the Comparison which Aristotle had established between the Anterior extremities of the Bird and Mammifer; and carries on the Analogy from the whole of these apparently different locomotive organs to their respective Component parts;—Showing that in both there existed the same Scapula, Humerus, the two bones of the fore-arm, Carpus, Metacarpus, and digits:—and he illustrates this unity of composition by introducing the figures of a Skeleton of a Bird and of a Man, side by side, and denoting the Analogous bones by the same Letters.

My learned Colleague has already adduced the remarkable expressions of the great Newton in reference to[11] this subject—And Newton sums up the series of instances by which he illustrates this unity of organization in the Vertebrate classes by the following remarkable expression—[12] "In corporibus animalium, in omnibus *fere*, similiter posita omnia."[13]

The great Newton perceived the general application of this primary point

---

1   that of their] *interlined over pencil.*
2   In. . .animals] *insertion from facing fol. 65[v] with insertion marked by caret to replace pencil deletion of* 'as, for instance, Animals having Pectoral Mammae which in others are abdominal.'.
3   having. . .abdominal] *restoring crossed out phrase from fol. 64 for grammatical sense [ed.].*
4   Mr. Macleay] 'Mr.' *inserted in margin and deletion of* 'W. S.'.
5   Κατ: αναλογιαν] *transposed from before* 'as for,' *with insertion marked by caret.*
6   of a Lion] *interlined in pencil with caret.*
7   the] *in pencil over over* 'a'.
8   of a horse] *interlined in pencil.*
9   , with. . .homologies] *insertion from facing fol. 66[v] with insertion point marked with caret.*
10   objects] *after deletion of* 'principal'.
11   to] *followed by deletion of* 'ye sameness of plan which is discernible in the position of ye parts of the vertebrate animals—'.
12   My learned. . expression] *all inserted from facing fol. 66[v] to be integrated with original footnote:* 'Newton. "In corporibus animalium, in omnibus fere, similiter posita omnia.".
13   "In. . .omnia."] *incorporated into text from footnote [ed.].*

in regard to the analogies of position in the corresponding parts of animals, &[1] the illustrious Harvey extends the field in which this fruitful principle may be applied; by instituting a Comparison between a simple Organ in an inferior Animal with the simple condition of the same organ in a higher Species; and this in not equivocal or obscure Language, as we shall hereafter show.

66   Hunter was guided[2] by the same principle when he determined the Optic lobes in the Brain of Fishes to be the analogues of the Nates in the Human Brain:—and when he compares a Congenital Malformation with a certain stage of the incompletely developed Embryo.

One of the most beautiful of the Essays of Vicq d'Azyr in Comparative Anatomy is a Philosophical Comparison of the bones[3] and muscles of the upper[4] & lower, or the pectoral & pelvic[5] extremities in the[6] vertebrate animals—in which many of these analogies, more recently insisted upon by Meckel are pointed out—

This essay[7] you will find in the Memoirs of the French Academy,—[8] of which the Series in the College Library is remarkably complete.[9]

Of late years the Organization of Animals has been subjected to more minute comparison with reference to the points of agreement in detail:—some Anatomists have almost exclusively studied the animal frame with the view of establishing the Analogies of its different parts:[10] &—Believing they were working out an intirely new principle; and viewing the facts of Anatomy through a new medium, have[11] *dignified* their speculations[12] with the Epithet Transcendental! or by Terms implying theirs' to be the only *philosophical* mode of considering the Subject.

But, when Cuvier (who, however, has more sober views) traced the Analogy between the small and insignificant coracoid process in Man, and the large posterior Clavicle in Birds and Reptiles, which he calls the Coracoid:—

---

1   The great. . .&] *interlined insertion in pencil at bottom of page over footnote quote from Newton and continued on to fol. 66ᵛ with insertion point indicated by pencil line.* 'The illustrious' *rendered in l.c. for grammatical sense [ed.].*

2   was guided] *with two illegible words interlined by RO in pencil above.*

3   bones] *restoring cross-out with stet marks below.*

4   upper] *restoring cross-out with stet marks.*

5   pectoral and pelvic] *interlined with caret.*

6   the] *reading uncertain [ed.].*

7   essay] *interlined above deletion.*

8   Academy,—] *before deletion of* 'in our Library'.

9   One. . .complete—] *inserted note facing fol. 67ᵛ with insertion point marked with lead line.*

10 parts:] *before deletion of interlined:* 'Those who have been guided in the determination of those analogies *by a priori views {continued on facing fol. 67ᵛ}'* *with* '& —' *inserted in margin.*

11 have] *after deletion of* 'they' *with deletion of second interlined* 'have' *before* 'dignified' *[ed.].*

12 speculations] *interlined above deleted* 'labours' *with additional deletion of* 'views' *written below* 'labours'.

and again, when Geoffroy Saint Hilaire pursued the interesting comparison of the Bones of the Head of a Mammal[1] in the Fœtal State with those of the Reptile in Adult age;—these estimable Observers only extended the System of Analogies which Aristotle had established, and Belon and others had followed:—they did no more than add to the Antient and acknowledged bases of Philosophical Zoology.[2]

67   But, from the number of unexpected correspondences which this kind of comparison has established within certain limits, some Anatomists, and especially those whose knowledge happened to be limited to a single great Division of the Animal Kingdom, were led to form Theories which are undoubtedly new, as regards their extravagance: assuming in[3] these Speculations, that[4] Nature is[5] restricted in the development of Animals to a supposed unity of Composition;—a unity of plan;—and a constancy of Connexions[6] & also to a certain number and kind of component parts, which are all determined by *a priori* theory;[7] they[8] have proposed the most extraordinary Analogies.

Thus, the parts of the Tegumentary Skeleton which are retained in Fishes to be subservient to a mode of Respiration peculiar to that Class; (I allude to the bones of the Operculum or Gill-cover) have been gravely stated to represent the Malleus, Incus, and Stapideus;—and even the little Orbicular epiphysis has had its representative.

These, and many similar views propounded by Spix & Geoffroy Saint Hilaire, and adopted by some of their[9] followers, are the result of an abuse of a sound and fruitful Principle, which has only suffered by the unwarrantable extent, and unjustifiable mode, of its application.[10]

68   I know not,[11] [/] if its importance will be better appreciated, by stating that it forms the very Foundation of the Philosophie Anatomique of M. Geoffroy Saint Hilaire:— But as M^r. Macleay, has well observed,[12] if that

---

1 Mammal] 'al' *interlined over deletion* 'ifer'.
2 Zoology] *followed by three vertical pencil lines.*
3 assuming in] *interlined with pencil caret above deletion of* 'and'.
4 Speculations that] 'ions that' *interlined above deletion of* 'ors assuming'.
5 is] *pencil insertion to replace deletion* 'to be'.
6 —and. . .Connexions] *faint pencil parentheses around this.*
7 &. . .theory;] *inserted from facing fol 68^v with* '& also' *and* 'certain' *interlined in insertion and final* 'and' *after* 'theory;' *deleted in pencil. Insert point marked with pencil caret.*
8 they] *interlined above pencil insert caret for insertion from fol 68^v.*
9 views. . .their] *phrase revised with* 'views' *interlined above deletion of* 'Ideas'; 'Spix &' *inserted over erasure; and* 'their' *over* 'his'.
10 application] *followed by vertical pencil scoring for deletion of* 'But this principle was established 2000 Years ago; and has ever since been operative in the advancement of Zootomical Science.'.
11 not,] *on page break.*
12 as. . .observed,] *on facing fol. 69^v with insertion point marked with a lead line and caret.*

Naturalist arrived at the first Idea of his "Principe des Connexions" by Inspiration, as he tells us (Anatome Philosophique, p. 30.)[1] — "We are certainly justified in believing that Aristotle when he laid down the Rules for studying the Animal Organization, Κατ' αναλογιαν, και, κατα τψν Θεσιν, must also have been inspired before him."[2]

Our[3] Library is peculiarly rich in the Editions of the Works of Aristotle—especially of the Treatise on History of Animals, which Cuvier justly pronounces to be one of the most admirable works that Antiquity has bequeathed to us, and one of the grandest monuments which the genius of man has raised to the natural Sciences[4]—Here is a rare & beautiful copy of[5] an *Editio Princeps* of the first[6] Translation of this History of Animals[7] by Theodore Gaza, a learned Greek to whom Western Europe was indepted for the first knowledge of this admirable Treatise—It was one of the singular events contributing to accelerate the revival[8] of letters which resulted from the Capture of Constantinople by the Turks—when Theodore Gaza, Angynophilus, & other learned men fled to the west, bringing with them many manuscripts[9] of the ancient Greek Philosophers & amongst this precious one for Zoology—the Περι Τοε Ζοον—The latin translation of Julius Scaliger—also in the Library—is much preferable however to that of Gaza.[10] In the Complete Collection of all Aristotles Treatises by Duval in the french Translation by Camus,[11] the readings &[12] disposition of the text is much the same as that of Scaliger.[13] And we have an[14] interesting translation of the History of Animals[15] into our own language by the indefatigable[16] Taylor—which is as correct as can be expected from a man who was neither Zoologist nor anatomist—[17]

---

1   30.] *Owen's footnote to text indicated by asterisk after* 'us'*moved into text [ed.].*

2   "We. . .him."] *In single quotes. Modified to double quotes for consistency with previous quote [ed.]. Followed by horizontal line to indicate pause.*

3   Our] *preceded by* 'note' *in lower case deleted [ed.]*

4   which. . .Sciences] *interlined in passage inserted from facing fol 69ᵛ with insertion point marked with lead line.*

5   a rare. . .of] *interlined in insertion.*

6   first] *interlined with caret.*

7   History of Animals] *interlined in inserted passage to replace deletion of* 'ιμπαρτον Ζοον' *with deletion of interlined* 'Treatise' *above and* 'History' *below.*

8   contributing. . .revival] *interlined below deletion of* 'tending to the advancement of'.

9   manuscripts] *following deletion of initial* 'p' *of an intended continuation.*

10   Note. . .Gaza] *All insert from facing fol 69ᵛ. Ink line beginning on 68 and running to 69ᵛ to separate this passage from next insertion.*

11   In. . .Camus,] *inserting interlined* 'Complete. . .Duval' *with comma inserted [ed.].*

12   readings &] *interlined with caret.*

13   Scaliger] *period inserted [ed.].*

14   an] *altered from* 'a'*[ed.[.*

15   of. . .Animals] *interlined with caret.*

16   the indefatigable] *interlined with caret.*

17   anatomist—] *followed by* 'Schniedr—best—30—years—.1811.'.

In taking leave of the Father of Natural History, and of Philosophical Anatomy, we shall find but little to detain us in relation to these Sciences from the[1] period of Aristotle[2] to the Revival of Letters and the establishment of the inductive method in Philosophy. But in this interval, the knowledge of Human Anatomy was greatly extended, and some important steps were made in Physiology.

Theophrastus, the Successor of Aristotle, in the School of the Lyceum, is best known by his *History of Plants*, which is a Work similar in design to Aristotle's History of Animals:—and as M[r]. Hunter has drawn largely from Vegetables for preparations in illustration of the laws of Organic Life; we may mention that the Work of Theophrastus[3] contains the earliest Account of the Anatomy and Physiology of Vegetables:

69　　　　—In it, the organs of Fructification of Plants [/] are[4] described, and the distinct Sexes of the Date-Tree are mentioned:—and[5] in treating of Leaves, he makes the very just remark, that the inferior surface of these Organs is more absorbent than the Superior. As[6] Aristotle is[7] the Father of Zoological Science so must Theophrastus be regarded as the founder of ye Botanical &[8] Mineralogical[9] Science—he[10] was the[11] first who[12] distinguished pearls from precious stones, and described them as an animal production formed by[13] an Indian species of shell-fish.[14]

From this period, Philosophy declined in Athens: henceforward[15] a subjugated dependency on the Macedonian Monarchy. But the Sciences took refuge in Alexandria, and found protection and princely encouragement in the Courts of the Ptolomies. These enlightened Princes,[16] emulous of contributing to the advancement[17] of Human knowledge,[18] founded successively in

---

1　the] *over erasure.*
2　of Aristotle] *interlined with caret.*
3　Theophrastus] *with pencil* 'X'*in margin.*
4　are] *in margin in pencil* '83– 09/74' *to indicate pagination of lectures to this point.*
5　and] *following ink caret then crossed out in pencil to indicate intended insertion point of deleted* 'The majority of' *on facing 70*[v].
6　As. . .shell fish] *insertion from facing fol. 70*[v] *with insertion point on fol. 69 marked by ink* '<'.
7　is] *interlined in insertion on facing fol. 70*[v] *above deletion* 'was'.
8　be. . .&] *interlined above deletion of* 'is the Father of'.
9　Minerological] *before deletion of* 'as well as Botanical'.
10　he] *after deletion of* 'and'.
11　was the] *interlined with caret.*
12　who] *interlined with caret.*
13　formed by] *interlined above deletion of* 'produced from'.
14　species. . .fish.] *interlined above deletion of* 'species of Testaceous animal.—'.
15　henceforward] *terminal* 'ward' *interlined with caret to revise* 'henceforth'.
16　Princes,] *comma inserted [ed.].*
17　advancement] *terminal* 'ment' *interlined with caret and in margin faint WC pencil* 'advancemt. ?'.
18　knowledge] *comma inserted [ed.].*

Egypt[1] a public Library, a Menagerie, and a Botanic Garden:—And in the Alexandrian Schools where the Sciences were taught, permission was first given to dissect the Human body. It was hither that Galen resorted, Centuries afterwards, for the purpose of acquiring Anatomical Knowledge.

It is chiefly[2] by the writings of Rufus Ephesius[3] that we become acquainted with the progress made in Anatomical Science in the intervening period under the advantages we have just been describing.

It was in the Alexandrian School that the great physiological discovery of the general Nature and Functions of the Nervous System was made. To Herophilus, a Physician, and one of the Asclepiadæ belongs the merit of having first recognised in the Nerves the organs of Sensation and Volition; and of having carefully and accurately[4] distinguished them from the other *neura* of the Ancients[5] i.e.[6] the Tendons, Ligaments, and other white Tissues with which they had previously been confounded. This must ever be regarded
70   as one [/] of the Æras in the History of Physiology and Anatomy.

Herophilus also described different parts of the Brain, i.e.[7] the curvature of the Corpora Striata; the choroid plexus; the calamus scriptorius;—and lastly, that vascular arrangement which to this day bears the name of The *Torcular Herophili.*

Herophilus was the first Anatomist who taught osteology from the Human Skeleton. He also speaks of the happy disposition of the Muscles for the movement of the limbs; whence we may presume that the function of the muscles was then known. According to Galen, this important discovery is due to Lycus of Macedon, who wrote a voluminous work now lost[8] on Myology, in which the four straight muscles of the Eye are first[9] described.

Erasistratus: on motor and sensory nerves, a contemporary of

---

1   in Egypt] *RO interlined in pencil.*
2   chiefly] *interlined above deletion of* 'only'.
3   Ruphus Ephesius] *interlined in ink over pencil with caret over apparent deletion of* 'Galen' *with pencil cross referring to note on facing fol. 70v:* 'Rufus Ephesius who wrote 'De Partium Corp[ora] Hum[ani] appellat. Ed. Clinch. London, 1726. 4 to p. 65. Thus observes: "Secundum Erasistratum quidem et Herophilum sensorsi nervi sunt. Asclepiades autem ipsos, sensu vocare testatus est: cæterum secundum Erasistratum cùm gemina nervorum *(νευρων) [from facing Greek}* natura sit, sensoriorum (αισΘητιξιον) videlicet, atque moventium (κινητινων), sensorsi, qui cavi sunt, in cerebri membranis originem habent, moventes in cerebro ac cerebello. Dixit autem Herophilus aliquosos esse voluntati obedientes nervos, qui & à cerebro spinalique medulla oriuntur, aliquos qui ab esse orientes in os inseruntur: aliquos à musculo in musculum transeuntes, qui articulos etiam copulant—". *Minor grammatical changes in quote inserted in accord with the published text [ed.].*
4   and accurately] *interlined in pencil above* 'carefully' *with double repeat of* 'accurately'.
5   of the Ancients] *interline in pencil with pencil* 'X?' *in margin.*
6   i.e.] *faint pencil underline.*
7   i.e.] *inserted in margin on the line.*
8   now lost] *interlined in pencil.*
9   first] *interlined below line in pencil with caret.*

Herophilus, and a Grandson of Aristotle, also devoted himself ardently to the Study of Anatomy. To him we are indebted for the capital discovery that all the Nerves terminate in the Brain, either immediately, or through the medium of the Spinal marrow,[1] but not as we now know them.[2] Erasistratus also[3] distinguished nerves of Sensation & nerves of motion.[4] According to him, the nerves of sensation (ν. αισθητικα) come exclusively[5] from the membrane of the Cerebrum; the nerves of motion (ν. κινητικα) from both cerebrum & cerebellum. Had he spoken of the sensitive and motor tracts of the spinal chord this Ancient anatomist & Physiologist would have almost expressed our present state of knowledge as established by a very recent discovery of the connexion of the anterior or motor tracts with the cerebellum.[6]

Erasistratus also[7] studied the Comparative anatomy of the Brain, and contrasted that of Man with the brain of a great number of Animals;—not only in a general manner, as Aristotle had done, but taking it part by part.

Erasistratus[8] was the first who asserted that digestion was performed by the *action* of the Stomach;—though without any suspicion of the nature and properties of the gastric juice. Before his time the stomach was supposed to be a mere[9] passive receptacle[10] in which the food was macerated and concocted—these processes being assisted by the heat of the liver, the spleen and the viscera lying in the neighbourhood.[11]

He also recognised the lacteal vessels in a kid which had just suckled but[12] as D^r. Barclay well observes, he neither traced their [/] connections, followed their course, or explained their peculiar functions.

He was acquainted with the internal valves of the heart, and described the tricuspid valve; but with respect to the functions of the vascular System these Alexandrian Philosophers were less successful.

Herophilus had recognised the isochronism of the pulsations of the heart and arteries; and had discovered the pulmonary veins, which he calls the Arterial veins. There wanted but a few more steps towards the true knowledge of the circulation; which, however, neither of these Anatomists arrived

71

---

1   marrow,] *comma inserted [ed.]*
2   but. . .them.] *inserted from facing fol. 71^V with a long lead line to insert point marked with symbol* 'a' *and followed by the long insertion beginning* 'Erasistratus'.
3   also] *interlined in insertion with caret.*
4   of motion] *interlined below line in insertion. Period added for grammatical sense [ed.].*
5   exclusively] *interlined in insertion with caret.*
6   Erasistratus also. . .cerebellum] *all inserted from facing fol 70^V with pencil horizontal line in margin to indicate pause.*
7   Erasistratus also] *interlined below line to resume original text after insertion with deletion of* 'He'.
8   Erasistratus] *with* 'sis' *interlined below over erasure.*
9   mere] *interlined with caret.*
10 receptacle] *interlined above deletion* 'organ' *with period inserted [ed.].*
11 Before. . .neighborhood] *note inserted from facing fol. 71^V with lead line.*
12 but] *interlined with caret.*

at. An interval of more than eighteen Centuries elapsed between those discoveries, and that crowning one which has immortalized our own great Countryman.

The determination of the organs of Sensation and Motion were, however, great steps in Physiology; yet little more was effected in this Science during the Empire of the Ptolomies[1];—we may mention, however, as a fact interesting in Zoology, among the Animals which graced the triumphal processions of these Monarchs the Giraffe and Rhinoceros are, for the first time, described.

The means of advancing Comparative Anatomy were, in fact, peculiarly rich, but the impulse given by Aristotle soon slackened. Some of the learned Men adopting the dreamy philosophy which already began to prevail in [/] the Capital of Egypt, wandered out of the proper direction:—others abandoned themselves to a certain indolence, which made them neglect direct observation.

The writings of the illustrious Galen hold a distinguished place in the history of Human Anatomy; being professedly destined to illustrate the Structure of Man, and the uses of the different parts of his frame. The foundation of the celebrated Works of this great Anatomist was, however the dissection of the lower Animals.

At the period when Galen resorted to Alexandria, although he enjoyed there the advantages of studying an accurate model in Bronze of[2] the Human Skeleton,[3] the practice of dissecting the Human Body had been discontinued. He employed himself therefore in investigating the structure of the Animals which bore the nearest resemblance to Man;—and it is known, that the Magot,[4] or Barbary Ape (Simia Pithecus, Linn.) and not the Orang, as Camper supposed, was the Species from which Galen acquired practically most of his Anatomical knowledge. Yet we cannot rise from the perusal of many parts of Galen's writings without a tendency to a conviction that he *had* dissected the human body.— & how minutely he pursued the comparison he himself tells us:[5] "Simia inter universa animantiam genera, tum visceribus, tum musculis, tum arteriis, tum nervis, simillima homini est" De administr: Anatom: lib.l.c.2.[6]

Galen, however[7] compared the structure of the Ape[8] with that of other

72

---

1 Ptolomies] *altered in pencil from* 'Ptolemies'.
2 an. . .of] *interlined in dark ink with caret.*
3 Skeleton,]*with erased pencil note opposite on facing fol. 73ᵛ:* 'de administrationibus anatomicus *[followed by some illegible words and continuing]* Simia ubi universa animantium genera tum visceribus, tum musculis, tum arterias, tum nervis, similima homini est—De. adm. anat. Lib 1, c.1'
4 Magot] *with eight illegible erased pencil lines alongside on facing fol. 73ᵛ*
5 us:] *colon inserted for grammar [ed.].*
6 Yet. . .lib.l.c.2.] *inserted in dark ink from facing fol. 73ᵛwith insertion point marked with caret.*
7 Galen, however] *inserted in margin to replace deletion of* 'He'.

Quadrupeds, & these[1] with the organization of Birds and Fishes. He states that he had never dissected Insects or minute Animals; but impelled by the de-
73   sire of acquiring just notions in Physiology, he had [/] frequently opened Animals alive.

Comparative Anatomy was, therefore studied by Galen with two objects; first—vicariously, to obtain analogical ideas of the Human organization; and secondly with the more legitimate purpose of determining the uses of the various parts of the Animal Body.

The knowledge which he acquired by this means, is arranged with all the correctness of a Critic, with all the abilities of the first and most accomplished scholar of his time.

The physiological part forms a separate Treatise, under the Title *"De usu partium"* the high merits of which, need no eulogium before this learned audience.

In the Anatomical descriptions, however, the anatomy of the higher Quadrumana is almost every where substituted for that of the Human Species; and the labours of the earlier Anatomists after the revival of Literature were chiefly directed to the correction of the errors with which the otherwise—estimable writings of Galen abound from this cause.

The leading Discoveries of Galen in Anatomy are considered by[2] my revered[3] preceptor in Anatomy, the learned D[r]. Barclay,[4] to be, the determination of the nature of the true contents of the Arteries;—his vivisections and Zootomical researches enabled him completely to overthrow the old and generally received opinion that they were air vessels.[5]
74   Galen flourished under the Emperors Trajan, Marcus Antoninus, and some of their successors; He is the only individual during the Roman Æera whose works claim the attention of the Historian of Comparative Anatomy and Physiology. And it is painful to contrast the unparalleled means for successfully prosecuting Comparative Anatomy which were afforded to the philosophers during the ages in which Rome was the Mistress of the World, with the paucity or rather the nullity of the results which were derived there-

---

(contd.)
8   the Ape] *interlined above deletion of* "this Species, however'.
1   & these] *interlined above deletion of* 'as well as'.
2   by] *preceded by deleted parenthesis.*
3   revered] *before deletion of* 'and lamented'.
4   Barclay,] *followed by deletion of second parentheses.*
5   vessels] *followed by deletion by vertical pencil score of paragraph:* {¶} An opinion that [l *fol. 73*] carried along with it, in its fall, Systems of pathology and Physiology that had lasted for ages, and become venerable even from their Antiquity. {¶}Next to Aristotle, Galen [*interlined over deletion of* 'he'} was the first who mentioned a communication between the branches of the Veins and Arteries, in the substance of the Lungs; and a passage of the blood by that communication from the right to the left Ventricle of the heart. He [*interlined over deletion of* 'Galen'} was the first who mentioned the peculiar structure of the fœtal heart; and the first who clearly demonstrated the larynx at the top of the windpipe and showed it to be the organ of Voice.'

from.

The passion[1] for Spectacles, consisting of the exhibition[2] of savage and rare Animals in the Amphitheatre of Rome, which originated in the politic slaughter, by Curius Dentatus, before the People, of the live Elephants cap-
75 tured at the defeat of Pyrrhus;—continued [/] through every subsequent pe-
riod of the Roman Empire[3].

Metellus having gained a great victory over the Carthaginians captured 142 Elephants which were all killed with arrows in the Circus.[4]
76 Then more probably the fossil remains of some of these gigantic extinct animals which it has been reserved for modern times to rightly interpret, and are interesting as being amongst the earliest[5] historical notices of these phenomena.[6]

In the year 55 before ye[7] Christian æra Pompey showed to the people a Lynx; an Ethiopian Ape (of the Genus *Cephus*); a single-horned Rhinoceros;— 20 Elephants fighting with men;—410 panthers, 600 Lions, of which 315 had manes.

Cicero, who was present at these games, speaks of them with just disdain, and says, that the people at last took pity on the Elephants:—This how-

---

1 The passion] *with pencil line along margin and across part of page presumably to indicate pause mark.*
2 exhibition] *before deletion of* 'and slaughter'.
3 Empire] *followed by deletion by vertical pen scores with bracket along side and* 'next page forwards' *in margin of paragraph:* ''The unbounded extravagance in procuring and transporting to Rome the rarest Animals of distant Climes, and their profuse and indiscriminate slaughter in the Amphitheatre, continued through the first four centuries of the Roman Empire:— We shall adduce only one more instance:— {¶}Commodus is related to have slain, with his own hand, a Tiger, a Hippopotamus, and an Elephant, He let loose in the Circus, a great number of Ostriches, and, as they ran about, shot off their heads with crescent-shaped blades fixed on the points of arrows. {¶}Herodian, who relates the fact, says, that the Birds after being decapitated, continued running about for some time: a phenomenon which has often been witnessed in other Birds in the course of Modern Physiological experiments.' *Compare with passage on fols. 77–78.*
4 Circus] *followed by pencil deletion by vertical scoring of* '{¶}Scylla exhibited more than 100 male Lions. {¶}Emilius Scaurus, during his Ædileship, *was {inserted by RO in ink in margin} distinguished not only by the number of his Animals, but still more by showing many which had never before been seen at Rome. At these spectacles the first Hippopotamus was exhibited {interlined by RO above deleted* 'seen'};—there were also five living Crocodiles 500 panthers; and what appeared more strange, the bones of the Animal to which it was said [l fol. 75] that Andromeda had been exposed. These bones had been brought from the Town of Joppa on the Coast of Palestine. There were among them vertebræ a foot and a half long, and a bone not less than six and thirty feet in length:—probably the lower jaw of a whale.'.
5 earliest] *followed by pencil deletion of* 'notices'.
6 Then. . .phenomena] *RO pencil insertion from facing fol. 77ᵛ with insertion point marked by pencil line.*
7 before ye. . . Æra] *RO alteration in ink from* 'before Christ.' *with pencil interline of* 'Æra' *and on facing fol. 77ᵛ WC pencil query subsequently deleted* 'Year of Romer? [sic]'.

ever was a transient feeling; and the passion for these spectacles arrived at an almost incredible pitch under the reigns of the cruel and debauched Emperors.[1]

77   Galba exhibited an Elephant which walked [/] on a tight-rope to the top of the Theatre, with an armed Roman on his back. These Elephants were taught when young; being born at Rome. Ælian, in whose works many interesting facts in Natural History are recorded, distinctly relates this circumstance, in speaking of the Elephants of Germanicus.[2]

M[r]. Corse, some years ago (1799) in his valuable papers in the Philosophical Transactions, proved in opposition to the opinion of Buffon that with certain precautions Elephants will breed in a state of domestication. But the fact was known in Italy, from the time of Columella.[3]

Commodus is related to have slain with his own hand, at one of these cruel games, a Tiger, a Hippopotamus and an Elephant. He let loose in the Circus a great number of Ostriches, and, as they ran about, shot off their heads with crescent-shaped blades fixed on the points of arrows. Herodian, who relates the fact, says, that the Birds after Being decapitated, continued

78   running about for some time:—A phænomenon [/] which has often been witnessed in other Birds during the course of modern physiological experiments.

Had Aristotle enjoyed the Opportunities thus afforded to the Roman Philosophers of observing the rare animals which have been ennumerated, we should, doubtless, have had the same just and philosophic description of their characteristic qualities, as of the animals whose history formed the wonderful[4] Treatise we have considered in the last lecture.[5]

---

1   Emperors.] *followed by deletion in ink of* '*Of {*inserted in pencil}* The unbounded extravagance in procuring and transporting to Rome the rarest Animals of distant Climes and their profuse and indiscriminate slaughter in the Amphitheatre continued during the first four Centuries of the Roman Empire, we shall adduce only one more instance:—{¶}Claudius, at the dedication of the Pantheon displayed four royal Tigers;— a mosaic pavement which has lasted till our time, represents these Animals of their natural size.'.

2   Germanicus.] *with deleted intended ink insertion by RO on facing fol. 78*[v] 'through every subsequent period of the Roman Empire, Hundreds of rare and curious animals, Lions, Royal Tigers, Leopards, Lynxes, Apes, Rhinoceroses[,] Elephants were sacrificed year after year in the blood-stained Circus. {¶}The Elephants were taught when young, (being born at Rome, to walk on the tight rope, counterfeit maladies, and enact parts like Horses at Astley's Amphitheatre—'.

3   Columella.] *followed by cross-out by vertical pencil scoring of* 'Of the unbounded extravagance in procuring and transporting to Rome the rarest animals of distant climes, and the profuse and indiscriminate slaughter in the Amphitheatre, which continued during the first four Centuries of the Roman Empire, we shall adduce only one more instance:—'.

4   wonderful] *interlined by RO in ink with caret over pencil.*

5   considered. . . lecture.] *RO ink insertion deleting* 'already' *before* 'considered' *and adding* in. . .lecture' *with period inserted [ed.]*

The Hippopotamus was known to Aristotle only through the medium of a vague description by Herodotus. The two-horned Rhinoceros and the Giraffe he had neither seen nor heard of:—Yet all these animals were, on more than one occasion, exhibited in the Roman Circus.

At one Triumph, we are told, that seven Giraffes were slaughtered, to gratify the vitiated[1] Taste of an enslaved and cruel people. But a knowledge of these facts is derived not from the descriptions of the Naturalist, but from the Satire of the Poet,[2] the records and medals of Political History;—or the Mosaic Pavements and ornaments of public Buildings. We can scarce perceive, in the works of the Roman Philosophers, a single spark of that love of Natural Knowledge, which in the time of Alexander the Great, prompted his

79 still greater Master to investigate and record the habits and structure [/] of the animals which a Monarch's conquests, and love of Science, placed at his disposal.

How marked the contrast between the description of the Elephant in the "Historia Animalium" and the crude and casual notice in the Works of *Pliny*! of those rarer animals which more extended conquests brought within his reach and observation.

Such Spectacles as I[3] have been alluding to, were continued uninterruptedly for more than 400 years, and must have afforded to the Roman Philosophers ample opportunities of making observations on the form and organization of foreign animals. Yet it seems, that these animals, once killed, were applied to no farther use.

The Writers of the first, second, & third Centuries of our Æra who have treated on Natural History, have borrowed every thing from the Greek Authors:—Even the Zoological works of the elder Pliny, the greatest Naturalist of the Roman Epoch, contain little more than a translation of Aristotle, interwoven with fabulous narrations which show how little he had imbibed of the true Spirit of his Master.

It is yet not without its use & interest[4] to reflect on the Ages which elapsed before the philosophical study of the Animal kingdom was again re-
80 sumed and advanced beyond the limits it had reached [/] through the labours of Aristotle. It might be naturally[5] supposed that the track of such delightful[6] investigations needed but to have been indicated, to have been eagerly followed; and we ask ourselves the reason of this long-enduring apathy, and of Man's tardy resumption of the healthy exercise of his faculties of research and discovery.

---

1   vitiated] *interlined with caret with illegible pencil erasure on facing fol. 79*v.
2   the Satire. . .Poet,] *interlined in ink by RO with caret.*
3   I] *interlined in pencil by RO above deletion of* 'we'.
4   yet. . .interest] *interlined in ink by RO with caret above deletion of* 'melancholy'.
5   naturally] *with second* 'be' *inserted by RO in ink with caret. Deleted for grammar [ed.].*
6   delightful] *replacing by stet marks original RO cross-out with deletion of interline of* 'exhilarating & remunerative'.

To me it seems evident that some of[1] the noblest motives which now guide the Naturalist[2] in the investigation of the Works of the Creator must have been wanting in men who entertained such erroneous notions of their own nature and relations, as are manifest in the doctrines[3] of the different Sects of the Greek and Roman Philosophers.

We have seen indeed, how much[4] a Prodigy of Intellect could singly effect;—but the appearance of such in the World are phenomena which may truly be compared to Angel's visits, "few, and far between."

Natural Science, however, is happily dependent on a high degree for its successful progress on the application of ordinary powers to careful and patient observation. But the motives for such humble, though essential labours, must have been[5] wanting, in Men who either believed with Plato that the Mind had been once united with, and would again form part of the Divinity, or with Leucippus[6] that its phænomena[7], were the [/] result of a fortuitous concourse of atoms.

81

The followers of one sect hardly deigned to stoop to the observation of natural phænomena, and if these had some little interest for the Materialists, the life-spring of research was wanting:— recognizing[8] no higher Cause than Fate and Chance, they resembled the Animals which wandering in the Woods are fattened with Acorns, but never look upwards to the Tree which affords them Food. Much less have they any Idea of the beneficent Author of the Tree and its Fruit.[9]

The soil was not prepared in those days so that the Seeds of Science which were sown therein[10] could germinate and flourish.

Another stage in the great scheme of the moral education of the World was to be attained; It was necessary for a just appreciation of the subjects adapted to his powers of investigation[11] that man should be taught that if he

---

1   some of] *interlined by RO in ink with caret.*
2   the Naturalist] *interlined by RO in ink with caret over deletion of* 'us'.
3   doctrines] *interlined in ink by RO below deletion of* 'dogmas'; *with* 'Philosophies' *interlined. On facing fol. 81ᵛ in RO pencil subsequently deleted:* 'We*can {*interlined with caret}* trace the *relation between the successful observation**cultivation {*interlined}* of nature & the improvement on the speculations of this greek philosophers as to the origin of natural objects—**successful cultivation of nature {*interlined below}* progress of the Science of Observation in' *with following deletion* '& when Socrates had thought that all *the {*deleted}* nature bespoke'.
4   how much] *interlined above deletion of* 'what' *by RO in ink.*
5   must have been] *interlined by RO in ink over deletion of* 'were'.
6   with Leucippus] *insertion by RO from facing followed 81ᵛ with insertion point marked with caret.*
7   phænomena] *before ink deletion of* 'like all others in the material world,'.
8   recognizing] *preceded by pencil* '/'
9   Fruit] *followed by pencil* '/'.
10 therein] *with RO ink insertion of* 'there' *with deletion of* 'it' *after* 'in'.
11 for. . .investigation] *RO ink insertion from facing fol. 82ᵛ within insertion point indicated by lead line and caret.*

were something less than a Deity he was something more than dust;[1]—it appears to have been essential to the highest and utmost application of those powers[2], that he should know, not only that an Intelligence was superadded to matter, and had[3] presided over its arrangements, but that the Omnipotence to whose Fiat both he and matter owed their existence; had in this world endowed him alone with faculties to appreciate the Works around him; and would, in another, require from him a strict account of their Application. And what wonderful atchievments in science have been effected by those whose minds have been regulated by[4] education based on these principles: but above all by those who have throughout[5] their high career acknowledged these principles as their rule of action!

What name greater than *Newton* can be cited in the annals of Physical Science?

Whose individual labours in the wide field of Natural History can be compared with those of Linnaeus in their extraordinary results?[6]

Shall we place Milton second to any who, in the walks of a higher wisdom have sought to inspire & elevate[7] the intellectual character of men? How does the Poet[8] speak of the high gifts of which he was conscious of the possession; not as one, wise only to his own ends, but as one who knew that they were entrusted gifts, the improvement of which would be required even to a strictness.—[9]

82   I cannot pass from the ancient periods of the History of Human Science[10]

---

1   dust] *interlined above deletion of* 'a Clock'.
2   appears. . .powers,] *RO ink insertion with* 'to. . .powers' *interlined with caret.*
3   had] *interlined with caret.*
4   by] *followed by pencil crossout of interlined* '*an*'.
5   throughout] *before deletion of* 'life'.
6   with. . .results?] 'with' *above deletion* 'to'; 'in. . .results' *transposed from after* 'compared' *with lead line and* 'with' *interlined above deletion of* 'to'.
7   sought. . .elevate] *interlined with caret above deletion of* 'ennobled'.
8   the Poet] *interlined with caret above deletion of* 'he'.
9   And what . . .strictness.—] *RO ink insertion from facing fol. 82ᵛ with insertion point marked with caret to replace deletion of:* '{¶}Contrast the Language of the great Linnæus the individual by whom above all others the [l *fol.* 81] Science of Natural History has been advanced, with the speculations of Lucretius; or of Cicero, even while adorning with his manly eloquence the subject of Final Causes;—or lastly of Socrates, crying out to have the darkness removed which hid him from the knowledge of his great Creator;— and asking When shall that time arrive, and who is he that shall so instruct us? Contrast those expressions with the noble confidence with which the Swedish Naturalist declares the Principles that guided and sustained him in his Investigations:—' *and deleting from next line* 'Linnæus says' *with line drawn to deleted insertion on facing fol. 83ᵛ:* 'He who remembereth this cannot but sustain a *saner {followed by illegible word}* of mind, and more pressing, than any *supportable {followed by illegible word}* or weight which the body can labour under now & in what manner he shall dispose & employ those sums of knowledge & illumination which God hath sent him to this world to trade with,'.
10 the History. . .Science] *interlined with caret with* 'Human' *interlined with caret above*

to those in which it[1] again revived without adverting to a great change in self-knowledge which characterizes the latter epoch— And I cannot contrast the speculations of Lucretius, or those[2] of Cicero even[3] while adorning with his manly eloquence the subject of final causes, or lastly the expressions of Socrates crying out to have the darkness removed which hid him from the knowledge of his great Creator, and asking "when shall that time come, and who is he that shall so instruct us;" I cannot, I say, contrast these expressions with the noble confidence with which Linnæus declares the principles that guided & sustained him in his investigations without a conviction that those principles, derived from the Christian revelation,[4] have been operative as a cause of success[5] in the modern progress of Science. Linnæus in the preface to one of his early[6] Works; says—

> The Maker of all things, who has done nothing without design, has furnished this earth by Globe, like a Museum, with the most admirable proofs of His Wisdom and Power; but since this splendid Theatre would be adorned in vain without a Spectator; and if he has placed in it Man, the chief and most perfect of all his works, who alone is capable of duly considering the wonderful œconomy of the whole; it follows that Man is made for the purpose of studying the Creator's Works;—that he may observe in them evident marks of divine wisdom. Were it otherwise, it would have been sufficient for that Wisdom which does nothing in vain, to have produced an indigested Chaos, in which, like worms in Cheese, we might have indulged in eating and Sleeping:—Food and rest would then [/] have been the only things for which we should have had an inclination; and our lives would have passed like those of our Flocks, whose only care is the gratification of their appetite—But, continues Linnæus[7] our Condition is far otherwise. He who has destined us for future joys, has at present placed us in this World. Whoever, therefore shall regard with contempt the Œconomy of the Creator here, is as truly impious as the man who takes no thought of futurity. And in order to lead us toward our duty, the Deity has so closely connected the study of his works with our general convenience and happiness, that the more we examine them, the more we discover for our use and gratification.

83

It is the same Spirit which pervades the Philosophy of Bacon,—The same principle is its foundation: he spoke to congenial minds regulated and at peace through convictions of its truth—Hence the[8] rules for observing and interro-

---

(contd.)
    'Science', *all interlined above deletion of* 'Scientific History'.
1   it] *interlined above deletion of* 'one'.
2   those] *after deletion of* 'even'.
3   even] *interlined with caret.*
4   derived. . .revelation,] *interlined in ink with caret.*
5   of success] *interlined with caret.*
6   early] *interlined with caret.*
7   Linnaeus,] *comma inserted as in printed text [ed.].*
8   foundation. . .Hence the] *RO insertion from facing fol. 84ᵛ and repetition of* 'Hence

gating Nature which Aristotle, under[1] a less happy dispensation had taught in vain, became[2] operative and acceptable to mankind at the period when they were again enforced by our own great Philosopher.[3]

(contd.)
the'.
1 under] *with* 'der' *interlined and WC query in margin in pencil* 'under?'.
2 became] *interlined in ink with caret.*
3 mankind. . .Philosopher] *over WC pencil* 'Something follows here but will not follow kindly' *and then at bottom of page by WC in pencil* 'End of Part 1. 31/4 a.m. Novr. 22. 1836. W.C.'.

$\mathcal{N}otes$
to
Lectures One and Two

1–9 This is a reference to the arrangement of the front benches of the new theater which were in the form of a semi-circle (see facing plate). In the front circle were seated the most illustrious members of the audience, including the President of the College in the central chair, members of Parliament and other Trustees of the College.

2–15 [William Clift and Owen, Richard], *Catalogue of the Hunterian Collection in the Museum of the Royal College of Surgeons in London*. Pt. I (London: R. Taylor, 1830), 98 pp; Pt. II (London, 1830), 96 pp; Pt. III (London, 1831), 266 pp.; Pt. IV (London, 1830), 144 pp. Hereafter cited as *CHM*.

3–4 The specification of these duties are contained in the initial charge to the College in December 7, 1799, with emendations on 7 January, 1800. Relevant documents are contained in the report of Thomas Warburton's select committee on medical education, in: United Kingdom. Report from the Select Committee on Medical Education, *Reports from the Parliamentary Committees, 1834: Vol. 13, Part II: Royal College of Surgeons, London* 1834, "Appendix" p. 62. Hereafter cited as *PCR 1834*.

4–5 For a summary of the prior lectures see V. G. Plarr, *List of Lecturers and Lectures at the Royal College of Surgeons of England, 1810—1900* (London: Taylor and Francis, 1900).

4–8 Owen's allusion is clarified by the draft in which he has a deleted page citing Everard Home on the mechanical, nutritive and generative structures, John Abernethy on the opinions and discoveries of Hunter and the physiology of muscles and nerves,

Charles Bell on the mechanical functions of the animals, Astley Cooper on digestion, and Benjamin Brodie on digestion. Particular discussion is given to Bell's discovery of the "difference of function in connection with a difference of origin" of the spinal nerves, underlying the so-called "Bell's Law" (RCS 67. b. 12. A, fol. 3^v–6). All this has been deleted from the final manuscript. Bell's pamphlet of 1819 on the afferent and efferent nervous systems was published in his *An Exposition of the Natural System of the Nerves of the Human Body* (London, 1824). As the Hunterian surgical lecturer for the years 1825–28 and comparative anatomy lecturer in 1832 and 1833, Bell had probably discussed this important discovery often. The 1832 and 1833 lecture series, both of which Owen had attended, dealt in their concluding section with the nervous system, including a summary of contemporary experimental work on the issue. See Clift's "Memorandum on Museum Lectures," RCS Clift Papers, 275. g. 45. On Bell's work, see P. Cranefield, *The Way In and the Way Out* (New York: Futura, 1974), and E. Clarke & L. S. Jacyna, *Nineteenth–Century Origins of Neuroscientific Concepts* (Berkeley and London: U. California Press, 1987), chp. 4.

12–1 Deleted material from the Lecture reads as follows in the RCS holograph draft beginning on fol. 4: "The various organs of the digestive System have recieved [sic] ample demonstrations from Sir Everard Home and Sir Astley Cooper, and following the steps of the great Founder of Scientific Surgery the results of Physiology Experiment [l *fol.* 4] *were combined {*crossed out with stet marks* *below*} with those of zootomical observation and *thereby established {*transposed with lead line*} a series of facts *were {*deleted*} with reference both to the relative powers of *digesting {*altered from* 'digestibility'} in Different parts of the Stomach, and also of the *different {*interlined above deletion of* 'relative'} degrees of digestibility of different alimentary substances which form an important *feature {*interlined above deletion of* 'section} in the history of Digestion in every Physiological Treatise which has been published since those lectures were given—{¶} The experiments of Profr. Brodie, which have thrown so much light on the more recondite processes of chylification are too well known to require more than *brief {*interlined after deletion of* 'this' *above deleted* 'are'} allusion to them in the theatre of this College where they were first *promulgated {*above deletion of* 'made known'} in connexion with a full and able demonstration of the Hunterian specimens relating to the Nutritive Functions. [l *fol.* 5] The extensive series illustrative of the Nervous System and the Organs of the Senses have successively exercised the Talents of Professors Carlisle, Brodie, Bell & Mayo. Many are doubtless now present who participated in the enthusiasm with which those discourses were recieved which announced the most brilliant discoveries of the present era, in connection with the exposition of *that Department of {*interlined with caret*} the Hunterian Collection relating to the Nervous System. And what, we may ask, what would have been the feelings of Hunter himself could he have listened to such extraordinary and unexpected proofs of a *principle {*altered from* 'principab'} in the physiology of the nervous system of a difference of function in connection with a difference of origin, which he had himself been led to suspect from observing the constancy presented by the nerves in their origins and destination. [l *fol.* 6] {¶} In other courses of Lectures delivered in the Theatre of the College, the trea-

sures of the Hunterian Museum have been brought before you and made subservientto the illustration of *particular {*interlined with caret*} subjects [.] *Departments of {*interlined with caret above deletion of* 'more'} Physiological Science *originating in the Professor himself {*deleted*} selected by *the {*deleted*} and as it were peculiar to the Professor Himself. Of this nature were the Lectures of Sir Anthy Carlisle on those substances which enter into the Composition of Organized bodies, yet seem as it were to vibrate between the organic & inorganic kingdoms, and *to {*interlined over deletion of* 'for'} which he *first applied {*interlined over deletion of* "concieved'} the happy term of *Extravascular,* * a term {*interlined with caret*} which as ever since been adopted in the language of Physiology in reference to the calcareous combinations forming the *hair {*interlined*} teeth *& horns {*interlined*} of the Higher animals, and the skeletons *or shells {*interlined*} of mollusca & Zoophytes; {¶} Under the same class of Lectures rank the Celebrated Discourses of Lawrence on the Natural History of Man. The numerous specimens selected from the different series in illustration of these Lectures have thereby acquired *new & extrinsic interest and {*deleted*} *extrinsic & {*interlined*} additional value—They are *asked {*interlined over deleted* 'inquired'} for and contemplated with a new interest acquired by association with the *eloquent {*interlined over deleted* 'luminous'} descriptions in which [l *fol.* 7] *in this synthetical {*deleted at top of fol.* 8} *the {*altered from* 'The' *ed.*} various systems of organs were considered, as they are combined together in the different *forms {*interlined above deleted* 'forms'} of the animal Kingdom; the application of comparative anatomy was in this instance *made apparently {*deleted, with* 'formed for' *deleted interline above*} in the direction of Zoology, *it was {*interlined*} to the illustration of the *philosophical or {*deleted*} organical characters *in physiological peculiarities {*interlined with caret with deletion of* 'manifestations'} of the different *groups and {*deleted*} Classes of the Animal Kingdom; and tended to elucidate their *true { *interlined above deletion of* 'natural'} affinities, the highest aim of the Philosophical Naturalist—{¶} *Now, although, this {*interlined above deletion of* 'Such, however'} is not the plan according to which *Mr Hunter has arranged his {*interlined over deletion of* 'the Founder of the'} Collection; *yet it was admirably adapted {*with interlined deletion of* 'he has taken the same organs from the different Classes'} to convey the preliminary knowledge requisite for entering upon the study of the Collection as it is— {¶} I would say, also, that *there could {*over deletion of* 'nothing'} be no severer test of the *completeness {*over deletion of* 'adequacy'} of the Hunterian Collection than *depending {*interlined over deletion of* 'founding'} upon *its resources for the illustration of a complete course {*interlined above deletion of* 'our deriving from it the materials for'} on the *plan and {*interlined with caret*} principles adopted, and so admirably *accomplished {*interlined below deletion of* 'fulfilled' *and* 'atchieved' [*sic*]} by Prof. Green— *In some respects the Collection undoubtedly fell short of the wants of the Professor, but these deficiencies were supplied in a manner {*interlined in darker ink over deletion of* 'Whatever deficiencies **some undoubtedly{ {*interlined in deletion*}} there were'} which entitles Mr. G. to the lasting gratitude of this College, *and especially of all **who have and may { {*deleted in interline*} his successors in this Chair.{*interlined*} Preparations *and especially those of the minute and complex organs of the smaller animals {*interlined*} cannot *be made available to the illustration of the {*below deletion of* 'always' *and* 'serve to'} [l *fol.* 8] several steps of a correspondingly complicated description— Thus to be *readily {*interlined with caret*} followed and understood must be accompanied by reference

to enlarged representations of the objects described *Upwards of Diagrams {*deleted interline*}. Now at the period when Prof. Green commenced his Course of Lectures the College did not possess the diagrams *which were requisite {*interlined over deletion of* 'necessary'} to illustrate so extended and complete a view of the animal organization as it was the object of the Professor to give— He, therefore *caused {*deleted*} engaged at his own expense artists *for this express purpose, {*interlined with caret over deletion of* 'whose skill in this'} and caused upwards of 300 diagrams *to be prepared in a style of art it would be difficult to surpass {*interlined with caret*} of which it is not the least of this value that the subjects were selected* with {*above deletion of* 'by'} the greatest judgment from *the {*deleted*} recent dissections, interesting preparations in the Collection, *and {*deleted*} or from authentic figures in the rarest & most valuable works on Comparative Physiology—{¶} This valuable collection of materials so essential to *the animal {*interlined over deletion of* 'the Lecturing department of the Museum'} Illustrations *of the museum in this Theatre—{*interlined below line with caret*} Prof. Green presented at the Conclusion of his labours, to the College— where they will *even {*interlined above deletion of* 'ever'} remain *as {*deleted*} a monument of his Science, Taste, and disinterested *liberality {*followed by deletion of* 'and **form as { {*deleted interline*}}a powerful aid to the Anatomical Lectures as must {*deleted with* 'the means of so considerable *lightening { {*deleted*}} aiding the delivery of the Anatomical Lectures as must the especial gratitude of all his successors—} [l *fol.* 9] {¶} *In the brief survey which I have just taken of the various modes in which the directions of the Public Expositor of the Hunterian Physiological Collections have been fulfilled, I must request the indulgence of the distinguished Professors now present for the adequate & cursory manner in which I have alluded to their

labours.{*interlined at top of fol. 10*}
When I reflect on how much has been
done in the Exposition of the *Museum
{*interlined above deletion of* 'Hunterian
Collection'}, when I remember *also
{*interlined*} how well, and by what eminent Masters in Physiology the
Hunterian labours have been *explained
{*interlined above* 'illustrated'} from this
chair, I can hardly hope to interest you by
the more humble duties which I feel
*myself limited {*above deletion of* 'alone
aspire'} to *endeavour to {*interlined with
caret*} fulfil— {¶} Yet when I percieve
[*sic*] that ably as the different departments
of the Physiological Series have been
illustrated by different Professors, a
connected *review [*interlined over second
'review'*} of the whole has never yet been
*accomplished {*interlined over deleted
'attempted'}—, and that the lectures
which have been published are not
available to the visitor of the Museum in
acquiring a precise or adequate knowledge
of the scope and nature of its contents—'
*MS now continues as beginning fol. 12
of final recopy* 'I am willing' *etc.*

12–5 The Hunterian collection in 1837
contained 10,000 specimens in spirit;
8,000 dry mounts; 2,000 skeletal
remains; 1,500 preparations of
pathological anatomy, and 1,300 fossils
(Owen and Clift, "Report to the Curators,
February 1837," Minutes of the Curators
of the Hunterian Museum, RCS Archives
5:16–7. Hereafter cited as "Curator
Minutes"). These different series were arranged for display in the new museum in
a specific order. The newly-enlarged
main hall housed osteological specimens,
large fossils, and similar hard materials
on the main floor. Cases on the main
floor also contained smaller specimens
for viewing. The elevated galleries above
the floor then displayed the Physiological
Series, and commenced from one end
with an initial series under the general
heading of "Parts Employed in
Progressive Motion"(see Synopsis above,
" Introduction", Table One). This series
commenced with fifteen displays on the
"Component parts of Vegetables and

Animals" and then followed eleven displays on "Sap and Blood: Their different
kinds". "Compounded Instruments for
Progressive Motion" formed the thirteenth series in this division, comprised
of preparations 279A to 288A which followed upon preparations on the skeletal
system. Specimens forming the Natural
History series were housed in the top
gallery and the adjoining smaller museum. The adjoining secondary museum
housed the Pathological Series, containing monstrosities and specimens demonstrating pathological anatomy. Cabinets
on the floor of this museum contained
specimens of abnormal osteology, calculi
and other pathological specimens. See
Owen's later *Synopsis of the Royal
College of Surgeons Physiological
Museum* (London: Taylor, 1845) for a
fuller description of the arrangement.
This full re-arrangement was not completed until May of 1838 ("Curator
Minutes," vol. 5, 3 May, 1838). For
Clift's summary of the preserved specimens in 1830, see his *CHM* Pt. 4.
(1830).

13–14 This series is the exclusive topic of
the five-volume [R. Owen and W. Clift]
*Descriptive and Illustrated Catalogue of
the Physiological Series of Comparative
Anatomy in the Museum of the Royal
College of Surgeons London*, (London:
R. Taylor, 1833–40). Hereafter cited as
DIC. Volumes 1–3 had been published
by the delivery date of the lectures, and
volume four (London: Taylor, 1838) was
in progress at this time. By 1840, the
physiological series numbered 3790 separate specimens and preparations.

13–16 This comment is compelling evidence that Owen's larger plan for the
lectures generally followed the outline in
the *DIC*. Owen's colleague, Edward
Stanley (1793–1862) had opened the
Hunterian series on 18 April with a set
of six lectures dealing with the skeletal
system, treated as part of the Hunterian
physiological series. Owen for this
reason did not deal with osteology until
his lectures in the 1840s. For a summary
of Stanley's lectures, see *London Medical*

Gazette **20** (1836–37): 342–5; 379–86; 421–4; 497–501; 577–80; 641–7. In the opening lecture Stanley developed the concept of the unity of type, using this as the basis for arranging the various organs and discussing their structure. (ibid., p. 343). Owen could therefore presume knowledge of this material by his audience. See also comments in my introductory essay above, p. 62. See also below, note 41–6.

14–13 The reference is most likely to Georges Cuvier's negative assessment of Hunter's importance in his *Rapport historique sur les progrès des sciences naturelles depuis 1789 et sur leur état actuel* (Paris: Imprimerie imperiale, 1810), p. 243. See also below, Lecture 3, note 39–15.

20–12 The reference is obviously to Plato.

20–13 Parmenides of Elea (fl. 450 B.C.)

21–4 Pythagoras of Samos (fl. 531 B.C.)

22–4 Alcmaeon of Croton (fl. 450 B.C.), member of the Pythagorean school, physician and author of a work on natural philosophy known to Aristotle.

22–7 Empedocles of Acragas (ca. 493–33 B.C.), author of poem *On Nature.*

22–11 See Aristotle, *HA* I. xi. 492a 17.

23–6 Epicharmus of Cos (540–430 B.C.), poet, playright, and member of Pythagorean school. Some minor writings on medicine are attributed to him known only by title.

23–10 Leucippus of Miletus (fl. 5th C. B.C.), the founder of the Atomic philosophical school.

25–1 Democritus of Abdera (fl. 5th C. B.C.)

27–10 Anaxagoras of Clazomenae (ca. 500–428 B.C.).

28–8 Plato *Phaedo* 97b 46-d 90.

29–10 See Hippocrates, *The Sacred Disease* 6. I am using the edition edited by G. E. R. Lloyd in *Hippocratic Writings* (Baltimore: Penguin, 1978.)

30–12 Hippocrates, *On the Heart* 4, in *ibid.,* p. 348.

31–15 Aristotle, *Metaphysics* 1.9 991a 10ff.

32–6 See Aristotle, *Posterior Analytics* 1. 31.87b28ff. Owen gives a distorted empiricist misreading of these statements

32–20 Ibid. In his quotations here and elsewhere, Owen is generally translating directly from the Latin of the Julius Scaliger edition of Aristotle. However, I have been unable to locate this exact quotation, which is an empiricist assembling of several Aristotelian maxims probably being taken by Owen from a secondary source. It somewhat resembles Aristotle, *Posterior Analytics* 1. 31.87b28ff, and *Nicomachean Ethics* VI. 3. 1139B 25–35. On the editions available to him at the College see below, Lecture Two, fol. 68.

33–2 See Georges Cuvier, *Histoire des sciences naturelles depuis leur origine jusqu'a nos temps* (Paris: Fortin, Masson et Cie, 1841) I: 146–7.

33–10 Aristotle, *Parts of Animals* 2.1. 647b10–22.

33–16 The first section of the Physiological Series in the Owen and Clift *DIC* commenced with preparations demonstrating the "Component Parts of Vegetables and Animals". See outline of the collection in "Introduction" table 1.

35–4 Owen is referring to the new histological classifications of distinct tissues set forth in François Xavier Bichat's *Traité de membranes* (Paris, 1800). He is also referring to the chemical analyses of bone, muscle and other tissues instituted by the French chemists.

35–21 This quote approximates that at Aristotle, *HA* 1.2.488b29–489a20–24.

36–8 This again resembles but does not reproduce, *HA* 489b23–490a25.

37–5 *PA* 1.1 642–65 ff.

37–7 Owen here is probably displaying to the audience an enlarged version of the drawing of the circulatory pattern of the Lobster printed in *DIC* 2.

38–4 G. Cuvier, "Discours préliminaire" to , *Le régne animale distribué d'après son organisation, pour servir de base à l'histoire naturelle des animaux et d'introduction à l'anatomie comparée,* 4 vols. (Paris: Deterville, 1817). On this basis Cuvier rejected Lamarck's division into Vertebrates and Invertebrates, and proposed in its place four autonomous *embranchements.*

39–3 J. Hunter, "Observations on the Structure and Oeconomy of Whales," *Philosophical Transactions of the Royal Society* **77** (1787), reprinted in: J. J. Palmer, ed. *The Works of John Hunter* 4 vols. (London: 1835–7), 4: 331–92. Hereafter cited as *Hunter's Works.* The fourth volume has extensive notes and a preface by Richard Owen, dated as November 1837. The preface incorporates verbatim some sections of these Hunterian lectures. See Introduction, p. 49.

39–15 Hunter's classification has been summarized by Owen in his preface to ibid.

40–9 Aristotle, *HA* 2.1.497b15–18.

41–2 Ibid. 4.1.523b8–12.

41–6 This is an allusion to Edward Stanley's surgical lectures which had preceded Owen's series. See above, note 13-16.

41–12 *DIC* 1. The first subdivision of this catalogue, covering series 1-12, was designated "Organs of Motion".

42–8 Aristotle, *HA* 2.1.501a8 ff.

42–11 Exact reference not traced. The same passage is found in the RCS holograph.

42–14 In William Clift's manuscript catalogue of 1816 the displays on the teeth are located between Section Ten "Coverings of Animals," and Section Twelve, "Parts of Generation and Reproduction."

42–16 Aristotle, *HA* 2.1.501a8–23.

43–5 Ibid. 2.5.501b30–502a2.

43–7 Reference uncertain. Possibly to his memoir on the fossil elephants. See 76–2 below.

43–10 Ibid. 2.3.501b1–5

43–14 Reference uncertain. He is possibly referring to Bacon's comments at *Sylva Sylvarum*, para. 688. I have used the *Works of Francis Bacon*, new ed. (London, 1826), I: 467.

44–3 *HA* 1.17.497a1–5

44–7 Ibid. 3.2. 511b 14–20.

45–13 Ibid. 20–25.

46–4 Ibid., 513a 27–35. See also *PA* 3.4. 66b22 ff. Extensive discussion on this issue exists in the Literature. See Harris, *The Heart and Vascular System in Ancient Greek Medicine* (Oxford:

Clarendon, 1973), chp. 3. D'Arcy Thompson's notes to the Oxford edition of *HA* discuss the possibility that Aristotle's claim was based on the dissection of a fetal heart.

46–15 *HA* 1.16.494b25 ff.

47–2 The reference is apparently to the opening discourse to Cuvier's *Régne animale*, vol. 1. See note 38–4.

47–5 *DIC* III. Pt. 3, p. 140 and Plate 42, fig. 2. This is a description of the eyes and brain of a fresh specimen of *Sepia officinalis* Linn. Owen is probably diplaying at this point a drawing of this preparation to the audience and passing around the bottled specimen.

47–7 *HA* 1.16.495a1–5.

47–12 See backdrop to Hunter portrait below, fig. 10, p. 164.

48–4 Ibid., 7–10.

48–6 Ibid., 1.16.495a12

48–10 Ibid. 3.16.515b15–27.

48–16 Ibid. 1.9.491b30.

49–9 See above, note 22–11

49–15 Reference uncertain.

49–17 Ibid. 3.7.516a20–5.

50–7 J. Hunter, "An Account of the Organ of Hearing in Fishes," *Philosophical Transactions of the Royal Society* **72** (1782), as reprinted in *Hunter's Works* 4: 297. Owen in a note to ibid. cites Guilio Casserio, *Pentestesion, hoc est, De quinque sensibus* (Venetiis, 1609), p. 224. A copy of this is at the RCS.

50–9 *HA* 4.3.534b17–20.

50–11 Hunter, "Observations on Bees," *Transactions of the Royal Society* **82** (1792) in *Hunter's Works* 4: 458.

51–2 *HA* 2.6.502a4.

51–3 Ibid. 503a15–28. Aristotle does not mention the Chameleon's prehensile tongue in this description.

51–6 Ibid. 4.9.536a5-10.

51–7 Ibid. 2.17.508a27.

51–8 Ibid. 8.13.597b26–9.

51–11 Ibid. 2.10.503a1.

52–2 *HA* 3.17.6-10.52–4 *HA* 3.2.517b8 ff.

52–5 Ibid. 2.1.499b33–500a2. The preferred manuscript reading of this passage refers to the "pardion." According to some editors, the hippardion is the giraffe. See D'Arcy Thompson's note to the Oxford edition. A. L. Peck, the translator of the Loeb edition of this text, has traced the hippelaphus to the short-horned Indian antelope, *Boselaphus tragocamelus*.

52–9 Ibid. 2.1.498b33–499a4

53–5 *HA* 2.1.499b17–20. Owen is interpolating discussion of the camel from the preceding 499a13–25. He is developing in this section of the lecture the claim that Aristotle held to something similar to Cuvier's principle of the law of correlation of parts.

53–10 *GA* 4.12. 694a115.

53–16 See Hunter, "Observations on the Torpedo," *Philosophical Transactions of the Royal Society* 63 (1773) in *Hunter's Works* 4: 409–21.

54–10 *HA* 4.1.524b17–20; 8.17.600b23–601a1; 2.17.508b7; 4.9.535b8.

54–15 J. Hunter, *Observations on Certain Parts of the Animal Oeconomy* (London, 1786).

55–2 *HA* 3.1.509a32 ff.

56–2 Alcmaeon of Croton. See Ibid., 7.1. 581a16 and note 22–4.

56–5 Ibid. 4.3.561a3 ff. Aristotle was obviously aware of the Hippocratic experiment which recommended the opening of eggs over a twenty-one day period to study development.

56–9 Ibid. 6.3.562a22.

56–15 The reference is to William Harvey's mention of having used a "speculum" in his studies on the generation of the chicken. See William Harvey, *Disputations Touching the Generation of Animals*, trans. G. Whitteridge (Oxford: Blackwell, 1981), Ex. 17, p. 96. Future citations to this text will be from this edition.

57–2 *HA* 6.1.588b 10ff.

57–7 Ibid. 5.33.558a 25.

57–11 Ibid. 6.10.564b 27–31.

58–2 Ibid. 8.30. 607b 18–21.

58–7 Reference untraced.

58–11 *GA* 1.15. 720b 17ff.

59–2 G. Cuvier, "Considérations sur les mollusques, et en particulier sur les céphalopodes," *Annales des Sciences naturelles* 19 (1830): 241–59. A reprint of this paper, in a volume with Owen's annotations, is found at the RCS.

59–8 G. Cuvier, *Mémoires pour servir à l'histoire et l'anatomie des mollusques* (Paris: Deterville, 1817). A copy of this work is at the RCS.

59–10 *PA* 4.9.684b15. The reference is to the major scientific controversy between Cuvier and Etienne Geoffroy St. Hilaire which reached its peak in 1830 and 1831 in the debate before the French Academy. Cuvier's line of attack is well-summarized in his G. Cuvier and A. Valenciennes, *Histoire naturelle des poissons* 22 vols. (1828–49), 1:545–51. For recent discussion of the debate see Toby Appel, *The Cuvier Geoffroy Debate: French Biology in the Decades before Darwin*. (Oxford: Oxford University Press, 1987).

59–19 *PA* 4.9.684b25–30. Reconstruction of the diagram from the indications in this text have produced several difficulties. See D'Arcy Thompson's notes to this passage in the Oxford edition.

60–3 *HA* 5.21.553a17–25.

60–5 Ibid. 5.23.554b21–2.

61–1 Ibid. 5.27.555a27 ff.

61–6 *GA* 2.9.758b ff.

61–10 *HA* 6.17.570a13–24.

61–13 Francesco Redi, *Experimenta circa generationem insectorum* (Amsterdam, 1671).

61–15 See William Sharpey, "An Account of Professor Ehrenberg's More Recent Researches on the Infusoria," *Edinburgh New Philosophical Journal* 20 (1835–6): 42–66 and earlier paper by Sharpey in ibid., 15 (1833): 287–308. Reference to "the other day" also appears in the draft of the lecture and is not a late addition.

62–8 At this point Owen apparently demonstrated a drawing or chart comparing Aristotle's and Hunter's arrangements. I have been unsuccessful in locating this diagram. See also end of Lecture One.

63–2 *HA* 8.26.605a23–605b5.

63–9 Ibid. 8.21.603b17.

63–21 See ibid. 1.1.486a20–25. In the passage to 65–2, Owen is closely paraphrasing W. S. MacLeay's discussion in "Remarks on the Comparative Anatomy of Certain Birds of Cuba With a View to Their Respective Places in the System of Nature or to Their Relations with Other Animals,"*Transactions of the Linnean Society of London* 16 (1823): 6–8.

63–25 Green also delivered a lecture series in 1828 which Owen attended.

64–2 Aristotle, *HA* 1.1.486b16.

64–13 Owen is commenting on the fact that the Hunterian displays typically presented isolated functional systems with examples drawn from several organisms. In such displays, the viewer was presented with systems running from their simplest manifestations to their most complex expressions.

64–20 MacLeay, "Remarks," p. 13.

64–22 Aristotle, *HA* 1.1.486b19.

65–4 This is the earliest usage of the distinction of these two terms I have found in Owen's work. It is to be noted that at this stage he has continued to follow the French and use "homology" and "analogy" as equivalent terms. Only in 1843 in his *Lectures on the Invertebrates* will he make a formal distinction of these.

65–13 Pierre Belon (1517–64), French anatomist and author of the *L'Histoire de la nature des Oyseaux* (Paris, 1555).

65–18 Isaac Newton, *Optice : sive de reflexionibus. . .* (Londini: G. & J. Innys, 1719), Qu. 30 (=31), p. 411. Owen has modified and rearranged the Latin of this passage. The full passage reads in its English version: "Such a wonderful Uniformity in the Planetary System must be allowed the Effect of Choice. And so must the Uniformity in the Bodies of Animals, they having generally a right and a left side shaped alike, and on either side of their Bodies two Legs behind, and either two Arms, or two Legs, or two Wings before upon their Shoulders, and between their Shoulders a Neck running down into a Back-bone, and a Head upon it; and in the Head two Ears, two Eyes, a Nose, a Mouth, and a Tongue, alike situated." *Opticks*, 4th ed. . edited D. Roller (New York: Dover, 1952), pp. 402–3.

In the deletion to this text, Owen makes reference to Edward Stanley's appeal to the same passage in Newton with which Stanley had defended the conception of a unity of type in his opening Hunterian surgical lecture of 18 April. See summary of Stanley's inaugural lecture in *Medical Gazette* 20 (June 3, 1837): 342. This indicates that Owen even at this early date was not opposed to a modest version of the theory of the unity of anatomical plan.

65–25 Reference is to Harvey's comparison of the heart in simpler forms to those in the vertebrates in his *De Motu Cordis* of 1628.

66–4 See Owen's introduction to *Hunter's Works* 4: xv–xvii and quote from Hunter in ibid., xxv–xxvi. Owen cites Hunter's endorsement of a mild form of recapitulationism here, in which the stages of an embryo represent completed forms of lower organisms.

66–9 Felix Vicq-d'Azyr. His papers are collected in his *Oeuvres de Vicq d'Azyr*, ed. J. L. Moreau, 6 vols. (Paris, 1805). A copy of this is in the RCS library. The reference to Meckel is probably to J. F. Meckel's *Mémoires d'anatomie et de physiologie humaines et comparée* (Halle, 1806)

66–19 This passage, also present in the draft in slightly altered form, is the first occasion where Owen comments on transcendental anatomy. The target of this comment originally was likely Etienne Serres. See Serres, "Anatomie transcendante, Quatrième mémoire," *Annales des sciences naturelles* 21 (Sept. 1830), 5–49. Serres (p. 48) there defines *anatomie transcendante* as an anatomy which employs his two rules of symmetry and conjunction of parts for its analysis.

66–28 This passage is important for assessing the degree to which Owen is concerned to attack Etienne Geoffroy St. Hilaire. Owen has not rejected Geoffroy's recapitulationism as expressed in the *Philosophie anatomique*, and he has only previously cited with approval John Hunter's endorsement of the same point (see 66–4). The following attack is directed only at the excesses of this posi-

tion. It is notable that the delivery ver
sion has actually weakened the critique
of Geoffroy found in the draft of the
manuscript at this point by including a
critique at 67–16 of the German transcendentalist Johann Spix.

67–8 MacLeay, "Remarks," p. 9n. quoting
Aristotle *HA* 2.1.497b12. Owen reveals
in these passages his attraction to the approach of MacLeay, who endorsed aspects
of the theory of analogies and the unity
of type against Cuvier. MacLeay also
attacked the excesses of Cuvier's
functionalism by making the distinction
between "analogy" and "affinity."

68–16 Aristotle, *De Animalibus interprete
Theodore Gaze* (Venetii: Colonia &
Manthem, 1476). This text is in the
RCS library.

68–25 Aristotle, *Histoire des animaux*,
trad. M. Camus, 2 vols. (Paris:
Desaint, 1783). A copy is in the RCS
library. Owen is using Scaliger's
translation in the edition edited by J. B.
Schneider (*De animalibus, Graece et
Latine edita J. B. Schneider* [Lipsiae:
Hahnensis, 1811]). See emendation to
the text. A copy is at the RCS.

68–27 Aristotle, *The History of Animals of
Aristotle and his Treatise on Physio-
gnomy*, trans. T. Taylor (London: R.
Wilks, 1809). A copy is in the RCS
library.

68–36 The only work by Theophrastus at
the RCS is *Theophraste des Odeum, mis
en Grec en nostre langue françoyse, avec
annotations. . . .par I. de l'Estade*
(Paris: Gilland, 1556).

69–22 Rufus Ephesius, *De partibus corporis
humani edit Guilelmus Glinch* (Londini:
Bettenham, 1726), p. 71. This edition is
in the RCS library. Owen has apparently
made an error on the page number. As
translated by Heinrich von Staden this
reads: "According to Erasistratus and
Herophilus there are nerves capable of
sensation, but according to Asclepiades
not at all. According to Erasistratus there
are two kinds of nerves, sensory and motor nerves; the beginnings of the sensory
nerves, which are hollow, you could find
in the meninges and those of the motor

nerves in the cerebrum [*enkephalos*] and
in the cerebellum [*parenkephalis*].
According to Herophilus, on the other
hand, in the *neura* that make voluntary
[motion] possible have their origin in the
cerebrum [*enkephalos*] and the spinal
marrow, and some grow from bone to
bone, others form muscle to muscle, and
some also bind together the joints."
(Heinrich von Staden, *Herophilus, The
Art of Medicine in Early Alexandria*
(Cambridge: Cambridge UP, 1989) p.
201, translating the Greek as in
Daremberg/Ruelle, pp. 184–5. Quoted
with permission of Cambridge University
Press.))

69–29 See ibid., pp. 200–3.

70–5 See ibid., pp. 224–5.

70–11 See Galen, *On Anatomical
Procedures* 14.1.232. See translation by
W. L. H. Duckworth (Cambridge:
Cambridge University Press, 1962), p.
184.

70–19 On Erasistratus, see text from Ruphus
of Ephesus in von Staden, *Herophilus*, p.
201.

71–4 See Harris, *Heart and Vascular
System*, chp. 4.

71–8 See Galen, *De pulsuum differentiis*
4.6 as in von Staden, *Herophilus*, p. 322.

71–12 Referring to William Harvey.

72–12 I have been unable to locate the
source for Owen's claim that Galen had
access to a bronze cast of the human
skeleton at Alexandria.

72–17 Modern scholarship has traced
Galen's simian material to uses of both
the Barbary Ape (*Macaca inuus*) and the
Rhesus monkey (*Macaca mulatta*) in his
dissections outlined in the *Anatomical
Procedures*. See discussion by Charles
Singer in notes to Galen, *On the
Anatomical Procedures* vol. 1, trans. C.
Singer (London: Oxford UP, 1956).

72–22 Claudii Galeni, *Opera Omnia*, ed. C.
G. Kühn (Lipsiae: 1821), 2: 219. The
only edition of Galen's works containing
this treatise at the RCS is his *Opera
omnia his accedunt nunc primum C.
Gesneri praefatio et prolegomena* 4 vols.
(Basiliae: Frohen 1568–62). There are
slight spelling and punctuation

differences from the Kühn text.
73–2 *On Anatomical Procedures* 6.1. 537
(Singer, p. 149).
73–6 Ibid., 6.1.536., p. 148.
73–19 Reference is probably being made to
John Barclay, *A Description of the
Arteries of the Human Body* (Edinburgh:
Bryce, 1812). Owen cites this work in
his notebook of extracts from scientific
works, BL Add MSS 34,406. See also
Barclay, *Introductory Lectures to a
Course of Anatomy,* ed. G. Ballingal
(Edinburgh: Maclauchlan & Stewart,
1827), pp. 82–3.
76–2 Allusion here is to G. Cuvier,
"Mémoire sur les espèces d'elephans vi-
vantes et fossiles," *Mémoires de l'Institut
national* 2 (1799): 1–22.
76–8 See Herodian, *History of the Empire
from the Time of Marcus Aurelius,* trans.
C. R. Whittaker, 3 vols., (Loeb Classical
Library; Cambridge, MA & London:
Heinemann, 1969) 1.15.1–7.
76–13 Letter of Cicero to Marius in: Cicero,
Epistulae ad Familiares, ed. D. R.
Schakleton-Bailey (Cambridge:
Cambridge University Press, 1984) I, 84.
77–5 Aelian, *On the Characteristics of
Animals,* trans. A. F. Sholfield, 3 vols.
(Loeb Classical Library; Cambridge, MA

& London: Heinemann, 1958) 2:11–12.
Aelian does not speak specifically of
elephants tightrope-walking.
77–8 John Corse, "Observations on the
Manners, Habits, & Natural History of
the Elephant," *Philosophical
Transactions of the Royal Society*
(1799), p. 31; and idem.,"Observations
on the Different Species of Asiatic
Elephants, & their Mode of Defense,"
ibid., p. 205
77–9 Reference is to Lucius Junius
Moderatus *Columella* (fl. 100 A.D.),
known primarily for writings on horticul-
ture.
78–3 Herodian, *History of the Empire. . .*
1.15.1–7.
78–10 *HA* 2.7.502a9–15.
82–5 See especially Cicero, *De natura
deorum* para. 47 ff. I am using the
translation by C. D. Yonge (London:
Bell, 1902).
83–11 C. Linnaeus, *Reflections on the
Study of Nature,* trans. J. E. Smith
(London: Nichol, 1785), pp. 13–16.
Owen has deleted almost a page from this
quote and made other minor
modifications. This is a translation of the
preface to Linnaeus, *Museum Regis
Adolphi Fridrici,* 1754.

Lectures Three and Four
Analysis of the Manuscript

These two lectures have been formed by the splitting of an original manuscript of one hundred and one pages, located at the Royal College of Surgeons (Owen Papers 42. d. 4). The division of this manuscript is clearly indicated by the crossing out on the cover sheet of the title "Lecture Two" in Clift's hand, with "3 & 4" and "May 6th & 9th" written in Owen's hand. The division point is clearly indicated at page forty-seven with a pencil line and then the crossing out of the first half of page forty-eight with an insertion of a new introduction in Owen's hand to the new fourth lecture, written on the back of folio forty-nine. As with Lectures One and Two, this lecture is written through page ninety-three on folded sheets of "J. Bune 1832" paper measuring 34.2 x 40 cm. folded in half. As discussed in the introduction, the copy-text of Lecture Four then deviates from the original draft to insert seven pages of new material on embryology, discussing controversies in comparative anatomy. This inserted section is on folded sheets of "J Green & Son 1837" measuring 19.7 x 33.2 cm. This is the same paper used in the final copy of the original third (now fifth) lecture of 11 May and for the main body of the manuscript of the sixth lecture of 13 May. The importance of these revisions to the end of Lecture Four is discussed in the Introduction.

Owen's draft of this manuscript is located in the Owen papers at the College of Surgeons (MS 67. b. 12.O). As with the draft of Lectures One and Two, the draft is primarily composed on sheets of "J Green & Son 1836" paper measuring 21 x 26 cm, to which have been added insertions, primarily in the opening section, on "J Whatman 1836" paper. A few sections of the latter part of the draft manuscript are also written on "Ruse & Turners 1834" paper and may represent incorporations from material prepared for Owen's lectures in the spring of 1835 at St Bartholomews' Hospital.

137

Fig. 9. Watercolor Drawing By George Scharf Depicting The Reconstruction Of The College Of Surgeons In 1835 Looking Northwest Toward Lincoln's Inn Fields. Depicted is the complete demolition of the main front structure, with only the facade over the entry left intact. The new museum and lecture theater are under construction to the rear. See also figs. 4, 8 above, pp. 46, 96. *(Courtesy President and Council, Royal College of Surgeons of England)* .

Lecture Three [1]
6 May 1837

1   Mr. President.

The Writings[2] of Bacon having taught and established the inductive method in Philosophy, and the same great Intellect[3] having wrought out a Classification of the legitimate objects of Human research, and having severally assigned to them their natural limits; each Science began to be cultivated by Men who devoted themselves exclusively to it with the whole energies of their minds, and by this judicious division[4] of labour, an unprecedented success was obtained.

In tracing the progress of Comparative Anatomy and Physiology from this period to the time of Hunter, we shall find that the Organization of Animals became the field of Inquiry to many ardent and indefatigable investigators of Nature who, bringing to the Task different capabilities and habits of thought, advanced the Science in different directions, and with various success.

Permit me, however, briefly to advert to the principal circumstances which
2   influenced the [/] progress of[5] anatomical knowledge during the dark ages[6] which preceded this epoch.[7]

The Crusades, while they carried abundant proofs of European valour[8] into the East, brought back, in return, some portions of Arabian Literature into the West: and we owe the revival of Anatomy, and of[9] Science in general to the singular mental contagion which overspread Europe at that period.

---

1   Three] *Lecture is entitled 'Lecture 2' in Clift's hand with pencil addition in Clift hand of '/3/'.*
2   Writings] *dark ink over erasure.*
3   Intellect] *inserted over erasure in dark ink.*
4   division] *interlined over pencil 'subdivision' above deletion of 'distribution'. On facing page in Clift's hand 'Subdivision of labour'.*
5   of] *interlined in pencil.*
6   ages] *deleting period for grammatical sense [ed.].*
7   which. . .epoch] *RO interlined in dark ink.*
8   valour] *followed by deletion of* 'and folly'.
9   of] *interlined in ink with caret.*

139

Among the first fruits of the acquaintance thus obtained[1] with the Writings of the Antients, was the Treatise of Mundinus, a professor of Bologna, entitled "Anatome omnium corporis humani partium"[2], written in 1318. In this work, besides so much of the Anatomy of Galen as was to be gained from the imperfect Translation of an Arabian Physician, there were added the results of actual observation, founded on the dissection of the body of a human female in the year 1306; (the first on record in the Christian Æra.) and of two other subjects in 1315.

Mundinus describes the three great venters or cavities of the Human Body, 3   and [/] their contained viscera; together with the Bones.

The seeds of Anatomical Science fell, however, on an unpropitious soil:— the human mind was trammelled by superstition and scholastic Metaphysics,[3] more baneful to the progress of Natural[4] Science than either the Atomic Philosophy of Democritus, or the Idealism of Plato.

The labours of Mundinus were, for a long while most ingeniously nullified by the misplaced zeal of his Patrons and Admirers. The statutes of Padua allowed no other System to be taught; for upwards of a Century Anatomists lost their time in commentating *on*, and explaining Mundinus:—nay so great was their[5] reverence of this Author that his descriptions were made the standard of Normal Structure:— so that whatever, in dissecting the human body, was observed to deviate from them, was held to be a malformation.[6]

4   Pope Boniface 8th forbad him to boil the bones of his subjects—but this appears to have been the only restriction under which he laboured—so that in some respects he was better off than the English Anatomists of the present day.[7]

---

1 obtained] *after deletion in dark ink of* 'singularly'.
2 partium] *interlined in dark ink with caret.*
3 Metaphysics,] *followed by erasure of* 'far'.
4 Natural] *interlined with caret.*
5 their] *modified from* 'the'.
6 malformation] *followed by deleted paragraph:* '{¶}It was in vain that Mundinus pointed out the deficiencies which existed in his own Work, and the difficulties superadded by Papal Ignorance {'authority' *in pencil by RO with pencil* 'authority'*on facing 4ᵛ* } to those which are inherent in the subject itself; [I *fol.* 3] {*with* 'itself:—'*repeated top fol. 4*}. He tells us that he would have described the *bones of {*interlined with caret*} the internal ear, but was forbidden to practise that which he considered to be the requisite means:—"Ossa autem alia, quæ intra basilare sunt, non bene ad sensum apparent, nisi decoquunter {*erasure of letter after* 'q'}, sed *propter peccatum* dimittere consuevi.".
7 Pope. . .day] *RO insertion from facing fol 5ᵛ to replace pencil deletion of paragraph* '{¶} In some respects, however, Mundinus was better off than English Anatomists of the present day. He appears to have laboured under one restriction only, with reference to the disposal of his subjects:—Pope Boniface the 8th. merely forbad him *to boil their bones* '. *The insertion is then followed by a comment by WC on 5ᵛ:* 'If he might do what he pleased with his subjects, what prevented him from examining the *internal Ear.* This is too absurd to be even mentioned as it cannot be believed; and therefore had better be omitted; unless the authority for the absurd *restriction* had some better foundation

Another event of War aided indirectly in the march of the human Mind. On the capture of Constantinople by the Turks in the middle of the 15th. Century many[1] learned men, fled to the West, and took refuge in Italy;—bringing with them a number of Greek Manuscripts, among which as we observed at the last lecture[2] was that most precious one for Comparative Anatomy, the Περι των ξωων of Aristotle, of which Theodore[3] Gaza published the first Translation.

The fabrication of Paper, the discovery of the Art [/] of Printing, and that of Engraving in wood, gave additional impulsion, and means of success to the returning spirit of Inquiry.

Berenger de Corpi, a professor in Bologna in 1502, (and the first who employed mercurial friction in Syphilis) published, in the beginning of the 16th. Century, an Anatomical Work, in which for the first time, the Text was accompanied by illustrative Wood cuts. We have seen, however that the Manuscript of Aristotle contained, also, Anatomical figures or Diagrams.[4] He first corrected the literal acceptation of the statement of Erasistratus that the nerves of motion arose partly from the Cerebellum; by showing that no nerves took their origin immediately from that part.[5]

Coiter (Observ. Anat. 1573) is the first who speaks of the fibrous structure of the brain: he says that the nerves are continued from the 'fibræ capillares' & recieve a sheath from the dura mater as they escape from the skull.[6]

In the Anatomical School of Paris, in which Gonthier[7] of Andernach was the first Professor of that Science, the works of Galen were preferred to that of Mundinus. Gonthier[8] applied himself to the improvement of Myology.

Michael Servetus, a pupil of Gonthier[9] following the track of inquiry which Galen had indicated respecting the course of the Blood in the lungs, describes, with much precision the lesser or pulmonary Circulation.

(cont)

than the authors *bare word*, which might have been only an excuse for his not having examined the part.[¶] Boiling the Bones *was unnecessary* —and Pope Boniface is not so absurd as apppears at first sight:[¶]Human fat thus procured was a hellish ingredient, for witches and others, long after this period.[¶]*Ask Shakespeare and even King Jemmy.[*with brackets* } Moreover the Italians were and *are* considered very expert poisoners.' *RO reply then follows:* "The fact of the restriction being admitted, it is worth comparing with the restrictions under which the pract. Anaty. labours at present.'.

1 many] *interlined above deletion of* 'Theodore Gaza, Argynophilus, and other'.
2 as. . .lecture] *interlined with caret.*
3 Theodore] *interlined with caret.*
4 Diagrams] *followed by pencil* '<' to indicate insertion point of facing text.
5 He. . .part.] *RO insertion from facing fol. 6ᵛ. Final* 'part' *followed by date* '1526'.
6 Coiter. . .skull] *second insertion from 6ᵛ to follow on above.*
7 Gonthier] *altered from* Gouthier'.
8 Gonthier] *second alteration from* Gouthier'.
9 Gonthier] *third alteration from* Gauthier' [sic}.

6   Sylvius, another of the pupils of Gouthier, [/] discovered the valves of the
Veins, and thus laid the germ of the subsequent discovery of the greater circulation.
          But the progress of Anatomical Science was still trammelled by the high
veneration in which the works of Galen were esteemed.[1] He still continued
the sole Dictator in Anatomy and Physic, until Andreas Vesalius boldly presumed to question his Authority.
          Vesalius, Fallopius, and Eustachius were the chiefs[2] of Anatomy in the
16th. Century;—Vesalius was successively professor of Anatomy at Paris and
at Padua. He first combined the dissection of the human body with
Comparative Anatomy, in order to determine the true peculiarities of the
Human Structure;—and established an able parallel between the bones and
muscles of an Ape, and Man.[3]

7   Vesalius examined, more minutely than his predecessors, the connections
of the Arteries and veins;—he distinctly mentions the valves of the veins, and
describes the valves of the Heart;—yet he saw them to no purpose. He
mentions the difference between the arteries and veins and demonstrates, by a
more direct experiment than Galen's, that the arterial pulse depends on the
Systole of the Heart.[4]

8   The disposition to generalize the observations furnished by Human
Anatomy in comparing the organs of Man with those of the lower animals,
which the Work of Vesalius had begun to spread abroad, was still more
strongly manifested in the labours of Fabricius of Aquapendente, his pupil

---

1   in . . .esteemed] 'in which'*inserted over deletion of* 'paid to' with 'works'*over erasure
and* 'were esteemed' *interlined with caret.*

2   chiefs] *Pencil X in margin and WC pencil query on facing* 7v*:* 'Archons—?Arbiters?
Archiatres? —Highest Authorities [?]'.

3   Man.] *followed by deletion of:* 'Nevertheless while he so unsparingly censured the
Errors of Galen, he *occasionally falls into {in pencil above* 'commits'} the same
abuse of Comparative Anatomy;—as where he describes the Heart:—since which period
the terms right and left, applicable only to the Heart in Brutes, has been applied to the
different cavities of the Human Heart.' *with erased WC five line note on facing fol.* 7v*:*
'This seems to be hypercritical of poor Vesalius; who was a fine old fellow and
probably only followed the *terms* right and left of those who went before him; knowing
his *before* from his *behind* perfectly well.'.

4   Heart] *followed by deletion by vertical pencil scoring of paragraph :* '{¶]Vesalius was
Physician to Charles the 5th. and followed that Monarch to Spain, where Anatomy was
at so low an ebb, that the great Anatomical reformer could not find a single Human
Cranium in the University at Madrid:— Nevertheless, Vesalius pursued his researches,
and at length fell a victim to his Zeal for Science:— On the accusation of having
dissected a subject of which the fibers were observed to palpitate, he was condemned by
the Inquisition to make a Pilgrimage to the Holy Land. On his return, he was
shipwrecked on the Coast of Zante; and it is related, he there perished {*over deletion of*
'died'} of Hunger. His [l*fol.* 7] great work is chiefly a refutation of Galen: it abounds
in beautiful Engraved Anatomical figures on Wood, it is said from drawings by Titian.'
with pencil 'X' *in margin and WC query in pencil on line* 'not Leonardo da Vinci?'.

and successor to the chair of Anatomy[1] in Padua. This Anatomist is best known by his supposed discovery of the valves of the[2] Veins. He undoubtedly[3] has the merit of having given the most exact description of these parts; and of having particularly pointed out their peculiar disposition in being all turned towards the Heart. Yet he could not raise himself to the true inference which was deducible from this anatomical observation. He said that these Apophyses Venarum were to the Veins what joinings or knots are to Vegeta-

9   bles; and that they served to moderate the current [/] of the Blood as it flowed backwards and forwards in their course.

At this time William Harvey arrived at Padua; and the glowing descriptions of Fabricius of the parts of which he deemed himself to be[4] the first discoverer, must have deeply interested the young Englishman. We know that subsequently Harvey acknowledged to the Honble[5] M[r]. Boyle that he was led, by reflecting on the use of these Valves to the knowledge of the Circulation; the noblest discovery in the whole range of Physiological Science.

This discovery, however, was established principally by observations and experiments on the lower animals. The works of Harvey, both on the motion of the Heart, and on Generation, abound with most interesting observations in Comparative Anatomy:—And here, the Modern History of that Science may be said fairly to have commenced.[6]

10   As the Progress[7] of Comparative Anatomy is here [/] traced with a view to determine the state of the Science, and the Physiological and Zoological deductions which had been derived from it, before M[r] Hunter began his great and[8] useful labours, I shall first consider the successive acquisitions of[9] a knowledge of the Structure of different Animals, *class by class*[10]:, and then the uses which the Discoverers made of their Observations:—lastly,[11] I shall contrast those with the amount of Facts established by Hunter, and the improvements in Physiology[12] which he founded upon them.

Among the first fruits of the establishment of the Royal Society of London, are the separate Anatomical Monographs of species[13] of Animals, which constitute so valuable a store-house of Facts and materials, for the use

---

1   Anatomy in] *interlined with caret.*
2   the] *over erasure.*
3   undoubtedly] *ink over pencil interline with caret above deletion of* 'certainly'.
4   to be] *over erasure.*
5   the Honble] *interlined with caret.*
6   commenced] *followed by horizontal ink line across middle of page to indicate pause.*
7   Progress] *interlined in pencil above deleted* 'History'.
8   great and] *interlined in ink with caret.*
9   successive acquisitions of] *ink over erasure.*
10   of the . . .*class:*] *in dark ink over erasure.*
11   lastly,] *after erasure of* 'and'.
12   Physiology] *before deletion of* ', and in their practical. . . .'
13   species] *after pencil deletion of* 'single'*with WC query on facing fol 11*[V] '*Monograph of single* Species?'.

of the Physiologist, the Philosophical inquirer into the natural Affinities of the Animal Kingdom; and the Physico-Theologist.

The Anatomy of the Bear and the Beaver appear in very early numbers of
11   the Philosophical[1] [/] Transactions. But the most valuable materials for the Anatomy of the Mammalia in this Collection are contained in the papers of Dr. Tyson.

It is here that we find the first account of the[2] complicated[3] stomach of the Pecari, and of the Marsupial organization as it is exemplified in the American opossum[4]. Tyson's description of the singular female organs of this animal[5] is more exact than many later descriptions; and his determination[6] of the Fallopian tubes[7] and Cornua Uteri, and his conjecture as to the seat of gestation in the complicated apparatus he describes, have recently been proved to be accurate notwithstanding the animadversions of Buffon & Geoffroy,[8] by an examination of the pregnant uterus of a Kangaroo.

Tyson's[9] Monograph on the Anatomy of the Pigmy, the *Simia Troglodytes* of Blumenbach, has rarely been surpassed, for the clearness of its descriptions, the justness[10] of the Analogies[11], or the profound learning with which it is dignified. By the labours of Tyson, and the previous observations of Ray and Rondeletius, many of the peculiarities in the organization of the Cetaceous Mammalia unnoticed by Aristotle, were first[12] made known.
12   The example of the Fellows of the Royal Society of England was followed at[13] the Institution of the Royal[14] Academy of Sciences at Paris[15], by some of its early Members. Dissections of 47 Animals, among which are many interesting Species of Mammalia are carefully recorded, with accurate illustrations in the collection of the French Academicians.[16] The singular stomach of the Camel, and the peculiar mechanism of the Claws of the Feline Animals are described for the first time in the Parisian Collection of Anatomical Monographs, which though published under the superintendance and name

---

1   The. . .Philosophical] *double vertical pencil line along margin.*
2   the] *RO pencil on facing fol 12ᵛ as if to continue, but without insertion point*: 'Anatomy of animals peculiar to the new World—'.
3   complicated] *inserted over erasure.*
4   as. . .opossum] *RO interlined in pen with caret.*
5   this animal] *inserted over erasure.*
6   determination] *over erasure.*
7   tubes] *terminal 's' inserted in pencil.*
8   notwithstanding. . .Geoffroy,] *interlined in ink with caret.*
9   Tyson's] *inserted in dark ink over erasure.*
10 justness] *terminal 'ness'over erasure.*
11 Analogies] *terminal 'ogies'over erasure.*
12 first] *interlined in ink with caret.*
13 at] *over erasure.*
14 Royal] *after deletion of* 'French'.
15 at Paris] *interlined in pencil.*
16 in. . .Academicians.] *inserted from facing fol. 13v in same hand as text with deletion of terminal period and* 'x'*after* 'illustrations' *[ed.].*

of Perrault, are principally due to the labours of the modest Duverney.

The descriptions and figures of the Digestive organs of various Animals by the learned Dr Grew;— the special study of the Stomach of the Ruminants by Conrad Peyer;— the observations of Cowper[1] on the Male Oppossum;— and the *Myographia Comparata* of Dr[2] Douglas are valuable[3] Contributions towards the Anatomy of the Mammalia.[4]

13   Much[5] interesting information in the Anatomy of Birds is contained in the Writings of Harvey, especially in his Classical Work *De Generatione Animalium*. He describes the peculiarities of the Gizzard, and anticipates Hunter in ascertaining its slow peristaltic motion, by applying the Ear to the Abdomen of the Bird. He first discovered the peculiar extension of the Air-Cells into the abdomen of the Bird; describes the fixed condition of their lungs to the ribs, and the nature of their rudimental diaphragm.

Here again also he anticipates Hunter in the philosophical comparison of the abdominal air-cells of the Bird with the abdominal lungs of the Reptile, and air-bladder of the Fish:— but he appears to have had no suspicion of the extension of the Air into the osseous System in the feathered Class.

Harvey is most rich, as might be expected, in the details of the Anatomy of the Generative organs, and is the first to describe those of the Cassowary,[6]
14   and Ostrich.

The crop, echinus, and gastric glands, and the gizzard of several species of Birds, together with the intestines and cæca, are well described and figured by D^r. Grew, as early as 1676; and the modifications of the different parts in relation to animal and vegetable food, are ably pointed out.

The[7] Brain of several species of Birds is figured in the System of Anatomy of Collins; who, in some of his explanations, rightly compares the Optic lobes with the bigeminal bodies; but further examination of his Work shows clearly that he had no just knowledge of the philosophical analogies of the different parts of the cerebral Organ.

We find the earliest attempt to elucidate the peculiar structure and formation structure[8] by which the lateral barbs of feathers are hooked together;[9]— also

---

1 Cowper] *changed from* 'Camper'.
2 Dr] *interlined with caret.*
3 are valuable] 'are valu' *inserted in dark ink over erasure.*
4 Mammalia] *replacing colon by period [ed.] before deletion of* 'and it is with these Works that the young Comparative Anatomist will most advantageously commence the [1 *fol.* 12] study of the Science.' *On facing fol. 13^v WC pencil note* 'X Douglis [sic], I think, preceded Camper many years? A D^r. Douglas was D^r. Hunter's preceptor— and Camper was, though possibly older, a contemporary of John Hunter; at least, he outlived D^r. Hunter.'.
5 Much] *faint vertical pencil line down margins of both fols. 13 and 14 from this point.*
6 Cassowary] *after deletion at end of fol. 13 of* 'Java'.
7 The] *following one-inch space as if to indicate paragraph division [ed.].*
8 microscopic structure] *in dark ink over erasure with* 'Microscopic' *in margin.*
9 ligniperdus together;] *with erased RO pencil above* 'is described & figured by the work

15 the air-cells of the Ostrich, and the curious mechanism of the Membrana nictitans are described and figured in [/] the Anatomical Memoirs of the French Academicians.

In the same collection there occur some of the earliest elucidations of the anatomy of Reptiles. It is here that we find the first account of the singular growth, shedding and replacement of the teeth of the Crocodile. Perrault & Duverney describe the anatomy of the Gigantic Land Tortoise.[1] The heart in the Testudinata is minutely described by Bouchier, Mèry & Duverney. The anatomy of the Turtle is admirably illustrated in Caldesi's celebrated monograph: The organization of the Rattle-snake is ably unfolded by Tyson. Swammerdam & Malpighi describe many of the peculiarities of the Batrachian Organization and first mention the mode of inspiring air by deglutition in the Toad & Frog.[2]

The[3] metamorphoses of the Batrachia, one of the most interesting discoveries in Animal physiology, are alluded to by Harvey in his 19th. Exercitation;[4] and[5] are fully detailed by Jacobæus[6], and are[7] beautifully illustrated, in the works of Roesel.

With respect to the Organization[8] of Fishes, much valuable information is contained in the Works of the Anatomists of the Italian School founded by Vesalius, and Eustachius. An anatomical description of a Shark [Squalus Mustela] is given by Fabricius ab Aquapendente;[9] who also enters into some details respecting the generation of Fishes.—

These details are repeated, and many are[10] added to them[11] by Harvey, who[12] describes the simple structure of the heart in this Class:—the single Artery dispensed to the Gills; and the course of the blood through that[13]

16 Artery. [/] "A general idea of the Structure of the Internal parts of Fishes," is

---

(cont)

of Duverney'.

1   Tortoise.] *period inserted [ed.].*

2   In. . .Frog] *RO ink insertion from facing fol. 16ᵛ to replace deletion of* ' ¶Anatomy of Reptiles:— By the labours of Bouchier, Duverney, and Mery, the structure of the Heart of Reptiles was made known:— and the Anatomy of the Chelonia is admirably elucidated in Caldesi's justly celebrated Monograph.'. *Three periods after* 'Organization' *deleted [ed.].*

3   The] *paragraph break inserted following deletion and insertion [ed.].*

4   are. . .Exercitation;] *RO interlined in ink with caret.*

5   and] *inserted [ed.].*

6   by Jacobæus] *interlined with caret.*

7   are] *interlined below line with caret.*

8   Organization] *over erasure with* 'tion' *interlined with caret.*

9   ab Aquapendente] *interlined in dark ink.*

10 many are] *interlined with caret.*

11 them] *interlined with caret.*

12 who] *followed by deletion in ink of* 'also'.

13 that] *inserted in ink over* 'the'.

given, by Preston, in the 19th Vol. of the Philos. Trans.: and our great Countryman Ray has consigned some interesting observations on the swimming bladder of Fish in the same Collection.[1]

The nature of the branchial circulation in Fishes, and the mechanism of their respiration by means of Gills, are very[2] fully described by Duverney.

Casserius, the pupil and successor of Fabricius ab Aquapendente[3] makes known many interesting details respecting the brain and organs of Sense of Fishes. He describes particularly their nostrils, eyes; and he[4] first discovered the membranous semicircular canals, vestibule, and calcareous concretions, or otolithes, of the internal Ear.

Malpighi, in dissecting a Sword-Fish, discovered the remarkable plicated structure of the Optic nerve, which we now know to be[5] peculiar to Fish.

Steno, a Dane by birth, and professor of Anatomy at Florence in 1667, is the author of some Anatomical and Physiological Works which abound in interesting observations in Comparative Anatomy. He treats of the brain, the eye, and the Teeth of the Shark, and gives a good description of the viscera of the Ray and Torpedo.

17   The illustrated Monograph of the Squalus Alopecias [/] in which the spiral valve of the intestine is figured, and which is due to the labours of the French Academicians, must not be passed over. But, the most remarkable contribution to the Anatomy of Fishes in the 17th. Century, occurs in our Countryman Collin's System of Anatomy before mentioned, in which the viscera and brains of upwards of 20 Fishes are represented in a style which was unequalled for a long time, and which still surpasses the illustrations appended to many of our Modern Compendiums.[6]

In the Anatomy of the Mollusca some interesting points are noticed in the works of Harvey;—as the heart and breathing organs in the Snail:—and the learned Dr. Willis gives a Monography on the Anatomy of the Oyster in his Work *De Anima Brutorum*. The anatomies[7] of the Snail and of the Cuttle-fish are[8] treated by Swammerdam; but those portions of the *Biblia Naturæ*

---

1   "A. . .Collection.] *interlined in light ink with* 'A. . .our' *at top of fol. 16, and* 'great. . .collection.' *inserted from facing fol. 17*[v] *and written over pencil* 'Ray consigned some interesting observations on their swimming bladder in the same Collection. {¶}Marinus Pisces—Rivinus, Acta Erud. Lips: 1687—p. 160–162-{¶} Schneider—Anatomia et Physiol. Piscium, {¶}Jacobæus, Gills of Lamprey—'.
2   very] *interlined in light ink above* 'subsequently more'.
3   ab Aquapendente] *inserted by RO in pencil.*
4   he] *inserted in margin in ink.*
5   we. . .be] *interlined with caret above deleted* 'is'.
6   Compendiums.] *followed by deletion by vertical scoring in light pen of* 'The letterpress *of Collin's {interlined in pencil with caret }* is more amusing, for its affected quaintness, than instructive.'.
7   anatomies] 'ies' *interlined in pencil above deleted* 'y'.
8   are] *interlined in pencil above deleted* 'is'.

18   are not[1] characterized by the same[2] detail and accuracy which [/] astonish us in reviewing his Anatomy of the Articulated animals.

    When we consider how well the History of the Mollusca was commenced by our countryman Lister, who combined the Study of the organization of the soft parts[3] with that of the forms of their calcareous coverings, and that these animals are more easy to dissect than Insects, it is difficult to account for the long neglect which they experienced, until the imperishable[4] works of Poli and Cuvier, & Della Chiaje[5] revealed the wonders displayed in this type of the Animal Organization.

    The poverty of the materials towards the Anatomy of the Mollusca which existed prior to the labours of Poli and Cuvier is, however, interesting to us in considering the Discoveries of Hunter:—for if it can be shown, as I shall be able to do, that the Collection affords evidence of Hunter having pushed his investigations as widely and as successfully into the organization of Mollusca, as into the Anatomy[6] of those Classes of the Animal Kingdom respecting which it was possible for him to have derived from the labours of

19   his predecessors a knowledge of much[7] [/] that he undoubtedly acquired by Dissection: then we possess strong grounds for claiming for him that rank in the History of Science which is due to[8] the acquisition of an universal knowledge of Comparative Anatomy by original labours guided by original views.

    Anatomical[9] preparations are ill calculated to convey an[10] idea of the Complex Organization of Insects;—yet, Hunter, notwithstanding, attempted to illustrate by this means all the important parts of their Anatomy. In this department, however, he is excelled in extent and minuteness of research by Swammerdam, Roesel, and Reamur, and especially by his contemporary Lyonnet whose monograph on the *Cossus ligniperdus*[11] Cuvier pronounces to be at once the Chef d'Œuvre of Anatomy and Design.

    We know from Hunter's published Memoir on the Bee, that he was acquainted with the labours of Swammerdam on the Anatomy of the same insect. We are willing to assign to him the praise due to a higher and more

---

1   are not] *omitting pencil interline above* 'by' *for grammatical sense [ed.]*.
2   the same] *interlined over pencil deletion of* 'that'.
3   the soft parts] *interlined in ink with caret with subsequent deletion of* 'of' *in the interline above deletion of* 'the Animals'.
4   imperishable] *interlined with caret.*
5   & Della Chiaje] *interlined in pencil with caret.*
6   into the Anatomy] *interlined in dark ink with caret.*
7   Classes. . .much] *pencil* '#' *symbol inserted alongside in margin .*
8   rank . . .due to] *interlined and inserted in dark ink with* 'rank . . .which' *interlined, and* 'rank' *above deletion of* 'positions'.
9   Anatomical] *after pencil deletion of* 'Articulata.'.
10   an] *deleting* 'y' *from* 'any'.
11   *ligniperdus*] *over erasure of* ' {illeg. word} which'.

physiological[1] application of the facts which he observed, than is to be found
20   in the [/] Treatise of the Dutch Anatomist; but we cannot subscribe to
Hunter's opinion of the inutility of the details in which he found himself so
far surpassed by Swammerdam.

Comparative Anatomy has other applications than to Physiology; and the
admirable harmony and unity of design traceable within certain limits[2]
throughout the Animal Organization, &[3] of which Hunter himself had arrived at the perception, but which is to be[4] developed chiefly[5] by attention to
details from which no physiological deductions immediately result;—this
high and philosophic application of the facts of Comparative Anatomy, demands an attention to the minutest character and variation of every organ, independent of considerations as to the function to which such variation is subservient.[6]

The materials for the Comparative Anatomy of the different classes of
Animals which have just been noticed, together with other less important observations too numerous to recapitulate; and scattered in different Collections
21   of Scientific Memoirs [/] of the 17th. Century, are for the most part collected
together and reprinted in the well-known[7] work of Gerard Blasius,[8]
published in[9] 1681. Or in the "Amphitheatrum Zootomicum" of Valentin,
which forms the indispensable Complement to "Blasius";—or lastly, in the
"Bibliotheca Anatomica" of Mangetus.

These compilations convey a favourable idea of the extent of Comparative
Anatomical knowledge attained in the 17th. Century, in the light of a simple
accumulation of Facts. Let us now briefly review the advantages gained by[10]
the Application of Comparative Anatomy to General Physiology, during the
same period.

The result of Harvey's experiments[11] and extensive researches into the
Structure of the lower Animals, was the discovery of the Circulation; and by
pursuing the same track of Inquiry, and associating with it observations on
the development of the Embryo, He laid the foundation of a true knowledge

---

1   physiological] *over erasure.*
2   within . . .limits] *interlined in pencil with caret.*
3   &] *inserted in pencil with* '&'*interlined.*
4   is to be] *interlined in pencil over deletion of* "has been".
5   chiefly] *followed by pencil deletion of* 'by the labours of the Continental Anatomists
of the present day; and'.
6   to which. . .subservient .] *inserted over erasure.*
7   well-known] *interlined with caret.*
8   Blasius,] *followed by deletion of* ' "Anatome Animalium terrestrium volatilium,
aquatilium, etc; ex veterum, recentiorum propriusque observationibus proponens"
Amsterdam'.
9   published in] *RO insertion in margin in pencil.*
10   review. . .by] *with* 'review' *in ink in margin and* 'advantages. . .by' *inserted over
erasure.*
11   experiments] *RO pencil comment on facing fol 22[v]* 'Circulation'.

22   of the more [/] mysterious function of Generation. It would have been strange
if the observation by such a man, of the progressive stages of Complication
presented by the different Organs of a Warm-blooded Animal during its development, should not have led to a perception of the Analogy which these[1]
different Stages present, to the various conditions of the same Organ in the
lower Animals. And we find accordingly that this Philosophical comparison,
so essential for appreciating the harmony of design in the Animal
Organization, is expressed in no equivocal or obscure language, in reference
to the development of the Heart in the Chick.

> For if you observe the formation of this part, say Harvey, first of all
> there is only a little vesicle, or auricle, or drop of blood, which pulsates;
> and increasing afterwards the Heart is perfected; so in some creatures; (as not
> reaching a further perfection) there is only[2] a certain little pulsating[3]
> vesicle, like a point, red, or white, as in Bees, Wasps, Snails, Shrimps,
> &c.—[4]

23   This just and philosophical comparison of the simple heart of Insects,
Crustacea and Snails, to an early condition of the heart of the Chick, is the
first indication I have yet met with of a Principle which is enunciated in[5] more
general terms by Hunter, but of which we are now only beginning to appreciate the importance.[6]

But to proceed with the examples of the influence of Comparative anatomy
on the progress of Physiology;—The lacteals originally observed by
Erasistratus in a Kid, were rediscovered by Asselius in 1622, while examining the mesentery of a living Dog. The importance of this Observation in a
physiological point of view, arises from the circumstance that the previous
discovery of the Antient Alexandrian Anatomist had not been productive of its
legitimate results. No deductions had been drawn from it; it was a barren fact,
and as such, had been forgotten.

Something more than Eyes are required to establish an Anatomical discov-
24   ery productive of real fame and honour to the Observer. At the [/] time of
Asellius it was universally believed that the Mesenteric Veins were the Agents
in absorbing the Chyle; and as they terminated in the venæ portæ, the office

---

1   these] *over* 'the'.
2   only] *interlined with caret.*
3   pulsating] *after deletion of* 'bladder'.
4   For. . .&c—] *RO insert from facing fol. 23ᵛ to translate Latin quote but without
    insertion point marked:* 'Sed si in ovo pulli conformationem advertas, primum inest, ut
    dixi, tantum vesicula, vel auricula, vel gutta sanguinis pulsans; postea incremento facto
    absolvitur Cor: ita quibusdam animalibus (quasi ulteriorem perfectionem non
    adpiscentibus{*with* 'en' *over* 'ea'}) pulsans vesicula quædam, instar puncti cujusdam
    rubri vel albi, dumtaxat inest; quasi principium vitæ; uti apibus, vespis, cochleis,
    squillis, gammaris, &c." (Op. Omnia p. 32.)'.
5   in] *followed by erasure.*
6   importance] *RO comment on facing fol. 24ᵛ* 'Absorption'.

of converting the Chyle into blood was assigned to the Liver. Asellius transferred the office of Absorption from the mesenteric veins to the white vessels; but as he still supposed that they terminated in the Liver, he calls them its Arms, by means of which, like so many leeches, he believed the Liver sucked up the Chyle.

Eustachius had seen the Thoracic duct, while dissecting a Horse; but like Erasistratus, had not suspected its important function. It was the good fortune of Pecquet in 1651, while dissecting a living Dog to perceive a white fluid mingling with the Blood, and flowing on with it into the cavity of the Heart in a constant stream. He traced the white fluid to its source, and successfully followed the Thoracic duct to the point of convergence of the lacteals;—and thus established the second important step in the discovery of the Absorbent System.

25   It was by the [/] dissection of the lower Animals chiefly that Olaus Rudbeck, a Swede, in 1651:—Bartholomis, a Dane, in 1652, and in the same Year Jolyffe,[1] an Englishman, discovered many other vessels which we now call Lymphatic Absorbents, terminating, in   addition to the Lacteals of Asselius, in the Duct of Pecquet.

Swammerdam, whose anatomical discoveries[2] (as unapplied facts,) we have already noticed, contributed an important Share towards the establishment of just Ideas on the Physiology of the Absorbents, by the discovery of their valves.

Lister and Musgrave anticipated Hunter in Experimenting on the Lacteals, by which they ascertained that fluids artificially coloured were taken up by the Lacteals from the Intestines. These experiments[3] are recorded in the 12th. (996) 13th. (6) 14th. (812) & 22^{d} (819)[4] Volumes of the Philosoph. Transactions.[5]

Notwithstanding, however, these observations,[6] the system of Lymphatic Absorbents was not known as it is now understood. The minute veins which accompany the capillary arteries were universally regarded as forming a part, and the principal part, of the absorbent System; and it is to these capillary veins *that*[7] Noguez in 1737 applied the term "conduits absorbans"[8] and, which form what he terms his second Class of Lymphatics.

26   These facts are necessary to be borne in mind [/] in considering the Merits

---

1   Jolyffe] *after erasure and pencil crossout of* 'by'.
2   discoveries] *interlined with caret.*
3   experiments] *interlined above deletion of* 'observat'.
4   12th] (819)] (996); (6) ; (812); (819) *interlined below volume numbers in pencil below line.*
5   Lister. . .Transactions.] *RO insertion in ink from from facing fol. 26^{v} with inserion point marked by* '<' *symbol.*
6   observations] *after deletion of* 'Anatomical'.
7   that] *pencil underline and in margin. WC query* 'That ? X'.
8   absorbans] *deleting period after* 'absorbans' *for grammatical clarity [ed.].*

of M^r. Hunter, as a Discoverer[1] in reference[2] to the absorbent System.

The physiology of Digestion was a subject to which Hunter expressly devoted much of his attention; and it will be seen that the nearest approaches to his views on this important part of the Animal Œconomy were made—not by the Chemist, who sought to reduce digestion to a fermentation, or interplay of elective affinities,—nor by the Mechanical Philosopher who calculated the pressure exerted by the muscular parietes of the Stomach in reducing the Ingesta to a pulpy Chyme;—but by these[3] Physiologists, who endeavoured[4] to elucidate the functions of the different organs by tracing their modifications of Structure through the different classes of the Animal kingdom.

Tyson was led to suspect the true nature and agent of digestion from observing the state of the Contents of the stomach in the Rattlesnake,—a cold blooded Animal in which there could be no concoction of the food by heat, nor any mechanical attrition capable of reducing the Ingesta to a pulp. He, therefore, very justly observes that he could not see how Digestion could be performed but by Corrosion: and the corroding or dissolving menstruum he concludes to be that which is poured out in Birds by the glands which are situated just above the Stomach, [/] and called the Echinus; or in other animals in the stomach itself, and called the glandular coat, and such he considers the inner coat of the Stomach of the Rattle-snake to be;—wherefore, says Tyson, the whole *Ductus alimentalis* from its uses may be divided into four parts; one, that which conveys the food, the œsophagus: 2^d. that which digests or corrodes it, the stomach: 3rd. that which distributes the Chyle, the intestines: 4th. that which empties the fæces, the rectum. But, having extended his researches to the lower animals, he observes that a Leech is all stomach.

The learned D^r. Grew,[5] a contemporary[6] of Tyson had described almost every modification of the alimentary canal in Mammalia and Birds; and the uses of the different parts are generally assigned to them with much judgment. He also attributes Digestion not to mechanical attrition merely, nor to coction; but to the action of a mucous secretion prepared from the blood of the stomach, which secretion he calls an animal corrosive: he considers besides that the nerves are essential to the process, but on this subject expresses himself obscurely, saying that they add a ferment which perfects the work. Elsewhere he proves that the solvent or corrodent does not act upon living

---

1   as a Discoverer] *interlined with caret.*
2   reference] *with erasure of nine lines in illegible RO pencil on facing fol 27^v and* 'See Note to Digestion' *in RO pencil below.*
3   these] *over* 'the'.
4   who] *before deletion of* ', prior to Boerhaave, sought' *with* 'endeavoured' *interlined in pencil over* 'sought'.
5   Grew] *with deletion of* 'Nehemiah'.
6   contemporary] *altered from* 'contemp' *[ed.].*

bodies [p. 23]. These were near approaches to the true doctrine[1], but obser-
vation being here uncombined with experiment the opinions of Tyson and
Grew were forgotten, until the idea of the dissolving fluid was revived by
Valisneri, Reaumur, & Spallanzani.[2]

30   The researches of De Graaf on the generation of the Rabbit, led him much
nearer to the true knowledge of the nature of the Mammiferous Ovum, than
did his carefully repeated dissections of the female organs of the human
Subject.

Had he been guided by a philosophical comparison of the facts which he
had established, he would have seen that the Vesicle of the human ovum
which now bears his name, was analogous to the follicle of the ovary of the
Rabbit, and not to the true Ovulum, whose independent existence and escape

---

1   true doctrine] *over erasure of* 'truth' *and on facing fol 28ᵛ* 'doctrine /'.
2   Spallanzani] *followed by deletion by vertical pencil and pen scoring of the following
    2ᵤₛ fol. pages reading:* '{¶} Among the Physiological works of the 17ᵗʰ century
    founded on *Comparative {*interlined with caret* } dissection, and experiments on the
    lower Animals that of Borelli "*De motu animalium*" is among the most interesting:
    **It is replete {*interlined with caret* }with new and important facts illustrative of the
    locomotion of Man, Mammalia, Birds, Insects, and Fishes: which are so perfectly
    demonstrated that they reappear, with little alteration in every System of Comparative
    Physiology. {¶} The Statical Essays of Dʳ. Hales, {*with deletion of* 'published in
    1773'} are {*over erasure of* 'is'} equally valuable in relation to the mechanics of the
    vascular System, and may still be studied with much advantage in connexion with the
    recent Hæmastatical experiments of Poiseuille, in which the principal objections in
    Hales's experiments are guarded against. {*With WC query in pencil margin:* 'Qu if
    here? See back of p. 26.'} {¶} In the works of this Author and of others of the same
    School, Physiology is reduced to a branch of Mathematics, and the different functions of
    the human body, the will excepted, are held to be susceptible of precise calculation, and
    to be explicable by means of Algebraic formulæ. While the Mathematician was thus
    attempting to explain every thing by means [l*fol.* 28]{*at top of page pencil* 'X'} of his
    Science, the Physiological Chemist was not less backward {*in margin second pencil*
    'X'} in resolving all the problems of Vitality into {*in margin third pencil* 'X' *with WC
    pencil note on the rear of this page* 'See forwards—out of place here Boerhaave p.'
    pages 29 and 30 held together by straight pin along left margin} an *interplay* of
    Chemical affinities. Thus by the abuse of two sciences, which, within certain bounds,
    are excellent auxiliaries, Physiology began to retrograde, and those laborious but
    essential labours by means of which the Circulation of the Blood had been established,
    and the first steps taken towards so many other important discoveries, were almost
    abandoned. {¶} {*in margin in WC pencil.* 'X If follow here?'} The Comparative
    Anatomist while sharing in the general admiration of the unexpected discoveries of the
    Natural Philosopher in relation to the properties of the Galvanic fluid, may look back
    with much pleasure to the circumstance in which the knowledge of that potent and
    universally operating agent originated. Perhaps it is one of the most remarkable among
    the accidental discoveries which have resulted from the dissection of the lower animals,
    that Galvani from observing, while dissecting a Frog, the tremulous contractions of its
    muscles, should have been led to a knowledge [l*fol.* 29] of a before unknown
    imponderable fluid, of which each day discloses some new *& astonishing {*interlined
    in pencil with caret*} powers.'.

after impregnation he so well describes in that Quadruped.[1]

The writings[2] of De Graaf also furnish us with another striking example of the inadequacy of Human Anatomy alone, to advance our knowledge of Physiology; nay more[3] of the Tendency of a knowledge so limited to cause Physiology to retrograde. Prior to De Graaf we find that the Testes, vesiculæ seminales, and prostate were considered to contribute three distinct fluids to-
31   wards the formation of the Semen, as we read [/] in the writings of[4] John Van Hoorne, in 1707.

De Graaf, from observing the communication of the outlets of the vesiculæ seminales with the ducts of the Testes, maintained that they acted as tempo-rary reservoirs of the Semen. His authority had great weight, and, notwith-standing that Tyson from his dissection of the Peccary, and other animals was enabled to support the old opinion by arguments drawn from the Glandular structure of the vesiculæ in these Animals, and the separate termi-nation of their ducts in the Urethra, it required the additional observations of Hunter, many years afterwards, to establish the correctness of these views which Comparative Anatomy had, and alone could, have originated.[5]

It is gratifying to find that the conclusions thus arrived at, have received recent corroboration by the Microscopic observations of Prevost and Dumas, who have been unable to detect the spermatozoa,[6] or Seminal[7] animalcules, in the fluid of the vesiculæ seminales, while they are traceable through the whole track of the Duct of the Testis.
32   While thus alluding to the microscope, I would observe that its discovery, and application to illustrate the Structure and Œconomy of organized bodies, about the middle of the 16th. Century, forms an Epoch in the History of Comparative Anatomy. This instrument appears to have been first used by Eustachius, in Anatomy; and afterwards by Grew and Malpighi, to unfold the intricate[8] structure of Vegetables;—and, as it has ever since been indispens-able to the Botanist, and productive of a thousand accurate results otherwise unattainable in that department of Natural History;—[9]Why, we may ask, made by means of this instrument by Swammerdam, De Heide, Lieuwenhoek, Baker, Ellis, Trembley, Roesel, Lyonnet, Spallanzani and Bonnet, have been established and confirmed by later observations and better Instruments;—and the causes of their errors have been satisfactorily explain-
33   ed and avoided in the modern Microscope, to the improvement of [/] which,

---

1   in. . .Quadruped.] *RO insertion in ink with* 'that Quadruped' *on facing fol. 31v* .
2   The writings. . .] *in margin pencil* 'X'.
3   nay more] *interlined in ink with caret above erasure.*
4   of] *interlined with caret.*
5   originated] *followed by WC pencil insert subsequently erased* '*See page 24 of text.'.
6   spermatazoa] *over erasure.*
7   Seminal] *inserted in margin.*
8   intricate] *over erasure.*
9   in. . .History] *RO ink insertion from facing fol. 33ᵛ with insertion point marked by line.*

the enlightened encouragement afforded[1] by our Society of Arts has greatly[2] contributed.

The capillary circulation, the form and structure of the blood-globules, the circulation in Insects, and the vibratile cilia which perform[3] so extensive and important a part in the Animal Œconomy[4], could never have been known[5] without the Microscope. Like other good aids to Anatomy, it has undoubtedly been abused, partly from the hope of establishing by its means, structures and arrangements of parts inappreciable to ordinary observers[6], but essential to the establishment of some favourite and preconceived hypothesis:— but above all, by its vicarious employment. We are justified in receiving with reserve[7] the descriptions of microscopic observations made under the latter circumstances; but when the unprejudiced or unprepossessed Observer relates what he has himself seen; and when others by means of instruments of the same power are able to corroborate his observations, we may then admit the details[8] with confidence.

34   Besides[9] the application of Comparative Anatomy to the discovery of the uses of parts, the numerous and striking instances it affords of the adaptation of means to an end in the different organs[10], and of their[11] relations of coexistence, had not escaped the notice of some enlightened Philosophers thus early engaged in the investigation of the Structure of Animals.

The Treatise[12] of Ray, entitled 'The Wisdom of God displayed in the Works of the Creation" is a well-known example of the application of Comparative anatomy just alluded to: but that Work is surpassed in the number and originality of the Anatomical and Physiological observations by the "*Cosmologia Sacra*" of Dr. Grew.

We have of late years[13] seen how admirably Cuvier has reconstructed, as it were, numerous animals of lost Species, by his profound knowledge and sagacious application of the laws of Coexistence of different Organs;—laws which he compares in their constancy to the Data of Mathematical Science.

Observe how clearly this fruitful principle was[14] enunciated by Grew so

---

1   afforded] *RO interlined with caret.*
2   greatly] *RO in pencil above deletion of* 'so much'.
3   perform] *RO insertion in pencil with* 'erform' *interlined above deleted* 'play' *and in margin WC query:* 'X perform)? '.
4   Œconomy] *RO interlined over deletion of* 'functions'.
5   known] *RO interline to replace* 'seen'.
6   observers] *RO pencil alteration of* 'observation'.
7   reserve] *RO interline in pencil over deletion of* 'due caution'.
8   admit the details] *RO interline above* 'receive the narration'.
9   Besides] *with pencil bracket at top of page in left margin.*
10   in. . .organs,] *RO interline with caret.*
11   their] *altered from* 'the'.
12   Treatise] *over erasure.*
13   of. . .years] *RO interline in dark ink with caret.*
14   was] *interlined in pencil over* 'is'.

early as the   close of the 17th. Century:[1]

35   In[2] the use of things is seen that relation which answers in some sort unto
Geometric proportions. So these Creatures whose motion is slow, are
blind: but those which have a[3] quick motion have eyes to govern or determine it: That is, as Blindness is to a slow motion, so is sight to a quick.
So those Animals which have Ears have also lungs, and *vice versa*, those
which have no Ears have no Lungs:—For as eyes are to motion, so are
Ears to Speech. So likewise, those Animals which have Teeth on both
Jaws, have but one Stomach; but most of those which have no upper teeth,
or none at all, have three Stomachs. As in beasts the panch, the read, and
the feck;—and in all graminivorous Birds, the Crop, the Echinus and the
Gizzard. For as chewing is to an easy Digestion, so is swallowing whole to
that which is more laborious.
          A man who hath a bigger Brain in proportion to his body than any
other Creature, hath also a better hand. A Monkey hath a hand, but with an

36   arm not so well fitted to a [/] hand as to a foot. Nor can he put his hands
and feet to their distinct uses at the same time, as a man whose posture is
erect. As Ears therefore are to Speech, or eyes to motion, so is Reason to
Operation.

There are a few errors, it is true, in these generalizations but the principle
or the law, as laid down by Grew, and compared by him, as afterwards by
Cuvier,[4] to Geometric relation, is not the less exact. Fishes have Ears,
though no true lungs, because their hearing is not exclusively for voice, or
for noises emitted by the Species exclusively, but for any kind of sub-aque-
ous noises.

Ray and Tyson subsequently showed that the Porpoise which has Teeth in
both jaws, has many Stomachs:—but then it swallows its food *entire*.[5]

After so many examples of the Influence of Comparative Anatomy on the
progress of Physiology, and a Philosophic perception of the general analogies

37   *or* [6] laws of the Animal [/] organization[7], it seems scarcely credible that any
hypothesis, with whatever acumen or eloquence supported, should have
succeeded in persuading the cultivators of Medical Science, that the Anatomy
of Animals was   foreign to the Art of Healing, and an inquiry of pure
Curiosity! Yet such appears to have been the effects of the powerful discourses and Writings of Boerhaave, at the early part of the 18th. Century.

---

1   17th . Century:] *below line on page break and not repeated at top of fol. 35.*
2   In] *at top of page in faint pencil* 'he likewise observes'.
3   a] *interlined with caret.*
4   Cuvier,] *underlined and comma inserted in pencil and in margin in WC pencil* 'X ?'.
5   entire] *underlined with pencil and modified from* 'intire'; *in margin WC pencil* : 'Qu.
    X' *with vertical pencil line in left margin. Entire page down to this point with faint
    pencil vertical score as if possible deletion.*
6   or] *underlined in pencil and in margin* 'X'.
7   organization] *in top margin by WC* '37 {*in ink* } *to 93 37/ 1{deleted} 56' {pencil} *to
    indicate the number of pages remaining to the original end of the second portion of the
    lecture.*

In the works of this Author, and of others of the same school, Physiology is reduced to a branch of the Mathematics, and the different functions of the human body,—the will excepted, are held to be susceptible of precise calculation; and to be explicable by means of algebraic formulæ.

It[1] is greatly to the Honour of the 1st. Monro—the Patriarch of a race of Anatomists who have successively & successfully devoted themselves to the advancement of anatomy—that at this period he endeavoured to recall the attention of Physiologists to the importance of the study of Comparative Anatomy, and his Essay still well repays a perusal from the philosophical views of the application of that Science which it contains.

The great Haller also[2] now arose to demonstrate how essential the study of the varieties of each[3] organ in different animals was to the attainment of just ideas of its function. While Daubenton & Pallas demonstrated the importance of Comparative Anatomy to the advancement of Zoological Science.[4]

And while[5] the Mathematician was thus attempting to explain every thing by means of *his*[6] Science, the Physiological Chemist was not *less* intent upon[7] resolving all the problems of Vitality into an interplay of Chemical Affinities:—Thus, by the abuse of two Sciences, which within certain bounds are excellent auxiliaries, Physiology at the beginning of the 18th Century[8] be-
38   gan to retrograde; and [/] those laborious and essential labours by means of which the circulation of the Blood had been established, and the first steps taken towards so many other important discoveries, were almost abandoned.

Comparative Anatomy, therefore, which had been cultivated with such ardour towards the close of the 17th. Century, became neglected in the first half of the 18th.[9]

To expatiate on the merits of such Men as Haller, Daubenton, and Pallas were as easy as it would be agreeable; but the influence which their example had in the restoration of Comparative Anatomy, and the important accessions
39   to that Science which resulted from [/] their labours have been so truly and

---

1   It] *paragraph break inserted [ed.].*
2   also] *in ink with caret over pencil.*
3   each] *over* 'an'.
4   It. . .Science.] *RO insertion from fol. 38v.*
5   And while] *RO insertion in dark ink over pencil with* 'While' *altered to l.c. [ed.].*
6   his] *underlined in pencil.*
7   less. . .upon] 'less' *underlined in pencil with* 'intent upon' *inserted by RO in ink above deleted* 'backward in'.
8   at. . .Century] *RO interlined in dark ink with caret.*
9   18th.] *followed by deletion of:* 'Linnaeus even contributed involuntarily to produce this indifference, by *applying {interlined in pencil above* 'carrying'} the *artifical method of {interlined in pencil with caret} Botanical *classification {interlined in pencil to replace deleted* 'Method into'} to the study of Animals; but Daubenton and Pallas*, however {interlined in pencil to replace deleted* 'opposed to this their example, and'} demonstrated the importance of Comparative Anatomy in Zoology; at the same time that Haller proved how essential it was to the advancement of Physiology'.

eloquently described in the Eloges of Cuvier, and also from this[1] Chair by Mr. Lawrence in the Introduction to his Classical Discourses on the[2] Natural History of Man, that I believe I may limit myself to this brief notice of the extent and scope of their works without being suspected of wishing to raise the Character of Hunter at their expense.

Cuvier, after paying a just tribute to the merits of Daubenton, Pallas, and Haller in his review of the progress of Science in the latter half of the 18th Century[3] goes on to state "John Hunter in England, the two Monros in Scotland, Camper in Holland, and Vicq D'Azyr in France were the first who followed their footsteps. Camper"[4], he observes, "cast[5], so to say, a passing glance of the Eye of Genius on a number of interesting objects; yet[6] almost all his labours were but sketches. Vicq D'Azyr with more assiduity, was arrested by a premature death in the midst of a brilliant Career; but their works inspired a general interest, which has ever since been on the increase."[7]

40   With reference to the nature of influence of the [/] labours of Hunter, the Historian of the Progress of the Physical Sciences, is silent:—he limits himself to an indication in a marginal note, of the Treatise on the Teeth, and "les autres ecrits de Hunter" inserted in part in the Philosophical Transactions.

Had a description of the Hunterian Collection at that[8] time been published, that astonishing result of Hunter's labours might perhaps have claimed a passing notice from one whose statements all Europe now receives, and all posterity will regard, with confidence and respect.

These "other writings" of Hunter to which Cuvier alludes, are devoted, indeed, more[9] to the development of general principles in Physiology, than to the detail of the Anatomical Observations upon which he founded them. Many of the facts ascertained in the course of his higher and more comprehensive enquiries, and incidentally alluded to in the Narration, are, however, fully as interesting and important, as those which other Anatomists have sometimes thought worthy of making the subjects of express monographs.

41   But Hunter had higher aims than the reputation [/] of a mere Discoverer[10] of Facts in Comparative Anatomy; and this he not only felt, but expresses, in an early period of his career. In a Manuscript of which Mr. Clift has

---

1   this] *in ink over* 'the'.
2   by. . .on the] *RO ink insertion above deletion of* 'in the Introductory Lectures to the' *with* 'Discourses on the' *continued onto facing fol 40*^{v}'.
3   in his . . .Century] *RO insertion in ink with caret from facing fol. 40*^{v}.
4   Camper"] *close quotes inserted [ed.]*.
5   "cast] *open quotes inserted [ed.]*.
6   yet] *RO interlined in pencil above deletion of* 'but'.
7   increase] *close quotes added [ed.]*.
8   that] *over* 'the'; '-at' *inserted above*.
9   more] *interlined in ink with caret*.
10 of. . .Discoverer] *RO inserted at top of fol. 41*.

preserved a copy[1] relating to a dissection of a Turtle, he says,

> The late Sir John Pringle, knowing of this dissection often desired me to collect all my dissections of this Animal, and send them to the Royal Society;— but the publishing of a description of a single Animal, more especially a common one, has never been my wish.

How much soever we may regret this feeling which has undoubtedly deprived the World of the results of much inestimable labour, and has operated in various ways disadvantageously to his own reputation, yet it[2] indicates the greatness of the mind which conceived it.[3]

Had Hunter published, seriatim, his notes of the Anatomy of the Animals which he dissected, these Contributions to Comparative Anatomy would not only have vied with the labours of Daubenton, as recorded in the "Histoire Naturelle" of Buffon, or with those Comparative dissections of Vicq D'Azyr, which are inserted in the early Volumes of the [/] "Encyclopedie Methodique" and Memoirs of the French Academy,[4] but they would have exceeded them both together.

It would be tedious to enumerate, name by name, the different Species of Animals whose organization was investigated and recorded by Hunter:— There is[5] evidence that he dissected and described the Anatomy of the following Mammalia:—

Quadrumana	21 Species	
Carnivora	51.	
Rodentia	20	
Edentata	5.	
Ruminantia	15.	
Pachydermata	10.	
Cetacea	6	
Marsupiata	10.	Total 138.

Of Birds	84 Species

---

1   of. . .copy] *RO interlined in ink with caret above deleted original interline of* 'of which a' *and deleting repetition of* 'which'.
2   it] *interlined in ink with caret.*
3   it] *with WC pencil note on facing fol. 42ᵛ with large bracket around it* ' See Life by Sir Everard, on this subject; of his regret that Mr. Hunter had not published more during his life, and *his reasons for it.* This was probably felt and meant by Sir Everard in 1794. See Hunter on the Blood, 1794. 4to. W.C.'.
4   and. . .Academy] *RO inserted in ink with caret at top of fol. 42 with erased pencil* '& in the Memoirs of the French Academy' *on facing fol. 43ᵛ. Comma inserted [ed.].*
5   There is] *RO interline in pencil to replace undeleted* 'We have'*with* ' had prepared notes of the dissections of which the Gerban {illeg} in the present' *on facing fol. 43ᵛ in same pencil.*

> Of Reptiles_____25.
> Of Fishes_____ 19.
> Of Insects_____29.

Of other Invertebrate animals, as Mollusca, Red-blooded-worms, and Radiata, say upwards of Twenty, so that from[1] the Titles of Manuscripts we

43   find the recorded dissections of 315[2] different species of animals.[3] [/] But in addition to these animals indicated in the titles of his manuscripts[4] Hunter's[5] preparations testify that he had dissected upwards of[6] 23 Species of Mammalia;—16 Species of Birds:—14 Species of Reptiles:—40 Species of Fishes:—42 Different Mollusca:—and about 60 species[7] of Articulate[8] and Radiate animals[9]: So that by adding these undescribed dissections to those of which we derive the evidence from the List of the Manuscripts, and of which described dissections the Collection in like manner contains evidences in the dissected and preserved[10] organs; there is proof that Hunter anatomized at least 500 different *Species* of Animals, independently of the Dissections of a considerable number of[11] Plants.

With respect to the rarer and less known Invertebrate animals, M^r. Hunter was not content with merely recording their Structure, and displaying their[12] leading anatomical[13] peculiarities in preparations; but he caused most elaborate and accurate Drawings to be made from the recent dissections; for which purpose he retained in his family *many years*[14] an accomplished Draughtsman.[15]

---

1   so that from] *RO insertion over erasure with* 'different. . .animals' *continued on facing fol. 43v.*

2   315. . .of] *RO ink insertion.*

3   we. . .animals.] *RO insertion in dark ink with* 'we. . .of' *at bottom of page and* 'different animals.' *continued on facing fol 43v.*

4   But. . .in] *RO ink insertion at top of fol. 43 with* 'But. . .these' *in dark ink and* 'animals. . .manuscripts'*continued in pencil with* ' the. . .manuscripts' *on facing fol. 41v with paragraph break and period before* 'Hunter' *inserted [ed.].*

5   Hunter's] *deleting paragraph break [ed.].*

6   upwards of] *RO inserted in pencil.*

7   species] *RO interlined in ink with pencil caret.*

8   Articulate] *altered from* 'Articulata'.

9   Radiate animals] 'Radiate' *altered from* 'Radiata' *with* 'e animals' *interlined in ink by RO and followed by pencil deletion of* '*All of them species of {RO interlined in ink with caret } Animals of whose Anatomy we have no evidence that he left written descriptions:—'.

10   and preserved] *RO interlined in ink with caret.*

11   a considerable . . .of] *RO interlined with caret above deletion of* 'to a considerable amount.'.

12   their] *RO insertion over erasure.*

13   anatomical] *RO insertion in ink with caret.*

14   many years] *underlined in pencil.*

15   Draughtsman] *followed by deletion in pencil of* 'who at the same time' *on page break*

44   I[1] allude to M[r]. William Bell, better known as the Author of *two*[2] papers in
the Philosophical Transactions, descriptive of the Sumatran Rhinoceros, and
the *Ecan Bonna* (*Chaetodon* (now *Platax*) *arthriticus*).[3]
        Several examples of these beautiful designs have already been published
by the Council of the[4] College in the illustrated catalogue of the Museum:—
they relate to the Anatomy of the Sepia and the Solen;—of the Ascidia and
Salpa; they illustrate the Circulation of the Blood[5] in the Crustacea, and in the
Class Annelidæ;—and here I may observe that[6] the figure which M[r]. Hunter
has given of the Circulation in the vessels of the *Chlosia*, a[7] red-blooded
Worm, far surpasses, in beauty and detail, any of those with which Cuvier
illustrates the Memoir devoted to[8] what he regarded to his latest breath as one
of his most interesting discoveries.
        Hunter also anticipated Cuvier in many points of the Anatomy of the
Cirripedes, of which dissection[9] it is to be lamented that, like many other researches, the preparations and drawings are now[10] the sole evidences. Hun-
45   ter's[11] illustrations of the Anatomy of the[/] Echinodermata, both of the
Spiny Species and of the unarmed Holothuria, have never been surpassed
either as to minuteness or Accuracy[12] and excepting the disputed article of the
Nervous System, little is added to the Anatomy of the Holothuria as displayed by Hunter, in the elaborate Monograph of Tiedemann.
        Let us now compare the Anatomical researches of Hunter with the works
of those Anatomists with whom Cuvier classes him in a secondary Category
of Contributors to Comparative Anatomy.[13]
        Daubenton's[14] anatomical labours were confined to that class of Animals

---

(cont)
        and continuing deletion at top of fol. 44 of 'was an able Anatomist.' *Period inserted*
        [ed.].
1   I] *Paragraph break inserted [ed.].*
2   *two] underlined in pencil and in margin pencil* 'X'.
3   *Ecan . . .arthriticus)] underlined in pencil and in margin pencilled* 'X' *with RO note*
        *on facing fol. 45*v *over erased* 'Platax': '*Chætodon* {*with faint pencil* 'Platax"
        above } (*now Platax*)'.
4   *the] inserted over* 'this'.
5   *of the Blood] RO interlined in ink with caret.*
6   *here. . .that] RO interlined in ink over pencil with caret.*
7   *Chlosia, a] RO inserted in margin with caret*
8   *devoted to] RO pencil interline above deleted* 'which recounts', *with* 'recounts' *inserted*
        in ink over erasure.
9   *dissection] RO insertion in ink in margin.*
10   *now] RO inserted in ink with caret.*
11   Hunter's] *RO interlined in ink above deletion of* 'The'.
12   Accuracy] *deleting period following [ed.].*
13   Let. . .Anatomy.] *RO insertion in ink from facing fol. 46*v *with insertion point*
        marked with '<'. *Paragraph break inserted [ed.].*
14   Daubenton's] *after deletion of* 'Now'.

whose structure most nearly resembles Man; he describes the position, and length, and breadth, and number of parts, with most praiseworthy Zoological precision; but never appears to raise his thoughts to the connection of the Structure with the habits, or its adaptation to function:—Hence he has been said to have made more discoveries of which he was unconscious, than any other Cultivator of Comparative Anatomy.

Vicq D'Azyr on the contrary adorns his descriptions with enlarged, &[1] beautiful philosophical views, but *he* [2] did not carry his Scalpel beyond the Vertebrate Series, while Hunter explored every modification of Animal Structure, from Man down to Zoophytes.—If Hunter surpassed his [/] contemporaries in the value and amount of the materials which he collected in Comparative Anatomy he rises far above them[3] in the application of his Facts.

46

By a profound and unintermitting meditation on the diversities of structure presented to his view, he derived[4] accurate ideas of the parts essential to the performance of the different functions of life[5] and every idea or doubt which was thus suggested he tested by the most[6] varied, &[7] ingenious[8] experiments.

> Many things he observes,[9] arise out of investigation which were not at first thought of[10]; and even misfortunes in experiments have brought things to our knowledge that were not, and probably could not have been previously conceived:—On the other hand, I have often devised experiments by the fire-side, or in my carriage, and have also conceived the result: but when I tried the experiment the result was different; or I found the experiment could not be attended with all the circumstances that were suggested.[11]

47   Few Physiologists[12] have made more experiments than M^r. Hunter[13];— still fewer have made such various experiments;—and none have made so many that were conclusive:—Yet Hunter says,[14]

> I think it may be set down as an axiom that experiments should not be often repeated which merely tend to establish a principle already known and admitted; but that the next step should be the application of that principle

---

1 enlarged, &] *interlined in pencil.*
2 on. . .he] *RO inserted in ink over pencil from facing fol. 46^v with insertion point marked with caret.*
3 them] *over erasure.*
4 derived] *followed by erasure of* 'more'.
5 of life] *RO insertion in ink with* 'life' *on facing fol. 47^v.*
6 most] *RO interlined in ink with caret.*
7 &] *RO interlined with caret.*
8 ingenious] *followed by deletion of* 'and accurate'.
9 observes,] *comma inserted [ed.]*
10 thought of] *over erasure.*
11 suggested.] *before RO insertion in pencil* 'On Bees'*at bottom of page.*
12 Physiologists] *RO interlined with caret in ink above deletion of* 'men'.
13 than M^r. Hunter] *RO interlined in ink with caret.*
14 says,] *comma inserted [ed.].*

to useful purposes.

This is an axiom that some late Modern Experimenters might have acted upon with advantage to Humanity, and without any detriment to Science. By this series of labours, of the mind and hand, prosecuted uninterruptedly from year to year, Hunter at length came to establish a body of Physiological doctrines whose influence on the Treatment of "the various ills that flesh is heir to," has been so frequently expatiated on[1], and with so much[2] eloquence and force, that aught that I could attempt to express, would be but a feeble echo.[3]

---

1 on] *followed by pencil deletion of* 'from this Chair,'.

2 so much] *RO interlined in pencil over* 'such'.

3 echo] *with pencil line across page and in RO pencil on facing fol. 48ᵛ* '/. End of Second' *to denote end of Lecture Three followed by deletion of* '{¶}If general acknowledgment of the beneficial influence' {'influence' *at bottom on page break; and continuing deletion in pencil at top fol 48:* } *'If general acknowledgments of the {*RO inserted in ink at top of fol.* } influence of Hunter's labours could have availed in raising him to his proper station in the history of Science, he would long ago have ascended from that secondary Category of Contributors to Comparative Anatomy with which he has been classed *by Cuvier. {*RO insertion in ink over pencil* } Let us endeavour, then, to define the grounds of our Eulogium;— let us consider the new views and improvements which resulted to General and Special Physiology from the labours of Hunter.'.

Fig. 10. 1798 Oil Portrait Of John Hunter By Henry Bone, After The 1786 Original By Joshua Reynolds. This copy more clearly represents the background referred to by Owen in the Second Lecture (p. 101) than does the Reynolds original. *(Courtesy President and Council, Royal College of Surgeons of England)*

Lecture Four
9 May 1837

49v Mr. President,

At this[1] last lecture we traced the progress of Comparative Anatomy and Physiology from the revival of Letters to the time of Hunter and considered the facts which had been[2] determined[3] in this period, and the inferences deduced from them, with especial reference to the Hunterian Discoveries and Doctrines.

We had proceeded to compare the inductive labours of Hunter—the data which he had acquired[4], with those collected by[5] his predecessors & contemporaries[6]—and we saw that he was the first of the Moderns who had embraced the entire range[7] of the Animal Kingdom in the scope of his anatomical[8] inquiries—This of itself is sufficient to raise Hunter from that 2dry Category of Contributors to Comp: Anat: with which he has been classed by Cuvier—[9]

It[10] remains for us now to compare the improvements in General & Special Physiology[11] and the Philosophical views of animal organization which Hunter[12] deduced from his observations, with the Doctrines previously entertained, and in vogue at the time of his[13] Enquiries—[14]

---

1 this] *interlined above deletion of* 'our'.
2 had been] *interlined above* 'were' *without deletion.*
3 determined] *word obscured by ink blot on MS [ed.].*
4 acquired] *interlined above deleted* 'collected'.
5 collected by] *interlined above deletion of* 'of'.
6 contemporaries] 'm' *over deleted* 'p'.
7 range] *interlined over deletion of* 'series'.
8 anatomical] *interlined in ink with caret.*
9 Cuvier—] *before deletion of* 'and'.
10 It] *Paragraph break inserted [ed.].*
11 Physiology] *before deletion of* 'which'.
12 Hunter] *interlined over deletion of* 'he'.
13 his] *interlined below deletion of* 'Hunters'.
14 Mr President . . . .Enquiries—] *all RO insertion from facing fol. 49v to replace deleted paragraph on fol 48. See emendations to Lecture III, fol. 47.*

48   First[1] as to the Vital powers: No physiologist would, at the present day, speak of[2] Hunter   as the Originator of the idea of a subtle imponderable principle operating in the fluids and solids of the Organism, and causing the phænomena of Life. Such a principle, under various names and with various attributes, has been assigned as the cause of Organization by Aristotle, Harvey, Willis, Cudworth, Grew, Van Helmont and Stahl.

           As both Harvey and Hunter had spent laborious lives in anxious inquiries, and repeated dissections and experiments on the lower animals,[3] to

49   ascertain relations between structure and function;—as both had [/] studied the changes which take place in[4] the form and structure of Animals from their Embryo state to that of maturity; and as both had carefully traced the successive phænomena which occur in the Egg during incubation, the approximation of[5] their opinions on the   Nature and powers of the Vital principle is correspondingly close.

           Both arrived at the conclusion that an animating principle exists and operates in the Ovum prior to the formation of any organ of the future animal.

           Both attributed the power by which the fecund egg resists putrefaction, while the unprolific one decomposes, to a principle of Life, which Harvey more precisely terms the "anima vegetiva."

50   Hunter, however, carries his researches a step farther, and submits the fecund egg to extreme cold; he ascertains new properties of the vital condition of the egg, of which Harvey was ignorant;—a power, viz. of resisting Cold; and, that when once frozen and killed by cold, the dead impregnated egg yields to putrefaction like the unimpregnated one.

           Both Physiologists observed that if the phænomena of a vital principle were manifested in one part of the Organization more than in another, it was in the Blood:

> For the blood says Harvey, is the first formed, and is the primary animate particle of the embryo;—it is generated prior to the *punctum saliens*;— before the first rudiment of the heart; and is endowed with the
51   vital heat or Principle, before it begins [/] to move, and from it does pulsation commence. For the thing containing, is made to be serviceable to the thing contained.
>
> Nor is the blood, therefore, to be called the primogenial part, because that in, and from, it the organ of pulsation is derived, but also because the Animal heat, and Vital principle are first implanted therein; and in it does life consist. For where heat and motion first begin there also life doth first arise, and last expire. Harvey on Generation, p. 274–275.

---

1   First] *inserting paragraph break after insertion from 49ᵛ [ed.].*
2   No.. .of] *RO interlined below deletion of* 'time is past when' *and deleting original interline* 'We can not longer regard' *with caret before* 'Hunter' *and after* 'Hunter' *deletion of* 'can be spoken of'.
3   on . . . animals,] *RO interlined with caret.*
4   in] *over erasure of* 'and'.
5   approximation of] *over erasure.*

   This explicit and beautiful enunciation of the pre-existence of the blood to
the Machine by which it is circulated, and of its endowment of Life, fell barren from the Pen of Harvey, (if we except the brief practice of Transfusion to
which it gave rise) and was forgotten, when Hunter resumed the inquiry.[1]
And why, it may be asked, was the Doctrine of the Vitality of the Blood
unproductive[2] as taught by Harvey?—Because, instead of establishing the
52   Doctrine by those observations and experiments from which [/] it was susceptible of deriving additional[3] proof;—instead of applying the Principle to
the explanation of the phænomena of Disease, and to a modification and improvement of remedial measures, Harvey obscures and forgets the conclusions of his cooler moments of observation, and excited by the discovery
which had extended his fame so widely over Europe, and had reflected such
lustre on his Name and country, he expatiates on the blood as something divine;—he has recourse to Hyperbole, and describes its properties in the ex-
travagant language of Romance.
   Hunter, on the contrary, carries a Series of calm and philosophical investigations on the vital properties of the Blood to an extent which has never
been surpassed:—he examines it under every condition, both in the vessels
and out of the vessels;—during circulation, and at rest;—in health, and in
disease: He seeks to define the respective importance of the component parts
of the blood by the way of experiment[4] by observing their relative constancy
in the whole animal series, and by watching the order of development[5] of the
Fibrine, serum, and coloured particles[6] in the higher animals in which they
all exist—In this profound & difficult enquiry[7] He not[8] only determined the
fact that colourless blood circulates in the Vertebrate embryo before the red
globules are developed, but draws all the legitimate conclusions[9] from that
observation—an observation which though more than once stated in the
Treatise on the blood, was reproduced in 1832 as a new discovery by MM.
Coste & Delpeck. Hunter more especially aims to establish the nature and re-
53   ality of the vital properties of the blood,[10] and he fully [/] details the changes
which it undergoes, and [/] the phænomena which supervene in the rest of the
Organism when these properties are lost:—lastly, he tells us how the Blood,
by means of its vital properties, assists in the restoration of parts when

---

1   inquiry.] *period added [ed.].*
2   unproductive] *over erasure.*
3   additional] *interlined over* 'further' *without deletion.*
4   experiment] *before deleted comma.*
5   development] *after deletion of* 'their'.
6   of the . . . particles] *interlined in ink with caret.*
7   In. . .enquiry] *interlined in ink with caret.*
8   not] *deleting repeat of* 'not' *[ed.].*
9   conclusions] *interlined above deletion of* 'deductions'.
10 He seeks . . .blood,] *RO insertion in ink from facing fol. 53ᵛ to replace deletion of* 'H(
     aims to establish the period in its formation at which it manifests the vital properties;'
     with insertion point marked by lead line. Comma inserted after 'Blood' *[ed.].*

injured or diseased.

Hunter carried his researches in the same spirit into the vital phænomena of the solids:—he endeavours to determine the specific powers and characters of those peculiar to[1] the nervous system, and[2] the Stomach:—he compares these important parts of the Animal body with reference to the degree of energy with which the vital[3] phænomena are[4] manifested;—he considers the influence which they reciprocally exert in maintaining the vitality of the Blood, and the relative dependance of the whole Organism on this[5] integrity.[6] He also dwells at great length on the Sympathies resulting from these mutual relations and dependencies.

        In short[7], instead of dogmatising on the powers and virtues of an abstract
54   essence, Hunter[8] [/] analyses the vital forces peculiar to each organic element, and classifies[9] the phænomena in which Life consists.

If we turn from Hunter's researches on Life to his investigations on another equally difficult and recondite subject in General Physiology;— vizt. Animal, or rather Organic Heat, we see the same exercise of the powers of the same great and original mind.

He first determines the relative extent to which the power of generating heat, or resisting cold is enjoyed, in the two grand divisions of organic Nature,[10] Plants and Animals:—he next investigates the degree in which that power is possessed by different Classes of Animals; and the relation subsisting between that degree, and the perfection and complexity of the organization with which the power is associated.

        He anticipates some Modern Physiologists in determining the different power of generating heat manifested by the same species, at different periods
55   of life; and goes further than D[r]. Edwards[11] in considering [/] the different powers of resisting Cold which different *parts* of the same organized body possess[12] in relation to their respective ages and periods of formation. He then analyses, so to say, the different functions, to determine the degree of dependance which the production of heat bears to their exercise; and reciprocally, the influence of the temperature of the body upon the active and healthy

---

1   peculiar to] *over erasure.*
2   and] *followed by erasure of* 'of'.
3   the vital] *RO interlined in ink with caret above alteration of* 'these' *to* 'the'.
4   are] *interlined in pencil with ink* 'are' *over erasure of* 'were' *with additional RO pencil* 'are' *interlined.*
5   this] *altered in pencil from* 'their' *with additional* 'ir' *interlined with caret.*
6   integrity] *before ink deletion of* 'of the vital powers of these parts'.
7   In short] *inserted in ink by RO over erasure.*
8   Hunter] *on page break with deletion of* 'endeavours' *at bottom of 53 replaced by* 'analyses' *inserted in ink by RO and with deletion of* 'endeavours to' *at top of fol. 54.*
9   classifies] *after erasure of* 'to' *and altering* 'classify' *to* 'classifies' *followed by deletion of* 'as it were,'.
10 Nature,] *inserted in left margin.*
11 Dr. Edwards] *RO interlined in ink with caret.*
12 possess] *altered from* 'possesses' *for grammatical sense [ed.].*

maintenance of the different[1] functions.

Throughout all this beautiful and justly celebrated inquiry, we see the Philosopher, conscious of the extent of His powers, and of the kind of knowledge which the right exercise of those powers was[2] adapted to acquire. We nowhere perceive a trace of a desire to establish a Theory of the nature of Animal Heat in the *abstract*.[3]

Let any one compare the language of Harvey or of Willis while expatiating on the *calidum innatum*, with the following just remark of Hunter's[4]:—

> I shall not, says Hunter, attempt to settle whether Heat is a Body or Matter, or only a property of matter, which appears to me to be merely a difference in terms, for a property must belong to something.

56

It is precisely in the same spirit that he conducts his researches on Life; and I would say, after a very careful Study of the Writings of Hunter, that he is the[5] Physiologist to whom a dogmatic Theory of abstract Life can least[6] be attributed.[7]

With the just notions which Hunter had acquired of Vitality and heat, he was enabled to explain many of the phænomena of digestion more satisfactorily than his predecessors Spallanzani and Reaumur were able to do. The following is a fair example of the different views and kinds of knowledge which these experimenters brought to the Inquiry.

Spallanzani had observed that Digestion did not go on in Reptiles below a certain degree of Temperature; he thought therefore that Heat was necessary to assist in the dissolving processes of the Stomach:—Hunter on the contrary[8] viewing the same fact from the lofty eminence of his vast induction, embraced[9] all the relative circumstances of this phenomenon[10] in his clear sighted intellectual[11] glance. He[12] showed that the [/] influence of heat in this

57

---

1 the different] 'the' *over erasure of* 'their' *with* 'different' *interlined with caret before* 'functions' *altered from* 'function'.
2 was] *over erasure with pencil* 'was' *interlined*.
3 abstract] *underlined in pencil with pencil* 'X' *in left margin*.
4 of Hunter's] *RO interlined in pencil*.
5 the] *before erasure of* 'last' *with pencil* 'X' *in left margin*.
6 least] *RO interlined in ink with caret*.
7 attributed.] *followed by deletion by vertical pencil score of paragraph* 'By {*with additional pencil* 'By' *in margin*} those, whose notions of Hunter's Doctrines are founded solely on a perusal of the posthumous "Treatise on the Blood," he is liable to be greatly misconceived and misrepresented.'.
8 Hunter. . .contrary] *following pencil deletion of* 'while' *and pencil interline with caret of* 'on the contrary' *to continue with insertion from facing fol. 57*[v].
9 embraced] *over deletion of* 'took in'.
10 phenomenon] *altered from* 'phenomen'[*ed.*].
11 intellectual] *followed by deletion of* 'vision'.
12 viewing. . .He] *RO pencil insertion from facing fol 57*[v] *inserted with caret after* 'Hunter' *on fol 56*.

case[1] was not merely Chemical, but that it[2] operated as a Vital stimulus[3] by first rousing the Sensitive Powers; these, the respiratory and circulating functions, upon which[4], the motive and other actions and faculties depended[5], and that the digestive Organs were necessarily excited[6] to corresponding activity in order to supply the waste occasioned by the working of the Machine,[7] which the heat had thus called into play.

Hunter more accurately determined, and first applied and rendered fruitful the fact which Grew incidentally mentions, vizt. That it is the property of a living body or part to resist the action of the Gastric juice; and his well known[8] paper on the Digestion of the stomach after Death, is a beautiful example of the application of his general Views in Physiology to the explanation of particular phænomena.

The published writings on Digestion convey a just idea of the extent of M[r]. Hunter's researches in Comparative anatomy, and the amount of his
58	knowledge in General Physiology. It would [/] be in vain to attempt to recapitulate, in a lecture, all the new observations which occur in these papers, with reference to the action of the Bile on the intestines, the digestibility of different substances &c[9]; but I cannot omit alluding to an instance of M[r]. Hunter's Method of considering a physiological question:—I mean, his Observations on the share which the Stomach takes in the act of Vomiting, and let any one compare the philosophical reasoning of Hunter, which led him to determine that the parietes of the Abdomen were the sole agents in Vomiting, with the bloody yet inconclusive experiments, by which Majendie arrived at the same result.

In attempting to form a correct estimate of M[r]. Hunter's merits as a Discoverer in reference to the absorbent System, it becomes necessary in the first place to distinguish between the discovery of the Vessels, themselves, whether Lacteals or Lymphatics, and that of their peculiar functions and anatomical relations to the other parts of the vascular System.
59	With reference to the Human subject, and the Mammalia generally, we have already seen that the existence both of Lacteals and Lymphatics had been determined long before the time of Dr William Hunter;—But the lymphatic absorbents had never been distinguished as a System, either Anatomically or Physiologically, from the capillary veins. Noguez, who of all Anatomists before the Hunters' had dwelt more particularly on the Lymphatics, divides them into Four Classes.

1	in this case] *RO interlined with caret after insertion of* 'of heat' *over erasure.*
2	it] *over erasure of* 'the Heat'.
3	as. . .stimulus] *RO interlined with caret.*
4	upon which] *interlined in pencil above* 'and lastly' *without deletion.*
5	depended] *interlined by RO in pencil above* 'and'.
6	excited] *over erasure with erased pencil* 'stimulated' *interlined above.*
7	Machine,] *light pencil* 'X' *in left margin.*
8	well known] *RO interlined in pencil above* 'celebrated'*without deletion.*
9	&c] *interlined in ink with caret.*

The first Class are the Minute exhalent arteries, which pour out Lymph or the Serous part of the blood upon the surface of the Membranes lining internal cavities, and upon the Skin.

The second Class of the Lymphatics are the Venous[1] branches corresponding to the exhalent arteries above mentioned. They open upon all the membranes and the Skin, and bring back the Lymph exhaled by the first Class of Lymphatics. These are the *Conduits absorbans* of Noguez[2]; but they are not described as forming a System like our Absorbents, and terminating ultimately in the [/] Thorascic Duct;—No, they gradually unite to form red veins: and we are informed by Noguez[3] that they lose the name of absorbent canals when they become large enough to be sensible to the naked eye.

60

The Third Class of Lymphatics correspond in reality to the Absorbent System in the Hunterian Sense, but terminate according to Noguez in the Receptaculum Chyli, the Thoracic[4] Duct, the Vena Cava, and the Vena Portarum.

The Fourth Class of the Lymphatics are the reflected terminations of the Arteries and the commencement of the Veins continued therefrom,[5] or the Capillary system of modern Physiologists.[6]

Such was the Doctrine which has been urged as anticipatory of the Hunterian Discoveries.[7] I need not say that the description of the Nature and functions of what Noguez denominates the *Conduits absorbans* is a gratuitous assumption.

John Hunter first proceeded to investigate the real relations and Anatomy of the Absorbents in the Human Body. D^r. Hunter describes one of his Preparations which showed the Lymphatic Vessels extending from the Ham upwards to the Thorascic Duct, as well as the Inguinal [/] and Lumbar glands:—the larger lacteals at the root of the Mesentery, and the receptaculum chyli, or what is so called[8], all finely injected with Mercury.

61

He also acknowledges his Brother's discovery in 1753 that the Lymphatic glands and the lymphatic vessels going from them, could be filled uniformly by pushing a pipe into their substance, and states it to have been M^r. Hunter's intention to have traced the Lymphatic vessels all over the Body, and to have given a complete description and figure of the whole absorbing System.

This work was unfortunately arrested by a very indifferent state of health, the effect of too much application to Anatomy, which obliged M^r. Hunter

---

1 Venous] *over erasure.*
2 of Noguez] *RO interlined in pencil.*
3 by Noguez] *interlined by RO in pencil.*
4 Thoracic] *altered in pencil from* 'Thorascic'.
5 therefrom,] *replacing period by comma for grammatical sense [ed.].*
6 Or...Physiologists] *RO interlined in pen.*
7 Discoveries.] *followed by pencil caret and on facing fol. 61^v incomplete RO insertion* 'in this by'.
8 so called] *over erasure of* 'called'.

then 25 years of age to retire into the country; when he took the opportunity of matriculating at the University of Oxford.

M^r. Hunter therefore[1], having, by anatomical Investigation,[2] determined that the true Absorbents were not continuous with the capillary veins, proceeded to examine their functions by experiment, &[3] resigned the task of describing the intire System of absorbents;—This was subsequently resumed
62   and ably accomplished by another [/] Ornament of the Hunterian School, the celebrated and learned Cruikshank.

When the question of the Office of the Lymphatics first began to be agitated, one of the Arguments against their being the exclusive agents in the absorbing processes was founded on their supposed absence in the Ovivarous Vertebrata;—

> for the Lacteal Vessels says D^r· Monro, have not as yet been certainly observed in Birds, nor in the more common Fishes; nor in general in the Animals called Oviparous; and from a considerable number of Experiments I have made, I am convinced they want the Lymphatics as well as the Lacteal Vessels.

Now the honour of having disproved this assertion and dissipated the error, by more exact and successful researches, unquestionably[4] belongs to John Hunter, to whom the discovery of the Lymphatics in Birds is assigned by M^r· Hewson who first published a connected and detailed account of those vessels in that class, and who is commonly regarded as the Original Discoverer
63   by such Anatomists[5] as derive their learning [/] from Indexes. Hewson, however, acknowledges in the Philos: Trans. 1768. that "it is but doing[6] justice to the ingenious M^r. John Hunter to mention here, that the Lymphatics in the necks of Fowls were discovered by him many years ago."[7]

---

1   therefore] *after deletion of* 'having'.
2   , by…Investigation,] *RO insertion in ink from facing fol. 62^v with lead line to insertion point. Below is WC pencil note:* 'See also Dr Hunters Medical Commentaries. {¶} See D^r. Hunter's Life by *Saml. {*interlined with caret*} Foart Simmons 1784—8vo. I think Hewson first succeeded M^r. Hunter as D^r. Hunter's Assistant, & afterwards declared off and published all that he had learnt from D^r. & M^r. Hunter as his own—Then followed M^r. Cruikshank— I'll find the Book. I have it. WC {¶} See also M^r. Hunter's Memorandum on Hewson.'.
3   proceeded…&] *interlined by RO in ink with caret.*
4   disproved…unquestionably] *with WC pencil comment on facing fol. 63^v:* 'See Hewson in Philos Trans? & Octavo{¶} See also John Sheldon.'.
5   Anatomists] *interlined in ink by RO with caret and pencil* '#' *mark inserted below caret.*
6   doing] *over erasure.*
7   ago.'] *before deletion by vertical pencil scoring of* 'And it further appears from D^r: Monro's reply to Mr Hewson, that this discovery of M^r. Hunter's had been communicated to Dr Cullen by D^r. George Fordyce, and had materially influenced his (D^r. Monro's) opinions respecting the Absorbent System.'.

M[r].[1] Hewson also informs us that prior to his own publications on the Absorbent System of Amphibia M[r]. Hunter had discovered and demonstrated to him the Chyle; and we must suppose, the Lacteal vessels, of a Crocodile. A copy of the Hunterian Manuscript detailing M[r]. Hunters discovery of the Lymphatic System in the Amphibia has been preserved by M[r]. Clift;—[2]And the date of the different dissections are given, by which the lacteal vessels of the Crocodile to which M[r].[3] Hewson alludes, were discovered.

With reference to the Absorbent System in Fishes the merit of that discovery must be awarded to M[r]. Hewson.

64   Having[4] made these observations on the discoveries of John Hunter, with reference to the Anatomy, Human and Comparative of the Absorbent System, I proceed briefly to allude to the share which his labours had in establishing their true function:—I say their *true function*; because although there may be many who now admit that the veins absorb, yet there are none, I apprehend, who do not believe that absorption if not the exclusive,[5] is the only[6] function which[7] the Lymphatic and Lacteal Absorbents perform.[8]

Prior to the Hunters, the Lymphatics were believed by Haller and most other Physiologists, to do little more than carry back into the circulation the serous or lymphatic part of the blood.

D[r]. William Hunter taught that the Absorbents were not the reflected capillary arteries, but originated in all the interstices and cavities of the body;—that they terminated exclusively in the Thorascic Duct; and were to the General System what the Lacteals were[9] to the intestinal canal:—the proper absorbing Vessels.

65   This Doctrine was confessedly founded on the well known experiments of John Hunter, which are 19 in number, made on the Dog, Sheep, and Ass. It would seem, however, that M[r]. Hunter himself did not consider the above as sufficiently conclusive to be submitted to the public; since he left them in Manuscript with his Brother who made use of them four years afterwards in the controversial Essay, with[10] the Monros, while John was abroad with the Army at Bellisle. The experiments are related in the "Medical Commentaries"

---

1   Mr.] *altered from* 'Mr:' *[ed.]*.
2   Clift;—] *with WC pencil comment on facing fol. 64v:* 'Qu Is not this account the Preface to the Old Gallery Catalogue? if so, it was not a copy of any thing destroyed.' *with RO reply:* 'No, it is a M.S. sent by the Capt— after Sir Evds decease.'.
3   Mr.] *altered from* 'Mr:' *[ed.]*.
4   Having] *on page break with erased* 'Hewson' *above* 'Having' *on fol. 63*.
5   if...exclusive,] *RO interlined in ink with caret*.
6   only] *RO interlined below deletion of* 'sole'.
7   which] *RO inserted over erasure*.
8   perform] *RO interlined with caret before deletion in pencil and ink of* ': allowing even the veins to participate with them in that function.'.
9   were] *before pencil deletion of* 'allowed to be'.
10 with] *interlined in pencil by RO above* 'against'.

without any additional proof or observation; being considered by Dr. Wm. Hunter as decisive in depriving the Veins of the power of absorbing, altogether.

Subsequent experiments appear to have shown that this Conclusion cannot be safely drawn from them:— but, without entering into this question, it is incumbent on me, (in regard to[1] the Character of an Author whose Works have exercised so salutary an influence over the Surgical profession,) to show that the charge of imperfection and negligence which has been cast upon these
66   Experiments;—whose want of [/] exactness, according to Majendie, can only be excused by the rude state in which the art of Physiological experiment was at the period when they were made: that this charge, I say, rests intirely on the culpable oversight[2] of the accuser.

In relating one of his experiments M[r]. Hunter states:

> Novr. 13. 1758. I opened the abdomen of a living Sheep, which *had eaten nothing for some days;* and upon exposing the intestines and mesentery, we observed the lacteals were visible, but contained *only a transparent watery fluid.*

It is conceivable that a Physiologist of M. Majendie's reputation, would have ventured, in his Commentary on M[r]. Hunter's Experiments, to object to them "because the Experimenter had neglected to notice whether the animal experimented on, was full or fasting; or whether the lacteals were or were not distended with Chyle." To give a colour to this objection all reference to the experiment above quoted, is avoided; but even in[3] the very experiment of which Majendie[4] gives a mutilated version, M[r]. Hunter expressly premises
67   that having exposed the intestines fully, he observed the [/] Lacteals filled

---

1   in regard to] *inserted over erasure of* 'to show'.
2   oversight] *RO interlined in pencil above deletion of* 'ignorance'.
3   in] *RO interlined in ink with caret and with WC pencil* '?X' *in margin.*
4   Majendie] *RO interlined above deletion of* 'he' *and WC pencil comment on facing fol. 68 ᵛ facing page:* '(Breakfasted with him at M[r]. Dusgate's){¶} In 1819 when in Paris I was taken to see M. Majendie's Travaux—, whom I found in a *dark* small room endeavouring to find the lacteals in the viscera of a very large Turtle—The Viscera were in a large washing tub of very dirty warm water. *With scalpel* in hand M. *Majendie {'j' inserted over* 'g'} was tumbling the parts backwards & forwards, and cutting unintentional holes wherever he went, through the mesentery, &c *and could not find a single lacteal.—{interlined }{¶}* All the apparel or apparatus was *in keeping*: vizt— a glass blow-pipe drawn to a point and snapt off, (an excellent tool of course to enter a delicate vessel;—) and a *wooden* quicksilver pipe split at the point. He asked me *to find* him an absorbent, and a few seconds I pointed out twenty as large or larger than Crow quills— He was quite *french-{with illegible small RO pencil insertion above }* frantic, capered about and showed to the lookers on what I supposed he told them he had discovered— He wished me to inject the parts with the quicksilver, but I said I could not think of depriving him of the honour of succeeding with such tools, which I could not think of attempting to use.W.C. {¶} After this, is it possible he could speak of M[r]. Hunter at all?—'.

with a white liquor at the upper part of the gut and mesentery; but in those that came from the ileon and colon, the liquor was transparent.

In the Herbivorous quadruped, the Sheep, in which M^r. Hunter employed Starch as the menstruum of the Indigo, the transparent watery contents of the lacteals was especially noted, before the coloured material was thrown into the intestine: they were afterwards observed to be filled with a fluid of a fine blue colour:—and yet M. Majendie [ibid. p. 210.] would have us believe that no alteration had been observed;—that the lacteals were of the same blue colour before the injection of the Indigo and Starch had been performed, as after.

Whether, however, the coloured matter had passed into the lacteals or not, it could not by the most careful and varied experiments, be detected in the veins.—Great precaution was taken to ascertain that fact:—Since the natural colour of the Blood rendered it difficult to perceive a change of hue, the contents of the Veins were collected, and suffered to coagulate, in the expectation
68   of the Serum manifesting the [/] presence of the Indigo, but it had not the least bluish cast.

Warm Milk was then made to circulate from the artery into the vein;— and it might surely have been expected, especially if the Modern Doctrine of Non-vital imbibition were true[1], to have then had a trace of the coloured contents of the intestine in the venal milk,— but no such result took place.[2]

It is by no means intended here to support the infallibility of M^r. Hunter: but a candid[3] and careful perusal of his experiments on Absorption recorded in the Medical Commentaries will not only exonerate him[4] from any charge of haste or negligence, but must impress the unprejudiced reader with the conviction that those experiments have rarely been equalled, and never excelled; either in the ingenuity and foresight manifested[5] in their contrivance;— in the skill and precaution against error displayed during[6] their performance,
69   in the fairness of the Conclusions [/] deduced from them;— and above all, in the minute accuracy and candour which pervades[7] their Narration.
Nerves.[8]

Mr. Hunter's *published* writings on the Nervous System bear no

---

1   true] *interlined in pencil by RO above deletion of* 'correct'.
2   place] *followed by deletion by vertical pencil scoring of paragraph* : '{¶}The experiment on the Ass in which the odour of Musk found in the Chyle, but not in the Venous Blood, is not referred to by Majendie.—'.
3   but candid] *with erased pencil interline above and below illegible erasure and continuing on to facing fol.* 69v.
4   him] *inserted by RO in ink over erasure of* 'M^r. Hunter' *with erased pencil interline of* 'him?' *above.*
5   manifested] *inserted by RO in ink over erasure of* 'displayed' *with erased* 'manifested' *interlined above.*
6   displayed during] *inserted by RO in ink over erasure.*
7   which pervades] *RO insertion in ink over erasure.*
8   *Nerves*] *in margin as if to indicate a topical division.*

proportion to the extent of his Anatomical investigations on this Subject, especially as manifested in the philosophical Series of Preparations in which the nervous System is traced through its progressive stages of Complication[1] from the simple filaments of the Entozoon[2], to the aggregated[3] masses which distinguish the organization of Man. The *fibrous Structure* of the Brain, which is displayed in Preparations made to show the fact, is stated by Hunter[4] in the Description of the Anatomy of the Whale Tribe.

In the paper on the Branches of the 5th. pair, which are distributed to the Nose, we cannot fail to observe that[5] Hunter had entered on that Track which has since led to such important improvements in this department in Physiology.

I do not know a more interesting paper in the intire range of Hunter's Works than this, taken[6] in connection with our present knowledge of the Subject.

70 He begins with stating his belief in the principle that particular Nerves have particular Functions, in relation to their Anatomical differences of Origin, Union, and Distribution.

I have no doubt, Hunter says,[7] if their Physiology was sufficiently known, but we should find the distribution and complication of Nerves so imme- diately connected with their particular uses, as readily to explain many of those peculiarities for which it is now so difficult to account. What natu- rally leads to this Opinion is, the origins and number of the Nerves being constantly the same; and particular nerves being invariably destined for par- ticular parts, of which the fourth and sixth pair of nerves are remarkable in- stances. We may therefore reasonably conclude, that to every part is allotted its particular branch; and that however complicated the distribution may be, the complication is always regular.

71 This general uniformity, in course, connection, and distribution will lead us to suppose that there may be some other purpose to be answered [/] than mere mechanical convenience. Whoever, therefore, discovers a new nerve, or furnishes a more accurate description of the distribution of those already known, affords us information in those points which are most likely to lead to an accurate knowledge of the nervous System: for if we consider how various are the origins of the nerves, although all arise from the brain, and how difficult the circumstances attending them, we must suppose a variety of use to arise out of every peculiarity of Structure.

---

1 Complication] *inserted by RO in ink over erasure.*
2 Entozoon] *followed by pencil deletion of* 'and Echinoderm'.
3 aggregated] *before deletion in ink of* 'and complicated'.
4 by Hunter] *RO interlined in ink with caret after* 'stated'.
5 that] *with WC note on facing fol. 70v in pencil with lead line :* 'Almost earlier than on the absorbent System Qu 1750 or 1755.'.
6 taken] *RO inserted in ink over erasure.*
7 doubt, . . .says] *double quotes after* 'doubt' *and before* 'if'. *Comma after* 'doubt' *inserted [ed.].*

It is scarce concievable that[1] a belief in the community of function of the different nerves could ever be attributed to Hunter after this explicit enunciation of the principle of a difference of function arising out of a difference of anatomical condition. It appears to me that Hunter needed only to have resorted to experiment in this, as he did so successfully in other fields of Physiological[2] Inquiry, to have established the nature and degree of the functional differences in these Nerves where[3] he describes the anatomical conditions on which he supposes the variety of uses to depend.

72 He limits himself, however, in his illustration of the grand Proposition, by anatomical examples [/] only:—He shows that Organs like the Eye and Nose which are endowed by means of one Nerve with a Special Sense, derive their ordinary sensation from a second nerve having a different origin:—this nerve, he determines, in the case of the Eye and Nose to be the Fifth pair. He says the same mode of reasoning is equally applicable to the Organ of Taste, and he traces the corresponding superadded[4] nerve to the Ear.

He says, with respect to this additional[5] nerve,

> I am almost certain it is not a branch of the seventh pair of nerves, but the last described branch from the fifth pair; for I think I have been able to separate[6] this branch from the Portio Dura, and have found it lead to the chorda tympani: perhaps it is continued into it.

The researches of recent Anatomists have confirmed the accuracy of Hunter's suspicion.[7]

73 The fact of the supplementary branches of the fifth pair being distributed to the eye, the ear, and the nose, it is well known, attracted the attention of a French[8] Physiologist of the present [/] but every Physiologist now adopts[9] the Conclusion of Hunter, that these Organs of Sense receive their endowments of common Sensation from the fifth pair, in preference to the well-known view which Majendie originally promulgated—[10]That the branches of the fifth were the Organs of the Special Sense. Hunter distinguishes the Sensations of the Stomach and of the Glans Penis as being peculiar; and

---

1   It. . .that] *RO inserted in ink over erasure with erased pencil* 'It is scarce concievable that' *interlined above. Paragraph break inserted [ed.].*
2   Physiological] *RO interlined in ink with caret over erased pencil interline of* 'departments of physiological'.
3   where] *RO interlined in pencil above deletion of* 'in which'.
4   superadded] *RO insertion in ink over erasure with pencil* '??X' *in margin.*
5   additional] *RO insertion in ink over erasure.*
6   separate] *over erasure.*
7   suspicion] *RO inserted in pencil to replace* 'conjecture'*without deletion.*
8   French] *in ink over erasure.*
9   but. . .adopts] *RO interlined in pencil to replace deletion in pencil beginning on page break reading* 'day;— but I apprehend [l *fol.* 72] there are few who do not adopt the'.
10 promulgated—] *RO insertion in ink over erasure with* 'promulgated' *interlined in pencil above.*

shows that as these peculiar sensations reside in particuliar[1] nerves; so, at whatever part of the nerve the impression is made, it always gives the same sensation as if affected at the common Seat of the sensation of that particular nerve.

In another place, Hunter makes the ingenious remark, that the Nerves which are specially designed[2] to receive peculiar impressions can convey only the ideas of such impressions to the Brain, in whatever way they may be affected or stimulated: Thus, he says,[3] "a mechanical impression on the retina produces an impression of light:—a blow on the ear, the sensation of

74   Sound."And [/] later experiments have only extended this principle, by showing, that whether the nerve be affected by mechanical, chemical, or electrical stimuli, it conveys the same sensation.[4]

Much[5] importance has been attributed to these Observations, on the Supposition that they were new;—and I have been induced to dwell thus long[6] upon Hunter's Contributions to the Physiology of the Nervous System, because in[7] the recent Works upon that Subject, he does not receive the Credit which is due to him for having made them.

The Physiological[8] discoveries of Bell, Majendie, Mayo, Bellingeri, Panizza[9] and Müller, have resulted from the combination of experiment with a philosophical consideration of the Anatomical peculiarities of the nervous System. It was the neglect of Experiment in this department of physiology which rendered Hunter unable to account for the peculiarities which equally struck his mind as being connected, in an intimate degree, with the functions of the Nerves.

Had Hunter combined experiment with dissection when he traced the

75   lateral branch of [/] the *Nervus Vagus* in the Cod, the Eel, and the Gymnotus; and wondered "that a nerve should arise from the brain to be lost in common parts, while there was a *medulla spinalis* giving nerves to the same parts":—this probably would not have remained to him as he expresses it[10] "one of the inexplicable circumstances of the Nervous System."

If I were to dwell on the new Views in Physiology which occur in the other published Works or papers "les Autres Ecrits" of Hunter, an equal space and time might be devoted to each;—and this is the less necessary, as

---

1   particular] *RO insertion in ink over erasure.*
2   designed] *RO ink alteration over original* 'organized' *with pencil* '?' *in margin.*
3   says] *with inserted* 'Parkinsons Lectures.' *in ink in margin.*
4   And. . .sensation] *Capitalizing* 'and' *[ed.]. with* 'it. . .sensation' *inserted by RO in pencil to replace deleted* 'they produce the same effect.'.
5   Much] *RO interlined in pencil above deletion of* 'Very great'.
6   thus long] *RO interlined in ink with caret.*
7   in] *interlined in ink with caret.*
8   Physiological] *RO interlined with caret in ink.*
9   Bellingeri, Panizza] *RO interlined in ink with caret.*
10 as. . .it] *RO interlined in pencil.*

all my Lectures will be devoted to the expostion of the Hunterian Labours. Let me, however, adduce a few instances of his Discoveries[1] in Comparative Anatomy; which on the supposition that they were original, have contributed to add to the reputation of subsequent Anatomists.

1.[2] The Organ of Hearing in the Sepia. The fact that this Cephalopod possesses the Organ in question is stated in the "Animal Œconomy," and it is said[3] to differ from that of Fishes. The discovery is attributed by Cuvier to Scarpa.

76   2.[4] The [/] semicircular canals of the Cetacea, described by Hunter[5] in the Paper on Whales[6]; a structure[7] which Cuvier rightly states that Camper overlooked, but incorrectly claimed the discovery as his own.

3.[8] In the latest sketch of the history of Comparative Anatomy prefixed to the French Translation of Carus's Comparative Zootomic[9] Anatomy, John Hunter is introduced as the impudent self-appropriator of Camper's Discovery of the Air-Cells in the Bones of Birds: and the Historian of the Science[10] does not honour him with any further Notice:—

Now[11] the facts with reference to this Subject are as follows: Campers account of the Air-Cells in the Bones of Birds[12] was first published, in the Dutch Language in the year 1774: the same year in which M^r. Hunter's discovery was published[13] in the Philosophical Transactions. The French Memoir of Camper was not published till the year 1776 in the 7th. Volume of the Memoires Etrang. de l'Academie des Sciences de Paris. Hunter gives the date of his discovery as being in 1758:— Camper his, in the year 1771.—

I believe both Physiologists to be equally worthy of Credit. The numerous observations and experiments, which their respective Memoirs detail are
77   not such as can hastily be got up for an unworthy [/] purpose; but as both Memoirs (which differ materially in their general scope, and the mode in which the Subject is treated) were published in the same Year, the honour of the Discovery must be attributed to both.

I might dwell on the philosophical comparison which Hunter makes between the abdominal air-cells of the Bird and those of Reptiles and Fishes, had he not in this Instance been anticipated by Harvey. Harvey, however,

---

1 Discoveries] *altered from* 'discoveries'.
2 1.] *inserting paragraph indentation [ed.].*
3 said] *RO interlined in pencil over deletion of* 'described'.
4 2.] *inserting paragraph indentation [ed.].*
5 by Hunter] *RO interlined in pencil with caret.*
6 Whales] *altered with deletion from* 'the Whale'.
7 a structure] *RO interlined in pencil above following* 'which'.
8 3.] *inserting paragraph indentation [ed.].*
9 Zootomic] *RO interlined in pencil.*
10 the Science] *RO insertion in ink in margin.*
11 Now] *paragraph break inserted [ed.].*
12 in. . .Birds] *RO interlined in ink with caret.*
13 was published] *interlined by RO in pencil above* 'appeared' *without deletion.*

was not aware that the respiratory system was extended into the muscular interstices, and the Cavities of the Bones:—the honour of which discovery must be assigned to Hunter and Camper, without the necessity of supposing that either had borrowed from the other.

The Preparations in the Hunterian Collection, and the Manuscript Catalogues, show that M[r]. Hunter had discovered the peritonæl canals or outlets in the Cartilaginous fishes; in the Crocodile; and the continuation of y[e][1] same peritoneal canals into the Corpus cavernosum penis in[2] the Tortoise and Turtle. The latter have been recently re-discovered by M. M. Isidore St. Hilaire and Martin St. Ange.

78

Hunter's Preparations, Manuscripts, and Drawings are the proofs of his discovery of the diffused venous receptacles of Insects and Crustaceans, in which latter Class they have been overlooked in some of the latest Treatises on the Subject.

Hunter first determined the bi-auricular structure of the Heart in the Batrachian Caducibranchiate Reptiles; and first described the circulating and respiratory Organs in the Perennibranchiate Batrachia, as the Siren, Amphiuma and Menopoma, which he called Pneumobranchiata: but he erred in supposing the Auricle to be single in this Division.

Hunter first discovered, by means of retrograde injections of the Tubuli uriniferi, that these essential parts of the kidney extended to the superficies of that gland, and were not confined to the medullary substance. Meckel who saw the beautiful preparations establishing this fact, while in London, described them on his return to Germany[3] to Müller; who in his recent elaborate work on the secreting organs[4] acknowledges how important this observation was in establishing true notions of the Structure of Glands.

79

Hunter extended his researches on the renal organ to the invertebrate classes, and shows "the kidney of the Snail" and the correctness of this ascription to the so called Mucous[5] Gland, has been recently established by the observations of Professors Jacobson and Blainville.[6]

I have selected a few facts from amongst the Multitude which Hunter ascertained in the progress of those investigations by which he sought in[7] the simpler modifications of[8] the structures in the lower Animals:—the true uses of the different Organs which are combined to form the complex frame of Man.

I have now to allude to the extent of Hunter's Labours in another Track

---

1   continuation. . .y[e]] *RO insertion in ink in margin.*
2   in] *with pencil* 'in' *interlined above.*
3   on. . .Germany] *RO interlined in ink with caret.*
4   in. . .organs] *RO interlined in ink with caret with* 'work. . .on the' *continued on fol.*
    79v in ink and 'secreting organs' *interlined in pencil above deletion of* 'glands' *on 79v.*
5   Mucous] *underlined in part in pencil with WC* query '?X Mucus ?' *in pencil in margin.*
6   Blainville] *followed by* 2-inch horizontal division line across center of page.
7   in] *interlined in pencil with caret.*
8   of] *over erasure.*

of Inquiry:—viz. That which unfolds the laws of Animal Development.—
Like his great Predecessors Aristotle and Harvey, Hunter endeavoured to
penetrate to the Mysteries which shroud the first formation of the parts of the
Embryo. We have from his own pen[1] a sketch of the general results of a long
continued series of researches on the development of the Chick.[2]

80   The astonishing materials which he has left to testify to the extent and
success of his inquiries, prove[3] that he had investigated the development of
the Embryo not only in the egg of the Bird, (which from the facilities it affords for observation, has been the subject usually selected for these researches[4]) but also in the Ovum of Insects, Mollusca, Fishes, Reptiles and
Mammalia.

These researches were to have formed the concluding part of a great
Work on Physiology, Human and Comparative; to which Hunter alludes in
some of his printed Memoirs; and on which he designed his reputation mainly
to repose. The "Treatise[5] on the Blood" may be regarded as comparatively an
imperfect[6] production, to the publication of which Hunter was reluctantly
compelled under the most unfavourable circumstances as regards health and
leisure. It contains a brief enunciation of many principles hardly intelligible
without the details of the progressive[7] observations and evidences on which
they were founded;—and which would have been fully detailed, with illustrations, in the great work on Physiology.

81   The Materials [/] which existed for the Chapters on Development[8] at the
period of his decease, were an extensive Series of Drawings; many of them
giving microscopic views of the earliest stages of the developement of the
Embryo, and all of them executed with a minute accuracy and elaborate finish

---

1 pen] *followed by erasure of* 'only'*which is also encircled in pencil.*
2 Chick] *before deletion in ink of* ', and it is greatly to be regretted that no descriptions
   exist of the' *with repeat of* 'the' *at top of fol. 80 altered to* 'The' *with period inserted
   and new paragraph break [ed.].*
3 prove] *following erasure of* 'His preparations' *originally beginning new paragraph.*
4 subject. . .researches] 'subject usually selected' *and* 'researches' *inserted in ink by RO
   over erasure with pencil* '? X' *in margin and WC pencil query on facing fol. 81ᵛ*
   'Chosen— Selected— some word apparently necessary.'.
5 Treatise] *after erasure of* 'imperfect'.
6 may. . .imperfect] *RO interlined in ink with caret after* 'Blood' *above deletion of* 'was a
   hasty' *with WC pencil comment on facing fol. 81ᵛ:* 'This work is said in his Life, or
   the preface, to have been the most laboured & for a greater part of his Life than any or all
   other of his works. {¶}*More than thirty years.* WC {¶} It had been begun to be printed
   during his life but the greater part was superintended through the Press by Sir Everard &
   Dʳ. Baillie after his death. I believe all the manuscript was considered complete: for it all
   went to the water closet in Sackville Street, piecemeal. {¶}I begged from thence part of
   the M.S. of the Work on the Venereal Disease— which I had bound in 3 Vols. 4ᵗᵒ.'.
7 details. . .progressive] *RO inserted in ink over erasure with erased pencil interline above
   with* '-ressive' *interlined with caret above* 'observations'.
8 Development] *followed by 5 ink periods and in margin pencil* 'X' *and WC query in
   pencil on facing fol. 82ᵛ* 'Qu Development of'.

which has never been surpassed and in this subject, never equalled.[1]

Among the facts observed may be mentioned the progressive separation[2] of the Serous, Vascular and Mucous layers of the germinal membrane, which are exemplified in the Drawings not by imaginary or diagrammatic sections, but by delineations of their exact appearance in Nature, as displayed by the most delicate and ingenious dissections.

The development of the nervous and vascular systems is traced from the earliest appearances;— the Heart's motion is described, whilst, as yet, it circulates only colourless Blood. [The situation where the red-globules first make

82   make their appearance, Hunter had described, as is well known, in the [/] celebrated Treatise on the Blood; and three figures were selected from the Series on Incubation to illustrate that work.] The gradual formation of the Lungs and Air-cells is traced.[3]

In describing the development of the generative System in the Bird,[4] the equality of the right and left Oviducts at an early period of their formation is expressly mentioned:—the formation of the Eye-lids;—and the growth of the horny knob on the Beak by which the shell is ultimately broken, are also described. But the most important passages in this valuable Manuscript are those in which some of the general laws of Development are enunciated, or at least definitely indicated.

We have seen that Harvey compared the simple heart of the Insect to an early condition of the heart of the Chick, beyond which stage[5] it could not ascend in the inferior organism.[6] Hunter rises to a more general expression in

83   his history of the development of the chick.[7] Hunter then states the Theories

---

1   equalled] *followed by deletion by vertical pencil scoring of paragraph* 'The Manuscript consists of a brief general account of the results of a series of examinations continued for 15 years.'.

2   separation] *RO inserted in ink over erasure with erased pencil* 'separation' *interlined above.*

3   celebrated. . .traced.] *RO pencil note on facing fol. 83ᵛ :* 'The relative importance of the proximate *fibrine & red globules {interlined above* 'proximate'} *constituents was inferred from the order of their development—' no insertion point indicated.*

4   in the Bird,] *RO interlined in ink with caret.*

5   stage] *RO inserted in ink in margin.*

6   in. . .organism] *RO interlined with caret.*

7   in. . .chick.] *period added [ed.]. This is RO pencil insertion from facing fol. 83ᵛ to replace deletion of original RO intended insertion in ink on fol. 83ᵛ:* 'of the analogy of condition presented by the embryonic phases of a *higher { interlined below deletion of* 'perfect'}*animal to the permanent *condition fo{deleted} states of the lower animals; in a {after deletion of* 'the'} *passage *which has {interlined with caret } already *been {interlined with caret } quoted by my learned Colleague {*'a' *interlined above* 'g'}' *All this to replace quotation on fol. 82 commencing after* 'expression': 'and says "If we were capable of following the progressive increase in the number of the parts of the most perfect Animal, as they were formed in succession, from the very first state [l fol. 82] to that of full perfection, we should probably be able to compare it with some one of the incomplete animals themselves, of every order in the Creation:— or in other words" says he, "in the fullness of his perception of this beautiful Law—{with pencil '&' over dash}

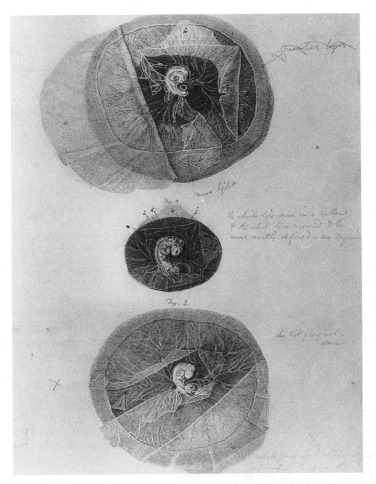

Fig. 11. Plate Depicting The Early Development Of The Chicken Prepared For
The *Descriptive And Illustrated Catalogue Of The Hunterian Museum,* Vol. 4
(1840). This illustrates some of the embryological researches Owen was
conducting near the time of the first Hunterian lectures. The plate is probably
his modification of an earlier plate by William Bell prepared for John Hunter.
(Courtesy President and Council, Royal College of Surgeons of England)

(contd)
*the legitimate extent to which it was applicable {*RO pencil continuation on facing fol
84v*}— If we were to take a series of Animals from the more imperfect to the perfect, we
should probably find an imperfect animal corresponding with some stage of the most
perfect."'.

of Evolution and Epigenesis[1] concerning which his predecessors and
contemporaries were warmly opposed to one another; and he justly observes
that he thinks he sees both principles together with a third, or that of
metamorphosis[2] introduced in the development of the Embryo. Hunter not
only points out the analogy of the transitory stages of the Embryo bird[3] to the
permanent structure of less complicated Animals but applies the conditions he
observed to the explanation of Monstrosities;—and reciprocally he reasons
from Malformations as to the probable formation of the parts of Animals
whose development he had not been able to trace by actual observation.

84   Thus,[4] he observes in the Chick [/] that

> every part is formed on the outside of the Animal, as the heart, then the
> lungs, and over the whole the Skin of the abdomen, which is not perfected
> 'till the animal is ready to hatch; and sometimes not even then. As this,
> however, he observes relates only to the Bird, it may be supposed to belong
> to it only; but there is reason to believe it is the same in other Animals; for
> in some Monsters in the Quadruped, we have no abdominal parietes. This
> state of deficiency in the parietes of the abdomen has all its degrees; some
> much more, others less.—

You will see these degrees of Malformation in the extensive series of
Monsters collected by Hunter and we can no longer wonder at the number of
specimens of this nature which Hunter had been at the pains to accumulate
when we find that he possessed the key by which so much that is Mysterious
in those productions, is explicable on definite & intelligible[5] Principles.

But I need not cite[6] this Manuscript to prove the Author's knowledge of
85   the [/] Application of the transitory structures of fœtal Life to explain congenital defects;—Have we not, at the commencement of "The Animal Œconomy"
a most beautiful illustration of the application of this principle. How? asked
the Surgeons of that day, How comes it that the Gut in Hernia Congenita is
in contact with the Testis?—Look, says Hunter, to the situation of the
Testis;—to its relation to the other abdominal viscera and to the peritonæum a
few months before the term of fœtal life has expired;—watch the progress of
the gland to the Scrotum, carrying along with it a peritonæal pouch like the
sac of a Hernia, and no longer wonder that this pouch must remain a hernial
sac, if its closure be prevented by the contemporary passage of a loop of

---

1   Epigenesis] *faint underline in pencil with pencil* 'X' *in margin and RO note on facing*
    fol 84v without insertion point 'His application of the same law where he compares
    tissue {after deletion of 'the'} of newly developed embryo of vertebrates to *Hydra*—'.
2   together. . .metamorphosis] *RO interlined with caret after* 'principles'*and with erased*
    pencil 'Together with a third, viz that of metamorphosis' *on facing fol. 84v.*
3   Embryo bird] *RO interlined in pencil with caret after* 'the' *before deletion of* 'Chick'.
4   Thus,] *inserting paragraph break for one-inch space to indicate sentence pause [ed.].*
5   & intelligible] *RO interlined in pencil above deletion of* 'Laws and'.
6   not cite] *inserted over erasure and in margin pencil* '?X'.

Intestine; a sac necessarily *common*[1] both to the part which has naturally, and the part which has præternaturally been extruded from the abdomen.

Then Hunter goes on to show how the early and transitory condition of the Tunica Vaginalis in the human fœtus, and also the still earlier abdominal position of the Testes are permanent structures[2] in the lower Mammalia.

86   With respect to Monstrosities in general, Hunter had drawn out a scheme for their classification, and had produced them by experiment. In the "Animal Œconomy" he states that every Species of Animal, and every part of an animal body is subject to malformation, but that this is not a freak of Nature, or a matter of mere chance;—for he observes that every species has a disposition to deviate from Nature in a manner peculiar to itself. It is this principle which forms the basis of the latest and most elaborate Treatise on Monsters, a Work,[3] which Geoffroy S[t]. Hilaire describes as being "the result of having established, by a great number of researches, that *Monsters* are, like the Beings called Normal, subject to Constant rules."[4]

Hunter differs from Geoffroy, in considering that in every Malformed part the Monstrous form was originally impressed upon it;—while the French Teratologist attributes the production ("l'ordonnèe") of Monstrosities, to the operation of exterior mechanical causes at some period of[5] fœtal development. The latter cause may occasionally operate; but the opinion of Hunter is that which is now most generally adopted, on this obscure and difficult question.

In expressing his opinion on this subject[6] Hunter again[7] enunciates another important[8] law of the aberration of the formative processes— "I should imagine," he observes, "that Monsters were formed Monsters from their very first formation, for this reason, that all supernumerary parts are joined to their similar parts, as a head to a head; &c. &c."[9]

---

1   *common] pencil underline with* pencil '?X' *in margin.*

2   structures] *inserted by RO in ink over erasure with erased* 'states' *interlined in pencil below and erased* 'structures' *interlined in pencil above.*

3   a Work,] *RO interlined in ink with caret.*

4   "the. . .rules"] *quote marks in pencil and underlined and in margin* 'X' *(2 times) to correspond to underlined quote marks.*

5   of] *RO interlined in ink with caret.*

6   subject] *after deletion of* 'intricate'.

7   again] *interlined with caret after* 'Hunter'.

8   important] *interlined below deletion of* 'beautiful'.

9   In...&c".—] *RO insertion from facing 87v with* 'In' *inserted over erased* 'Such' *on page break and vertical pencil deletion of entire fol. 87 except for last line:* '{¶} Such are the general laws *and Doctrines* {interlined with caret by RO in ink } which Hunter had established, and such are the Examples he has left, *even{deleted}* of their Application. {¶} Hunter receives, however, a sad reward for the devotion of a life to the discovery of Principles *which {inserted over erasure}* the supineness and neglect of his Countryman have suffered *so many of them{deletion }* to be forgotten until their proper application by Foreigners recalled their importance and transfer the discovery to other names.— {¶} Had the means and time been granted to Hunter to have made publick all the great results of his unexampled labours;— or had similar researches been conducted with the same spirit by his immediate Successors :— would our *Teachers {faint pencil underline and*

87   The Museum of Hunter is the principal but not the sole[1] Depository of
88   the [/] depictions[2] of these labours by which he established the general principles to which we have alluded.[3]
     Hunter had passed his 30th. year before he had collected a single prepa-
ration for himself. All that he had made before that time were added to his
Brothers' collection which is now the ornament[4] of the University of
Glasgow. In commencing his independent labours in Anatomy[5] he conceived
the Idea of a collection in which the illustrations of the Human Organization
should form a part only of a general display of all the Types and
Modifications of Animal Structure, and, practically, he was the first who reduced the scattered facts of Comparative Anatomy to a connected system.[6]
     When Hunter had brought his Museum to an approximative degree of
Perfection, he then set apart certain days in which he exhibited and explained
to some chosen Minds which could respond to the conceptions of his own,
his great Scheme, embracing the demonstration of all the leading
Modifications of every organ of the Animal Body, and of every stage which
each organ undergoes in its development to fulfil the functions demanded
from it in the highest organisms.
89   We know how far he had [/] advanced in the Completion of the first
Department of this philosophical Series[7] which embraces the permanent or
perfect[8] forms of Organs. That part of the second division, which demonstrates[9] the[10] transitory condition it is true[11] is but a Sketch;—yet it is the

---

(contd)
     in margin pencil '?X'} now, after a lapse of half a century have but begun to develop to
     their students those beautiful laws of Animal Development for *their* {'ir' *inserted* }
     knowledge of which they are indebted to the labours of the Professors and Conductors of
     those noble Schools of Continental {*after deletion of* 'continental'} Europe, where the
     Spirit of Hunterian Inquiries seems so long to have exclusively resided?' *followed by
     horizontal ink line across 1/5 of page.*
1   principal. . .sole] *RO interlined in pencil with caret and* 'sole' *after deletion of* 'onl'.
2   depictions] *RO interlined in pencil to replace pencil deletion of* 'evidences' *on page
     break and interlined again above deletion of* 'evidences' *on fol. 88.*
3   to…alluded] *interlined above deletion of* 'we have been endeavouring to explain.'.
4   ornament] *corrected from* 'ornament' *[ed.].*
5   In…Anatomy] *interlined and inserted by RO in ink with* 'labours in Anatomy he' *over
     erasure.*
6   and, practically…system] *interlined to replace deletion of* 'and be subservient to that
     End' *with RO pencil note without insertion point marked on 89v facing:* 'He first
     practically reduced {interlined above deletion of 'brought'} the scattered facts of
     Comparative Anatomy to a connected System—'.
7   of the…Series] *RO inserted in ink with* 'of the. . .Department' *over erasure and*
     'of…Series—' *interlined with caret.*
8   or perfect] *RO interlined in ink with caret.*
9   That…demonstrates] *RO alterations of* 'That' *from* 'The' *in pencil with* 'at part of the'
     interlined in pencil and with 'which demonstrates' *inserted on line over erasure in ink.*
10   the] *altered in pencil from* 'their'.
11   it is true] *interlined by RO in ink over pencil with caret.*

Sketch of a Master, and possesses all that charm and force which the traces of a Master-hand and mind never fail to impart.[1]

Future labours must complete this outline,[2] and a few years will I trust see that Noble Apartment destined by the munificient spirit of this College to the display of the Hunterian Treasures, occupied in orderly arrangement facilitating every inquiry & research, with specimens supplying each[3] vacant link, and harmonizing with those chosen examples which Hunter has left to guide the Comparative Anatomist in his[4] future investigations.[5]

Amongst the enlightened men who enjoyed the inestimable advantage of listening to the Explanations which the Founder of our Collection gave of his own Labours, and of their scope and tendency were Camper, Poli, Scarpa[6] and the now venerable Blumenbach.

90   Camper, as a contemporary, and in some respects a rival of Hunter, may have been less influenced in [/] the general tenour and success of his investigations in Comparative Anatomy than the last named and younger Naturalists and Physiologists. We cannot but suppose that the spectacle of the organization of so many rare Marine Animals so beautifully displayed by so consummate a practical Anatomist as Hunter, must have had a lasting Influence on the mind of Poli, and we may perhaps trace to this Source the numerous & minute dissections[7] of the[8] Mediterranean Mollusca, and the Magnificent Illustrations of their Anatomy[9] which have justly immortalized their Author.

In contemplating the gradational &[10] connected[11] series of the Organs of Animals which Blumenbach must have witnessed for the first time, in the Museum of Hunter, that learned and accomplished Physiologist doubtless vividly appreciated the cumulative force with which Comparative Anatomy urges the onward progress of Physiological Science, when all its scattered facts are concentrated into one orderly System. In his[12] subsequent publica-
91   tion[13] of the first Systematic Treatise of Comparative Anatomy the [/]

---

1   impart] *RO interline in pencil above deletion of* 'give.'
2   outline] *inserted over erasure with erased pencil* 'outline' *interlined above.*
3   each] *RO interlined in pencil above deletion of* 'every'.
4   the Comparative…his] *RO inserted in ink with* 'Comparative Anatomist' *interlined with caret and* 'in his' *inserted over erasure.*
5   investigations] *pencil pause mark* '#' *inserted at this point.*
6   Scarpa] *RO interlined in ink with caret.*
7   numerous …dissections] 'numerous & minute'*interlined by RO in ink above deletion of* 'perservering'*with* 'dissections'*followed by deletion of* 'which formed the foundation of'.
8   of the] *RO insertion in ink in margin following deletion of original* 'of the' *following* 'Anatomy'.
9   Magnificent…Anatomy] *RO transposed in ink by lead line before* 'of the' *from preceeding line.*
10 gradational &] *RO interlined in ink with caret.*
11 connected] *before deletion of* 'and complete'.
12 his] *RO insertion in ink over pencil erasure with pencil* '?X'*in margin.*
13 publication] *RO insertion in ink over erasure with erased pencil* 'publication' *interlined.*

erudition of Blumenbach supplied many of those links in the Series of animal[1] Structures which Hunter derived from Nature's original Sources.

The object of our previous Lectures[2] has been[3] to indicate the steps by which Comparative Anatomy advanced to this crowning Epoch in its History. We have traced it to the period when *the* [4] Individual arose whose intellect and energies were equal to the Task of grappling with the whole circle of the Science; and to whom[5] Life, alas! was too short to consummate the grand object of the labours in which all the disposable moments of that active[6] life had been expended:—the leisure to[7] which Hunter looked forward,[8] in order to lay before the Public his Opinions at large upon the Operations of the Animal Œconomy[9]—I use his own words—"That leisure never arrived."—

But the grand impulse was given; the published writings of Hunter, full of profound and original views, combined with the spectacle of his wonderful dissections, raised the Science[10] of Comparative Anatomy in the scale of Human Knowledge.

92   The precision and facility introduced into[11] the study [/] of Natural History by the ingenious arrangements of Linnaeus, had the happiest influence in the prosecution of the Collateral Science of Zootomy, not only by the habits of nice discrimination which a Study of the Linnæan System induced, but[12] by the number of additional objects presented to the Scalpel of the Anatomist as the boundaries of Zoology were by the same influence augmented.

The Eloquence and Genius of Buffon had given[13] a new charm to the Study of Zoology, while the numerous and accurate dissections, and the exact descriptions of Daubenton & Pallas, had contributed to[14] a sure foundation for its future superstructure.

Can we wonder at the unexampled[15] progress of[16] the Science of Zoology

---

1 animal] *RO interlined in ink with caret.*
2 previous Lectures] *with 'previous' interlined by RO in pencil above deletion of* 'present' *with* 's' *added in pencil to* 'Lecture'.
3 been] *before deletion of* 'chiefly'.
4 the] *underlined in pencil.*
5 whom] *retaining original reading and deleting pencil interline by RO of* 'whose' *[ed.].*
6 active] *RO interlined in ink with caret.*
7 to] *RO interlined in ink with caret.*
8 forward,] *before erased* 'to'.
9 Œconomy] *before erased quotation marks.*
10 Science] *interlined in ink by RO above* 'importance' *without deletion* .
11 into] *RO insertion of* 'to' *with erased pencil* 'to' *interlined above with pencil* '?X' *in margin.*
12 but] *RO insertion in ink over erasure.*
13 had given] *RO interline in ink above deleted* 'gave' *with deletion of* 'al' *between* 'had' *and* 'given'.
14 &. . .to] *RO interlined in ink with caret and altering* 'laid' *to* 'lay'.
15 unexampled] *interlined in ink by RO with caret.*
16 of] *restored after original deletion in ink with stet marks below.*

in its most extended sense, and most varied application, when urged forward at this auspicious period by all the energies[1] of Cuvier, one[2] of Nature's most favoured Interpreters;—[3]whose master-mind combined the precision of Linnæus, the patience of Daubenton, the eloquence and unquenchable zeal of Buffon;—who possessed Anatomical Skill equal to that of Hunter, and pow-
93   ers of design which rival those of Hunter's [/] most accomplished artist. The discoveries of Cuvier, and those of the distinguished Anatomists of the Continent, and of Great Britain who have contributed to the advancement of Science since the time of Hunter, we shall have frequent occasion to mention[4], in the exposition of the Treasures of the Museum.[5]

At present our time will scarce permit us to do more than advert[6] to the general theories or views by which it has been endeavoured[7] to simplify the[8] expression &[9] facilitate the[10] intelligence of the vast assemblage of facts in Comparative Anatomy—[11]

The great object of Cuvier was to determine the laws[12] which govern the co-existence[13] and condition of the different organs in the Animal body, so as to be[14] enabled from a part to predicate of the structure of the whole.

Cuvier was stimulated to this investigation from[15] finding himself sur-rounded by the remains of extinct animals respecting the nature of which he
94   could only form an [/] opinion after having previously determined the rela-tions subsisting between[16] the general habits of existing species and the structure of their Skeleton.[17] To answer the question What bone is this?

---

1   energies] *interlined by RO in ink above deleted* 'powers'.
2   Cuvier. . .one] *RO interline in ink above deletion of* 'one'.
3   ;—] *followed by deletion of* 'of one'.
4   mention] *inserted in ink by RO over deletion of* 'allude to'.
5   Museum.] *followed by double pencil slash marks to indicate pause or end.*
6   advert] *inserted by RO over erasure.*
7   endeavoured] *followed by pencil deletion of* 'to reduce the vast assemblage of facts in Comparative Anatomy to order'.
8   the] *altered in pencil from* 'this'.
9   &] *followed by pencil deletion of* 'to'.
10   the] *altered from* 'their'.
11   it has. . .Anatomy—] *all RO insertion from facing fol. 94v with lead line to insert point and* 'of the. . .Anatomy' *in pencil All replacing deletion from fol. 93* 'the vast assemblage of facts in Comparative Anatomy accumulated by their united labours, *have [inserted over* 'has'}, in the present day, been sought to be reduced to order,— their expression simplified, and their intelligence facilitated.'.
12   laws] *before deletion in pen* 'of co-existence'*with* 'co-' *interlined above with caret.*
13   co-existence] *RO interlined above ink deletion of* 'presence'.
14   so. . .be] *inserted by RO over erasure with deletion of following* 'being achieved, the Anatomist is' *with erasure of* 'is'.
15   from] *over erasure.*
16   relations. . .between] *RO interlined in ink over deletion of* 'influence which'.
17   and. . .Skeleton.] *RO interline over deletion of* 'had over the osseous System: and reciprocally. He was equally necessitated in this inquiry to determine the homologies of the* various {*over deletion of* 'different'} *parts which enter into the composition of*

or[1] of what bone is it a part? equally obliged him to determine the homologies of the parts which compose the skeletons[2] of different[3] Animals—and he was not less necessitated to consider the[4] affinities by which the different animals themselves[5] are naturally grouped together;—but, as many striking and unexpected correspondencies between different parts of different animals[6] were brought to light in the rapid succession of new facts which were now discovered[7], some Comparative Anatomists began to[8] devote themselves almost exclusively to the determination of these Analogies.

Now the inquiry into the homologies or signification of the different parts of the animal frame is replete with beautiful results when pursued in accordance with those rules of investigation, by which alone the Human Understanding can attain to any certain truth. But the remarkable neglect of the *Inductive method* displayed in some of the most elaborate Treatises on this subject constitutes perhaps one of the most striking incidents in the history of modern Science.[9]

One Inquirer starts from the metaphysical dogma that each part, and every part of a part must represent the whole:

95   He sees, therefore, in the Head a condensed representation of the intire[10] body:—the cranium is the head of the head; the nose is the thorax of the head; the jaws represent[11] the extremities; the Teeth, the nails; &c. &c.

The true analogues of the Cranial bones are pronounced by another Anatomist from an *a priori* determination, that the skull in every vertebrate animal consists of seven[12] vertebræ, each vertebra being composed of nine[13] elements. But the determination in this case is equally arbitrary with the pre-

---

(contd)
      different animal bodies; and the' *with pencil line in left margin alongside.*
1   or] *before deletion of* 'to'.
2   skeletons] *after deletion of* 'various'.
3   different] *interlined with caret.*
4   To. . .the] *RO insertion from facing fol.* 95ᵛ *to replace deletion on* 95. *No insertion point marked.*
5   themselves] *RO insertion in ink.*
6   different animals] 'different' *interlined with caret by RO in ink above deletion of* 'the'*and* 'animals' *altered from* 'animal' *and followed by deletion of* 'body'.
7   were now discovered] *with* 'were' *and* 'discovered' *interlined by RO with carets in ink and deletion of* 'brought to light' *after* 'now'.
8   some. . .to] *RO interline in ink over deletion of* 'speculative investigators of the*Science {underlined in pencil}'.
9   But. . .Science] *RO insertion from facing fol* 95ᵛ *below deletion of earlier insertion,* 'there is perhaps no branch of modern Science in which Induction has been more disregarded than' *to replace deletion in ink on fol.* 94 *of* 'But this has hitherto been rarely the case.'.
10   intire] *in ink over erasure of* 'whole'.
11   represent] *RO interlined in ink with caret.*
12   seven] *RO inserted in ink over erasure.*
13   nine] *over erasure.*

ceding; it is wholly[1] unsupported by an examination into the primary formation of the cranial bones, to determine[2] how many are actually developed
from the circumference of a gelatinous[3] *Chorda dorsalis;* the only true embryonal condition of a Vertebra.

A third Transcendentalist goes further; and limiting the means at the disposal of the Demiurgus[4] employed in[5] the construction of the framework of
an animal to modifications of a single element, sees a Vertebra not only in
every bone of the vertebrate body, but in each ring of the Worm, and in every
joint of the Lobster.

But what is really gained by propositions such as we have adduced? or
by stating that all animal organization is reducible to one primitive form or
96   plan[6]? What would be thought of the [/] Geometrician, who, because all
figures are modifications of the simple mathematical line, should gravely advocate their unity of composition, and put forth the observation as one of high
philosophical importance?

It was to obviate the retrograde[7] tendencies of these[8] metaphysical or
transcendental Theories of Animal Organization that Cuvier devoted his latest
energies; and the abuse of the doctrine of Analogies perhaps led him by a natural reaction[9] to under-rate its value.

The general laws of Animal Organization can never be developed from a
consideration of the perfect or matured structure alone: and such *was* the
general[10] character of the knowledge from which Cuvier deduced his inferences.

The laws of Coexistence;—the adaptation of structure to function; and to
a certain extent the elucidation of natural affinities may be[11] legitimately[12]
founded upon the examination of fully developed species:—But[13] to obtain
an insight into the laws of development,—the signification or *bedeutung*, of
the parts of an animal body demands a patient examination of the successive
stages of their development, in every group of Animals.

97   This truly philosophic inquiry is now in [/] progress. Tiedemann,
Purkinge, Baer, Rathke, Wagner, Valentin, Müller, and the disciples of their

---

1   wholly] *RO interlined in ink with caret.*
2   determine] *RO interlined in ink over deletion of* 'see'.
3   the circumference. . .gelatinous] *RO interlined in ink with caret with* 'a' *deleted before* '*Chorda*'.
4   at. . .Demiurgus] *interlined below line by RO in pencil.*
5   employed in] *RO interlined in ink over deletion of* 'for'.
6   or plan] *RO interlined in pencil with caret.*
7   retrograde] *RO interlined in ink above deletion of* 'baneful'.
8   these] *altered in ink from* 'the'.
9   by. . .reaction] *RO interline in ink with caret.*
10 general] *over erasure.*
11 may be] *interlined by RO over deletion of* 'can'.
12 legitimately] *followed by deletion of* 'be'.
13 But] *RO interline in ink.* 'To' *modified to* 'to' *for grammatical purposes [ed.].*

School, are cautiously but steadily laying the foundations of a just and true theory of animal development and organic affinities.

But care must be taken to distinguish the inductive labours of these Physiologists from the *a priori* speculations of the Transcendental school— and not to accept literally, or without mature consideration the exaggerated expressions, sometimes indulged in, of theories, which when simply enunti- ated, do in truth accord with Nature.[1]

It is thus that the beautiful observation of the resemblance of imperfect conditions of the organs of a higher species to the perfect conditions of corre- sponding organs in a lower organized species is misrepresented when it is stated that the Human Embryo *repeats in its development* the structure of any part of another animal; or that it *passes through the forms* of the [/] lower classes;—or when it is asserted that a Fish is an overgrown Tadpole. Such propositions you will at once, Sir, perceive, imply that there exists in the Animal Sphere a Scale of Structure differing *in degree* alone:—nay, they imply the possibility of an individual, at certain periods of its development, laying down its individuality, and assuming that of another Animal;—which would, in fact abolish its existence as a determinate concrete reality.

98

Individualities, however, manifest themselves at very early periods of development, and cannot be laid aside. There exist permanent structures among the lower animals not met with in the embryonal phases of any of the higher:—while there are many phases of the higher animals corresponding to which we do not find any permanent structures among the lower. And, again, No structure which peculiarly characterizes any one animal, or group of Animals in the perfect state, makes its appearance even in the Embryonic life of any other.

The doctrine of Transmutation of forms during the Embryonal phases, is closely [/] allied to that still more objectionable one, the transmutation of Species. Both propositions are crushed in an instant when disrobed of the figurative expressions in which they are often enveloped[2]; and examined by the light of a severe logic.

99

Even when it is asserted that Man, in the progress of his development is upon that grade upon which the several classes beneath him remain stationary in the progressive development of the intire animal kingdom:—the proposi- tion thus qualified, is by no means true. Man, in the progress of his[3] devel-

---

1   But. . .Nature.] *RO ink insertion in ink over erased pencil from facing fol. 98ᵛ to replace deletion by vertical pencil scoring of paragraph on fol. 97* '{¶} The progress of knowledge will, of course, be impeded in proportion to the influence and popularity of those Teachers who for the sake of the small and transient reputation, gained by exciting the wonder of their hearers, or readers, advocate the baseless speculations of the transcendentalists, and indulge in exaggerated expressions of views to the development of which they are unable or unwilling to lend the co-operation of honest and unbiased labours.'.
2   enveloped] *interlined in ink by RO above deletion of* 'concealed'.
3   progress. . .his] *in ink over erasure.*

opment, is not upon that grade on which any other animals remain stationary, unless the latter belong to the same *type*[1] as Man.

But can the various structures which Comparative Anatomy now unfolds[2], be referred to one, or do they manifest[3] different types? This is a question which is now in progress of Solution: and can only be retarded[4] by too[5] hasty generalizations;—by the seeking for analogies[6] on *a priori* conceptions.

The knowledge of the varieties of Structure presented by Animals in their perfectly developed states;—a knowledge so limited;—for this expression must be applied to it; is, and [/] ever will be inadequate to solve that question, or to establish a just doctrine of the homologies of animal structure, or even to ascertain the true affinities of the different groups or species of Animals.

Already the Entomologist acknowledges the indispensibility of studying the transient conditions of Insects during the various stages of their different Metamorphoses, in order to obtain an insight into their natural affinities:—and the study of the development of each organ in the other classes of animals is not less essential to their true[7] classification. What should we know of the zoological relations of the Barnacle, if we were acquainted only with its organization in the last & fixed stage of its existence?[8]

Nor[9] can the functions of a part in all cases be sufficiently determined even after a knowledge of all the varieties of form & structure[10] which it presents in the fully developed species of the animal kingdom. How many organs are there which manifest their activity at early stages of existence and relapse into a state of quiescence & atrophy at a later period of life.

The use of the thymus gland as an active manufactory and magazine[11] of the cerebral globules, while the brain is attaining its full development could never have been suspected had not its condition as an extensive receptacle,[12] and the period of its increase been first determined by the well known labours of our President—& had not[13] microscopic[14] structure and progressive

---

1 *type*] *in heavy bold ink.*
2 unfolds] *altered from* 'unfold'.
3 do. . .manifest] *interlined by RO in ink.*
4 and. . .retarded] *RO interlined in ink above deletion of* 'but that progress is also interrupted'.
5 too] *interlined by RO in ink with caret.*
6 analogies] *altered from* 'analogues'.
7 true] *interlined in ink by RO with caret.*
8 What. . .organ] *RO insertion in ink from facing fol. 101ᵛ with insertion point marked with* '<'.
9 Nor] *inserting paragraph break in inserted material following slash mark on MS [ed.].*
10 & structure] *interlined with caret.*
11 and magazine] *interlined with caret in ink.*
12 receptacle,] *replacing semicolon with comma [ed.].*
13 condition. . .not] *interlined with insertion point marked by lead line.*
14 microscopic] *interlined with caret.*

development been afterwards traced by Ehrenberg[1] in conjunction with observations on the corresponding phases of the Cerebral organ.[2]

In short,[3] Comparative Anatomy, to be productive of its full results, must embrace not only the knowledge of the differences of forms and structure of all organs in all animals, but also the varieties of form which each organ presents at every period of its development.

That Hunter fully appreciated the necessity of studying the Animal organization in both these directions, is evident from the general tenour of his physiological writings, and from[4] the fact that a great proportion of his Series of Comparative anatomy is devoted to the elucidation of the transitory conditions [/] of Animal Organs. His most elaborate Series of investigations had for their object the Metamorphosis of the organs of the vertebrate embryo;— and the facts which he thus ascertained enabled him to enunciate Propositions which each step that has since[5] been taken in the same track of inquiry, tends to establish.[6]

101

1   by Ehrenberg] *interlined in ink with caret.*
2   What should we. . .Cerebral organ] *All inserted in ink by RO from facing fol. 101ᵛ with insertion on 100 marked with '<'.*
3   In Short] *deleted in ink and then restored by stet marks below.*
4   and from] *interlined in ink by RO with caret.*
5   since] *interlined in ink by RO with caret.*
6   establish.] *with ink horizontal line across center of MS to mark end of lecture.*

Notes
to
Lectures Three and Four

2–9 Mondino da Luzzi (ca. 1275–1326), author of the *De omnibus humani corporis interioribus membris anathomia* (Bologna, 1316), first printed at Padua in 1478.

3–8 On the history of Paduan anatomy and these statutes, see E.A. Underwood, "The Early Teaching of Anatomy at Padua, with Special Reference to a Model of the Padua Anatomical Theatre," *Annals of Science* 19 (1963), pp. 1–26.

4–4 This is reference to the Bull issued by Boniface VIII of 1300 which was issued to prevent practice of boiling bones of eminent persons dying abroad. On this see C. Singer, *A Short History of Anatomy from the Greeks to Harvey* (New York: Dover, 1957), p. 85. Owen is alluding to the Anatomy Act of 1832 which established tight controls over anatomical dissection to control the grave-robbing practices.

5–7 Giacomo Berengario da Carpi (ca. 1460–1530), author of the first illustrated anatomical textbook (*Isagoge brevis*, Bologna, 1523).

5–14 Volcher Coiter (1534–1576). Reference is to his *Externarum et internarum humani principalium corporis partium* (Noribergae, 1573).

5–17 Johannes Guinter (ca. 1505–1574), known for his translation of Galen's *De anatomicis administrationibus* (Paris, 1531), and a teacher of Andreas Vesalius.

5–20 Michael Servetus (1511–53) one of co-discoverers of the lesser circulation and author of *Christianismi restitutio*

(1553)

6–3 Sylvius= Jacques Dubois (1478–1555), an ardent Galenist and professor of Medicine at the Collège de Tréquier in Paris. One of the teachers of Andreas Vesalius.

8–6 Hieronymous Fabricius ab Aquapendente (1537–1619), author of *De venarum ostiolis* (Padua, 1603).

9–8 This is reported in Robert Boyle's interview with Harvey. See Robert Boyle, *A Disquisition About the Final Causes of Things* (London, 1688), quoted in William Harvey, *Anatomical Disputation Concerning the Movement of the Heart and Blood in Living Creatures*, trans. G. Whitteridge (Oxford: Blackwell, 1976), p. xxvii.

11–1 The reference is apparently to the descriptions carried out in the papers of Samuel Collins (1618–1710), published as his *A Systeme of Anatomy* 2 vols. (London: Newcombe, 1685). A copy of this is at the RCS. See Cole, *History*, pp. 156 ff.

11–6 Edward Tyson, *Carigueya, seu Marsupiale Americanum, or Anatomy of the Opposum* (London: Walford, 1698) (with William Cowper).

11–11 Owen is referring to his own debates with Geoffroy St. Hilaire over the site of gestation in the monotremes and marsupials. Geoffroy had claimed that monotremes produced a true egg like the birds. Owen's counter was that the reproduction in the marsupials and monotremes was as in other mammals, with conception taking place initially in the oviduct. On this conflict see Owen's note to Hunter's "Descriptions of Some Animals from New South Wales," *Hunter's Works* 4: 482–4. See also A. Desmond, *The Politics of Evolution* (Chicago: Chicago UP, 1990), chp. 7.

11–15 E. Tyson, *Orang-Outang, sive Homo sylvestris: or the Anatomy of a Pygmy compared to that of a Monkey, an Ape and a Man* (London, 1699).

11–17 Idem. *Phocaena, or Anatomy of the Porpesse* (London: Tooke, 1680). Owen is also referring to John Ray's "An Account of the Dissection of a Porpess," *Philosophical Transactions of the Royal Society* **6** (1671); and Guillaume Rondelet's discussion of the whale in his *De Piscibus marinis* (1556).

12–9 [Claude Perrault], *Description anatomique de divers animaux dissequez dans l'Académie royale des sciences*, ed. 2 (Paris, 1682). An English translation of this was made in 1688. A copy of the French edition is at the RCS.

12–11 Nehemiah Grew, *Musaeum Regalis Societatis, or a Catalogue and Description of the Natural and Artificial Rarities Belonging to the Royal Society. . . whereas is subjoined the Comparative Anatomy of Stomachs and Guts* (London: Printed for the Author, 1681) (collection of several of his papers from the *Philosophical Transactions*),

12–14 Johann Konrad Peyer (1653–1712), *Exercitatio anatomico-medica de glandulis intestinorum earumque usu & affectionibus* (Schaffhausen: Waldkirch, 1677); Edward Tyson and William Cowper, *Carigueya, seu Marsupiale Americanum, or Anatomy of the Opposum* (London: Walford, 1698) ; James Douglas (1675-1742) *Myographia comparata specimen; or a Comparative Description of all the Muscles in Man and in a quadruped...* (London, 1707).

13–3. William Harvey, *De generatione*

animalium (London, 1651). Owen typically cites either the English translation of Harvey's works of 1653, or the Latin of the *Opera omnia: collegio medicorum Londinensi edita* (Londini, 1766). See *Hunter's Works* 4:176n. and below, n. 22–15. Copies of both are at the RCS.

14–4 Grew, see *Stomachs and Guts...*

14–10 Collins, *Systeme...*

14–12 François Poupart (1661-1708), French anatomist and surgeon.

15–9 Giovanni Caldesi, *Osservazioni anatomiche intorno alle Tartaughe Marittime d' Acqua dolce e Terrestri* (Firenze, 1687); Edward Tyson, "Anatomy of the Rattlesnake," *Philosophical Transactions of the Royal Society* **13** (1683), reprinted in idem., *Orang-Outang, sive Homo sylvestris* 2nd ed. (London, 1751). This work is in the RCS library.

15–12 On Swammerdam's work on the frog see Cole, *History*, pp. 296–7.

15–15 See William Harvey, *Disputations Touching the Generation of Animals*, trans. G. Whitteridge (Oxford: Blackwells, 1981), Ex. 19, p. 115.

15–15 Oligerus Jacobaeus, *De Ranis* (Paris: 1676).

15–16 August Johan Roesel von Rosenhof, *Historia naturalis ranarum nostratium* (Nurnberg: Fleischmann, 1758).

15–21 Hieronymus Fabricius ab Aquapendente, *Opera omnia anatomica e physiologica* (Lipsiae, 1687).

16–2 C. Preston, "A General Idea of the Structure of the Internal Parts of Fish," *Philosophical Transactions of the Royal Society* **19** (1697).

16–4 John Ray, "Some Considerations... About the Swiming [sic] Bladders in Fishes," *Philosophical Transactions of the Royal Society* **10** (1675).

16–11 Guilio Casserio, *Pentasthesion; hoc est de quinque sensibus* (Venetii, 1609). A copy of this is at the RCS.

16–13 On the interest of several anatomists in the peculiarities of *Xiphias* see Cole, *History*, pp. 353–4.

16–18 The reference is to *Nicolai*

Stenonis Elementorum myologiae specimen (Florentiae, 1667), containing his "Canis carchariae dissectum caput," and other shorter treatises. This is reprinted in Nicolai Stenonis, *Opera Philosophica* ed. Wilhelm Maar (Copenhagen:Tryde, 1910), vol. II. See discussion in Cole, p. 375.

17–3 C. Perrault et. al, *Memoires pour servir a l'histoire naturelle des animaux* (Paris: Imprimerie royale, 1671). On the complex history of this text see Cole, *History* 393–442.

17–8 Collins, *Systeme of Anatomie* (London,1685).

17–10 William Harvey, *Movement of the Heart*, pp. 46–7.

17–12 Thomas Willis, *De anima brutorum* (Londoni, 1672). See Cole, pp. 223–5.

18–2 Jan Swammerdam, *Biblia naturae, sive historia insectorum* 3 Vols. (Leiden, 1737–8)

18–5 Martin Lister, *Exercitatio Anatomica, in qua de Cochleis, maximè terrestribus & Limacibus agitur* (Londini, 1694). See Cole, *History*, 231–7.

18–7 Giuseppe-Saverio Poli (1746–1825), *Testacea utriusque Siciliae eorumque historia et anatome* 2 tom. (Parme, 1791–5).

18–8 Stefano Delle Chiaje, *Instituzioni di anatomio e fisiologia comparata* (Naples, 1832).

19–12 Pierre Lyonnet, *Traité anatomique de la chenille qui ronge le bois de saule* (La Haye: De Hondt, 1762). A copy of this is at the RCS.

19-15 John Hunter, "Observations on Bees," *Philosophical Transactions of the Royal Society* **82** (1792), reprinted in *Hunter's Works*. 4:422–66.

21–3 Gerard Blaes, *Anatome animalium* (Amstelodami, 1681). A copy of this is at the RCS.

21–4 Michalis B. Valentini (1657–1729), *Amphitheatrum zootomicum* (Francofurtii: Moenum, 1720).

21–5 Jean Jacques Manget, *Bibliotheca*

anatomica sive recens in anatomia inventorum thesarus 2 Vols. (Geneva: Chouet, 1685).

22–15 Harvey, *Opera omnia* 1:32. Owen is translating directly from the Latin rather than using an English version for this quote. The Latin quotation deleted from the MS has been slightly altered from the printed text.

23–9 Gasparo Aselli (1581–1625), *De lactibus sive lacteis venis quarto vasorum mesaraicorum genere novo in vento. . .*(Milan, 1627; Basel, 1628). Copies of both editions are at the RCS.

24–10 Bartholomeo Eustachi, *Opuscula anatomica* (Venetiis: Luchineus, 1564). A copy is at the RCS.

24–16 Jean Pecquet, *Experimenta nova anatomica quibus incognitum hactenus chyli receptaculum . . .*(Parisiis: Cramoisy, 1651).

25–2 Olaus Rudbeck (1630–1702), *Insidiae structae Olai rudbeckii sueci ductibus hepaticus aquosis* (Lugduni Batavorum, 1653); Thomas Bartholin, *Anatomia ex Caspar Bartholin; parentis institutionibus* nov. ed. (Lugduni: Hackium, 1660). This has appendix *De lacteis thoracicis & vasis lymphaticis* (Hagae: Vlacq, 1660). A copy is at the RCS.

25–5 See Owen's discussion of the discovery of the lacteals in *Hunter's Works* 4:307-8n. I have been unable to find more information on Joliffe.

25–12 Martin Lister, "Experiments for Altering the Colour of the Chyle in the Lacteal Veins," *Philosophical Transactions of the Royal Society* **13** (1702) p. 9; and William Musgrave, ibid. **12** (1701), p. 996. These are cited and quoted by Owen in *Hunter's Works* 4:303n and 309n.

25–20 Pierre Noguez, *L'anatomie du corps de l'homme en abrégé* (Paris,1723), p. 155. Owen quotes this work on a similar point in *DIC* 2:11. I have been unable to verify the existence of an 1737 edition of this work. No copy is at the RCS.

26–2 See Owen's further discussion and

references concerning Hunter's role in the discovery of the so-called absorbent system in Lecture Four below, p 169.

26–16 Tyson, "Anatomy of Rattlesnake," (see note 15-9).

27–13 Nehemiah Grew, *Stomachs and Guts.*

27–22 Antonio Vallisnieri, *Considerazioni ed esperienze intorno al creduto cervello di Bue impietrito, vivente ancor l'animale* (Padova, 1710); Réné Antoine de Réaumur, "Sur la digestion des Oiseaux," *Memoires de l'Académie royale des sciences de Paris* (1752), pp. 266–307; 461–95; Lazzaro Spallanzani, *Dissertazioni de Fisica animale e vegetabili: Della digestione* (1782), translated in *Dissertations Relative to the Natural History of Animals and Vegetables*, 2 vols. (Edinburgh, 1784).

30–4 Reigner de Graff, *De virorum organis generationi inservientibus...* (Lugduni: 1668); idem., *De mulierum organis generationi inservientibus* (Lugduni, 1672).

31–2 Johann Van Horne, *Opuscula anatomico-chirurgica...* [Lipsiae, 1707). A copy is at the RCS.

31–9 Edward Tyson, "Anatomy of the Mexico Musk-Hog," *Philosophical Transactions of the Royal Society* **13** (1683), p. 370. This is quoted by Owen in *Hunter's Works* 4:26 note.

31–11 John Hunter, "Observations on the Glands Situated Between the Rectum and Bladder, Called Vesiculae Seminales," *Hunter's Works* 4: 20–33.

31–16 See J. L. Prevost and J. B. Dumas, "Sur les animalcules spermatiques de divers animaux," *Memoires de la societé physique de Génève* **1** (1824), 180–207; see also *Annales des sciences naturelles* **1** (1824): 10–29;162–82; 274–93; **2** (1824): 100–20; 129–49; **3** (1824)113–38; **12** (1827): 415–43.

33–3 Owen is referring to the recent development of the achromatic microscope which was supported by

the Society of Arts. Owen's friend, Joseph Jackson Lister, was a major pioneer in the development of the achromatic scope.

34–8 John Ray, *The Wisdom of God Manifested in the Works of His Creation*, first ed. (London, 1681).

34–10 Nehemiah Grew, *Cosmologia Sacra, or a Discourse of the Universe as it is the Creature and Kingdom of God* 5 books (London: Rogers, Smith, Walford, 1671). Owen is using the 1701 edition in the RCS library.

36–4 Ibid. Bk 1, chp. 5, p. 29. Owen's quote slightly alters the punctuation in the original.

36–8 See Georges Cuvier, *Leçons d'anatomie comparée*, 2d ed. (Paris: Grochard, 1835) I, 50.

36–11 J. Ray,"Dissection of a Porpess," and E. Tyson, "Phocaena, or anatomy of the Porpesse," (London: Tooke, 1680).

37–10 Owen's allusion is to the so-called iatro-mechanical school of physiology which sought to develop a purely mathematico-physical analysis of physiological function. Boerhaave's advocacy of this is contained particularly in his *De usu ratiocinii mechanici in medicina oratio* (Lugduni Batavorum, 1709). A copy of this is in the RCS library.

37–16 Alexander Monro (*primus*) (1697–1767), *An Essay on Comparative Anatomy* (London: Nourse, 1744). A copy is at the RCS.

37–20 Louis Marie Daubenton (1716–99), the comparative anatomist who collaborated with Buffon on the monumental *Histoire naturelle générale et particulière, avec la description du Cabinet du Roi* 15 Vols. (Paris: Imprimerie Royale, 1749-67); and Pierre Simon Pallas (1741–1811), German physican and anatomist, and author of the *Miscellanea zoologica* (The Hague, 1766), and *Novae species quadrupedum* (Erlangen, 1778).

37–24 The reference is to the

iatrochemical school, which was concerned to explain all physiological function by fermentations and chemical interactions.

39–4 This favorable reference to Lawrence's controversial Hunterian lectures of 1817-21 possibly suggests that Lawrence was expected to be in the attending audience.

39–15 Owen is translating from G. Cuvier, *Rapport historique sur les progrès des sciences naturelles depuis 1789 et sur leur état actuel* (Paris: Imprimerie Imperiales, 1810), 2: 243. See also *Hunter's Works* 4:v.

41–9 Clift had made hand copies of several Hunterian manuscripts, many of which were only published in the *Essays and Observations of John Hunter* 2 vols, ed. R. Owen (London: Van Voorst, 1861). I have been unable to locate the source of this quotation.

42–2 Felix Vicq d'Azyr, *Traité d'Anatomie et de physiologie avec les planches colorées* (Paris: Didot et aine, 1786); idem, *Oeuvres de Vicq d'Azyr*, ed. J. L. Moreau, 6 vols. (Paris, 1805).

44–3 William Bell, "Description of the Double Horned Rhinoceros of Sumatra," *Philosophical Transactions of the Royal Society* **83** (1793): 3-6; idem., "Description of a Species of Chaetodon, Called, by the Malays, Ecan bonna," ibid., pp. 7–9. Bell was Hunter's accomplished anatomical illustrator.

44–12 Owen is referring to William Bell's plates prepared for Hunter while he was his illustrator. Many of these were published for the first time in the *DIC* with slight revisions made by Owen.

45–3 These plates are published as plates 14,16 and 21 in *DIC* 2 (1834).

45–5 Friederich Tiedemann, *Anatomie der Röhren Holothurie, des Pomeranzen Farben Seesterns und Seeigels* (Landshut, 1816).

46–15 John Hunter, "Observations on Bees,"*Hunter's Works* 4:424. Owen has slightly altered this passage from

original.

47–8 I have been unable to locate the source of this quotation.

49–6 John Hunter, "On the Vital Principle," *Hunter's Works* 1:223–5. Hunter's results were published in his "Experiments on animals and Vegetables, with Respect to the Power of Producing Heat," *Philosophical Transactions of the Royal Society* **65** (1777), reprinted in *Hunter's Works* 4: 131–55.

49–11 Manuscript has asterisk after "vegetiva" with following quotation in a note at bottom of page from Harvey's *Opera omnia* 1, pp. 278, 286 and 2:446. The quote has been synthesized from disconnected passages: "Plurimum itaque mecum ipse reputavi, qui fieret ut ova improlifica gallinæ supposita, ab eodem calore extraneo corrumpuntur, putrescant et fœtida *evadant* ovis autem fœcundis idem non contingat.— Ovum itaque est corpus naturale, virtute anomali proditum; principis nempe motus transmutationis, quietis et conservationis.—Cum enim in ovo macula prius dilatetur [l*fol.* 49] colliquamentum concoquativo et præparetur plurimaque alia, (non sine providentia) ad pulli formationem et incrementum instituantur, antequam quidpiam pulli vel ipsa primogenita ejus particula appareat, quidni utique credamus calorem innatum animamque pulli vegetativam ante pullum ipsum exsistere."

51–8 Ibid., pp. 274, 276. Owen is translating directly from the Latin in this quotation.

52–23 Owen's reference appears to be to Jean Jacques M. Coste and J. M. Delpech, "Recherches sur la formation des embryons," *Annales des sciences naturelles* **28** (1833): 158–80.

54–2 Owen seems to be alluding here to the functional classification of the Hunterian museum.

55–3 W. F. Edwards, *De l'Influence des agens physiques sur la vie* (Paris, 1824). A copy is at the RCS.

56–1 J. Hunter, "Experiments and

Observations on Animals," *Hunter's Works* 4:137. This is a synthesis by Owen of Hunter's two treatises in the *Philosophical Transactions* for 1775 and 1777. See references above, 49–6.

56–9 Hunter, "Some Observations on Digestion," *Hunter's Works* 4: 81–121; On Spallanzani and Réaumur, see above, note 27–22.

58–6 Ibid., 4:92.

58–10 For comments on Magendie, see Owen's notes to ibid., pp. 91–2.

59–7 Noguez, *Anatomie.* See also Owen's comments in *Hunter's Works* 4:308. Clift has consistently spelled his name as Noquez.

61–9 J. Hunter, "Of Absorption by Veins," *Medical Commentaries* Part I, reprinted in *Hunter's Works* 4: 229–314, esp. pp. 308–9. "Mr. Hunter" refers to John's brother William.

62–2 William Cruikshank, *The Anatomy of the Absorbing Vessels in the Human Body* (London, 1786). A copy is at the RCS.

62–12 Alexander Monro (*secundus*), *Observations Anatomical and Physiological* (Edinburgh: 1758), p. 57. This is also quoted in Owen's note to Hunter's, "Of Absorption," *Hunter's Works* 4: 308n. Owen has inserted some slight grammatical changes into this text.

63–4 William Hewson, *Experiments on the Blood, with some Remarks on its Morbid Appearances* (London, 1771), reprinted from *Philosophical Transactions of the Royal Society* 1768. See also Owen, *Hunter's Works* 4:308n.

65–2 Compare this passage with Owen's account in ibid., 309n.

66–4 This again reflects Owen's low opinion of Magendie. See Clift's comments on 66–18.

66–10 J. Hunter, "Of Absorption," *Hunter's Works* 4:303. There are no italics in the printed version being quoted.

66–16 François Magendie, *Precis élémentaire de physiologie* 3d. ed. 2 vols. (Paris: Mequinon-Marvis, 1833),

2:199, 201. This is quoted by Owen in his note to Hunter's "Of Absorption," p. 310n.

67–3 Ibid. The phrase 'having exposed... transparent.' is a quotation from Hunter.

69–11 J. Hunter, "Observations on the Structure and Oeconomy of Whales," *Philosophical Transactions of the Royal Society* 77 (1787), reprinted in *Hunter's Works* 4:331–92. Reference is to p. 373.

69–15 J. Hunter, "A Description of the Nerves which Supply the Organ of Smelling," ibid. 4:187–94.

71–8 Ibid., 187. Slight punctuation changes have been made from the original and a long sentence is silently omitted by Owen.

72–12 Hunter, "A Description of Some Branches of the Fifth Pair of Nerves," ibid. 4:94.

73–10 Hunter, "Organs of Smelling," ibid. 4:191n.

74–1 I have been unable to trace this quotation. The same general point is made in "Lectures on Surgery," chp. 4, *Hunter's Works* 1:262.

74–8 This allusion is to Magendie's criticisms of Hunter in Magendie's *Journal de physiologie expérimentale* **4**. See Owen's note in "Organs of Smelling", *Hunter's Works* 203 4:190.

75–5 Hunter, "An Account of the Gymnotus Electricus,"*Philosophical Transactions of the Royal Society* **65** (1775) in *Works* 4:414–21. Quote on p. 419.

75–8 See reference to Cuvier in Lecture 3, note 39–15.

75–13 J. Hunter, "An Account of the Organ of Hearing in Fishes," *Philosophical Transactions of the Royal Society* 72 (1782), 379, reprinted in *Hunter's Works* 4:292–4.

75–16 G. Cuvier, "Mémoire sur les céphalopodes," *Mémoires pour servir a l' histoire et a l' anatomie des mollusques,* (Paris : Deterville, 1817), p. 2; Antonio Scarpa, *Anatomicarum annotationem liber secundus: de*

organo olfactus praecipio deque nervis nasalibus interioribus (Ticini, 1785).
76–2 Hunter, "Oeconomy of Whales," *Works* 4:331–92.
76–8 A. J. L. Jourdain, "Preface" to Carl Gustav Carus, *Traité élémentaire Jourdain* (Paris: Baillière, 1835) 1:xxxi. These comments are quoted by Owen in his note to J. Hunter, "An Account of Certain Receptacles of Air in Birds, which Communicate with the Lungs and Eustachian Tube," *Hunter's Works* 4:177n.
76–11 Ibid. The original discussion appeared in Peter Camper's *Verhandeling van Bataafsche Genootschte*, (Rotterdam, 1774), cited in Owen's note to ibid., p. 176n. The French edition of this, probably the text being used by Owen, appeared as "Mémoire sur la structure des os dans les oiseaux," in *Mémoires de mathématique et physique* **7** (1776): 328–35. Also cited in this note. A copy of this is at the RCS.
77–7 Owen cites the 1653 English translation of Harvey, *On the Generation of Animals* (London, 1653), p. 7 elsewhere in his notes to "Certain Receptacles," *Hunter's Works* 4: 176n.
78–1 Hunter had not published on this issue and is being cited from the manuscripts. Reference is to Gaspard J. Martin Saint-Ange (1803–1888), probably to his *Analytic Anatomy: The Circulation of the Blood as Found in the Human Foetus, and Compared with that of the Four Classes of the Vertebral Animals*, trans. T. W. Jones (London, 1833). A copy of this is at the RCS.
78–17 Johannes Müller, *De glandularum secernentium structura penitiori earumque prima formatione* (Lipsiae, 1830), p. 95. A copy of this is at the RCS and is cited again in *DIC* 2 (1834), p. viiin and in Owen's preface to *Hunter's Works* 4:xxiv. Müller's report of Meckel's visit is on p. 95. J. F. Meckel had been in England during his travels of 1805–6. He was

offered the first professorship of comparative anatomy at the new University of London in 1827, but declined the offer. His refusal led to Robert Grant's appointment. See Desmond, *Politics*, pp. 82–4.
79–2 This does not seem to be a reference to a published work of Hunter's, but is probably derived from the displays in the collection. Owen makes the same comment in his preface to *Hunter's Works*, 4:xxiv.
79–4 Henri Ducrotay de Blainville, *Cours de physiologie générale et comparée*, 3 vols. (Paris, 1829–33). A copy of this is at the RCS.
80–10 Hunter's manuscripts on the development of the fetus and egg were published by Owen and Clift in the *DIC* (1841) 5: viii–xxiii. This volume was in the early stages of preparation as these lectures were being composed. The plates to accompany these manuscripts were originally prepared by Hunter's illustrator William Bell.
81–10 Owen's close study of Hunter's detailed writings on generation provided him with an independent source of discussions of developmental embryology in addition to those he absorbed from the German tradition represented by Von Baer or Martin Barry. It is probably Owen's study of Hunter's manuscripts in preparation for the *DIC* that led him to the work of Von Baer.
81–13 See Hunter's manuscript discussion in *DIC* 5:xxv.
82–4 Ibid., pp. xxx–xxxi.
83–6 Ibid., pp. xiv–xv.
83–11 See Hunter manuscript quote in *DIC* 5: xiv: "If we were capable of following the progress of increase of the number of the parts of the most perfect animal, as they first formed in succession, from the very first to its state of full perfection, we should probably be able to compare it with some one of the incomplete animals themselves, of every order of animals in the Creation, being at no stage different from some of the inferior

orders." Owen notes that the same
point is made by Hunter in his earlier
Animal Oeconomy, reprinted in
Hunter's Works 4:268. See below,
note 98–2.

84–10 I have been unable to locate the
source of this quotation. It may be an
unpublished Hunter manuscript.

85–15 Hunter, *A Treatise on the Animal
Oeconomy*, in *Works* 4:6–7.

86–10 Isidore Geoffroy St. Hilaire,
*Histoire générale et particulière des
anomalies de l'organization chex
l'homme et les animaux, ou
Traité de teratologie* (Paris, 1832).
This passage is quoted in French by
Owen in *Hunter's Works* 4:45n.

86–22 I have been unable to trace this
quotation. It is probably from a
Hunterian manuscript.

89–4 Owen's allusion is to the second
division of the collection on
reproductive organs and the process of
generation. See schema of the
Hunterian series in Introduction, Table
One. Owen was himself begining
work on this material as these lectures
were being written.

89–16 All of the named individuals had
paid visits to the Hunterian Museum.

90–9 Giuseppe-Saverio Poli (1746–1825),
*Testacea utriusque Siciliae eorumque
historia et anatome* 2 vols. (Parme,
1791–5). Owen's discussion in this
section of the lectures is reprinted
almost verbatim in his preface to
Hunter's Works 4:xxxvii-xxxix.

91–3 Johann F. Blumenbach, *Handbuch
der vergleichenden Anatomie*, 3rd. ed.
(Göttingen, 1824). The RCS has
Owen's annotated copy of the English
translation of the *Handbuch* by
William Lawrence (*A Short System of
Comparative Anatomy* 2nd. ed., trans.
W. Lawrence [London, 1827]).

93–5 At this point the new insertion of
material commences, probably added in
April of 1837. The original RCS
draft reads from this point to the end:
"*To enumerate their names alone
would occupy much time, had it even
formed part of my present design to

have pursued the progress of
Comparative anatomy to the present
day {*all crossed out* }To enumerate
their names *even {*interlined*}would
occupy much time— To have dwelt on
their *labours & Discoveries
{*interlined*} on the present occasion
would be almost superfluous once
their scope & tendency has been *so
often already{*separately deleted*}in the
Metropolis so frequently & {*all interlined*} so luminously explained, by the
learned and eloquent Professor of Comparative Anatomy at the London University—whose *beautiful&{*interlined*}original discoveries *in the more
recondite paths of Comp. Anat.
{*interlined with caret*}and *whose
{*interlined with caret*}unwearied
endeavours to diffuse the knowledge of
the Science which he adorns, will
place him in a high & honourable
station in it's history—" RCS Owen
MSS 67. g. 12 fol. 74. The reference
is to Robert Grant, holder of the
comparative anatomy position at
London University. This passage
should indicate with some decisiveness
that Owen and Grant were on good
terms at this time.

94–11 On this distinction see Lecture
Two, note 65–4 above, p. 133.

95–16 Owen's arguments should be read
carefully. The denial of a *single* unity
of plan, in the form claimed by
Etienne Geoffroy St. Hilaire and
attacked by Cuvier, still leaves open
the option of a more limited group
unity in terms of several different
plans, which was to be Owen's mature
view. The organization of the
Hunterian series must also be
considered in the origins of Owen's
views. This collection on one hand
suggested relations of homology
deeper than Cuvier's functionalism
through its displays of isolated organ
systems from several different animals
together. The attraction of Owen to a
branching arrangment is directly related
to this, and the same schema was
adopted by Joseph Henry Green

earlier. See Introduction, pp. 37 ff.
96–8 See especially Cuvier's
"Considérations sur les mollusques, et
en particulier sur les céphalopodes,"
Annales des sciences naturelles **19**
(1830), 241–59. See also Lecture
Two, fol. 59.
96–12 This displays Owen's decision to
utilize embryonic, rather than mature,
characters as the key to classification.
Again, his departure from Cuvier at
this early date is clearly in evidence.
97–4 Friederich Tiedemann, *The Anatomy
of the Foetal Brain* trans. W. Bennett
(Edinburgh, 1827); Johann Purkinje,
*Symboli ad ovi avium historiam ante
incubationem* (Lipsiae, 1830); Karl
Ernst von Baer, *Die Entwicklungsgeschichte der Thiere* (Königsberg,
1828); Heinrich Rathke,
*Abhandlungen zur Bildungs-und
Entwicklungsgeschichte des Menschen
und der Thiere* 2 vols. (Leipsig, 1832-
33); Rudolph Wagner, *Prodromus
historiae generationis hominis atque
animalium* (Leipsig: Voss, 1836);
Gabriel G. Valentin, *Handbuch der
Entwicklungsgeschichte des Menschen
mit vergleichenden Rücksichts der
Entwicklungs des Saügethiere und
Vögel* (Berlin, 1835); Johannes
Müller, "Ueber die Entwicklung der
Eier im Eierstock bei den
Gespenstherschrecken," *Nova Acta
Physico-Medico Academia Caesarea
Leopoldina-Carolina Naturae
Curiosorum* **12** (1825), 555–672.
Copies of each of these works are in
the RCS library.
98–2 See quote above, 83–11. Owen is
referring here to his own general
sympathy with Hunter's claim that the

fetus represents imperfect stages of
higher stages of development. Owen
admits in this passage a mild form of
recapitulationism in the sense of Von
Baer while rejecting the Meckel-Serres
variety.
98–7 Owen's targets in this critique
would appear to be Johann Spix and
Etienne Serres. See comments in his
later "Report On the Archetype and
Homologies of the Vertebrate
Skeleton," *Report of the British
Association* **16** (1846), pp. 244 ff.
99–2 The linkage of these two issues was,
of course, the principal thesis of the
Meckel-Serres law. Owen's
opposition to transmutation is here
given its clearest statement. See also
Lecture Five, fols. 34–5.
100–10 Reference is probably to
Hermann Burmeister, *Beiträge zur
Naturgeschichte der Rankenfüssen
(Cirrhipedia)* (Berlin, 1834). A copy of
this is at the RCS.
100–20 Reference is to Astley Cooper,
the current president of the College.
100–22 He is referring to Christian
Ehrenberg's Habilitationsschrift, *De
globularum sanguinis usu* (1833).
This is attested by a quotation in an
unpublished manuscript: "Ehrenberg
has recently proposed a new and very
ingenious theory of the use of the
Thymus Gland founded on the
observation of the close resemblance
not to say identity, of the globules in
the brain, retina, &c. with the nuclei
of the blood-disks, and the globules in
the thymus gland." Richard Owen
manuscript (undated, 1832 WM), RCS
Owen Papers, "Miscellanea," MS box
3(17).

Lecture Five
Analysis of the Manuscript

The manuscript utilized for the transcription of this lecture is located at the BMNH (O. C. 38). This bears on the cover page "Lecture III, V" with "May 11th" then written in by Owen. The opening page of the lecture is entitled "Lecture III. Organized Beings: Nature and Characters of." This manuscript was originally intended as the third lecture in the series, but was moved into its position of the fifth by the late splitting of the final manuscripts of Lectures One and Two.

In view of the important connection in theoretical content between this lecture and the late revisions added to the end of Lecture Four of 9 May, it is to be observed that Lecture Five is written entirely on the same "J Bune 1832" paper in folded sheets used in the body of previous four lectures except for the ending of Lecture Four. The constancy of the paper type throughout this lecture is further evidence that the final section of Lecture Four was inserted very late, contemporaneous with or after the final copy of Lecture Six.

The holograph original of this lecture is located at the RCS as Owen MS 67. b. 12. E. This is drafted on "J. Green & Son 1836" paper measuring 21 x 26 cm. The draft has been heavily reworked in sections with new insertions written on the rear of facing pages. As discussed in the Introduction, the holograph manuscript shows an insertion of several pages at folio twenty, corresponding to final manuscript pages thirty-four to forty-four transcribed below. Since this lec-ture was originally the third lecture, as indicated by the revision of the title, this new material would appear to form the section rewritten by Richard Owen on the evening of April 19, as recorded by Caroline Owen in her diary, following the Geological Society meeting at which he first presented his conclusions on the *Toxodon platensis* skull. The reflections within this section on the life and death of species and other issues generally bearing on the issues of the duration of forms has a special interest in light of this context. For further discussion of this see Introductory Essay, page 58.

Fig. 12. Lithograph Of Johannes Müller By P. Rohrbach, 1858, After Photograph By S. Friedlander, 1857. (*Courtesy Wellcome Institute Library, London*)

Lecture Five
11 May 1837

1   Mr. President,

Having in previous Lectures endeavoured to define the point, in the course of Science, from which Hunter commenced his investigations in the Physiology of Organized beings,[1] I now proceed to the proper business of this course,[2] the eludication of the Physiological Series of Comparative Anatomy in the Gallery of the Museum.[3] The great Founder of this Collection teaches us that before entering upon the subject of[4] the Comparative Organization of Plants and Animals;— it is necessary, in order to acquire just notions of the general nature of those beings, to examine and compare them with reference to the other objects of the Material World; and to determine their essential characters.[5]

The terraqueous Globe, and all that it supports, are endowed with properties which impress our Senses, and our Senses are expressly designed to receive those impressions.

If we can conceive the first-created Rational Being to have speculated on the first act of Sensation which revealed to him his own existence and the World around him, it might be questioned whether he would first have considered the nature of the things impressing, or that of the subject impressed:—
2   but as the Sensorium appears to be passive while receiving an impression, the reflective power which is intimately connected with the recipient of sensible impressions would be led from obvious causation to consider the matter impressing.[6]

---

1   beings,] *before interlined erasure of pencil* 'Mr. President'.
2   course,] *comma inserted [ed.]*.
3   Museum] *inserted before deletion of* 'Before entering upon the immediate'.
4   The great. . .subject of] *interlined above and below deleted* 'subject of this Course,'.
     Entire passage 'Mr. President. . .the subject of' *written in over erased section*.
5   ; and . . .characters.] *inserted over erasure*.
6   but. . .impressing.] *with note on facing fol. 2 by W. S. Bennett* 'I think it was the blind youth operated on by Cheselden who thought at first that all objects *touched* his eyes and for a long time had no ideas of comparative distance or size, visually, saying that he

207

Mr. Hunter defines Matter to be "everything that is capable of making such impression as to give us sensation." But we receive sensitive impressions from within by the acts of thinking, of reflecting, of understanding, and of willing;—and these acts are, by common consent, referred to a principle distinct from matter; and called Mind, or Spirit: and as the right Contemplation of any object of the Universe leads to the conviction that it bears the impress of thought, whosoever[1] signifies the principle which thinks, knows, wills, &c by the term Spirit; must, while investigating the properties of Matter, be irresistibly led, (with Newton) to acknowledge the existence of Infinite Spirit, with Infinite Power and Wisdom. Material objects (thus understood) were ear-

3   ly divided into the Animal, Vegetable, and Mineral [/] kingdoms, characterized according to the ideas which were most obviously excited by each:—the first, as enjoying[2] Life and Sensation:—the second as having[3] Life without Sensation;—and the third, by the absence of both; the last,[4] in a comparative sense were termed *inert*.

The Founder of the Collection observing that Animals and Vegetables had many properties in common considered them as members of one grand division of material objects, now understood by the term *Organic*.[5] He attracts our attention to this division by placing at the commencement of the Physiological Series a few simple examples of Plants and Animals: and we find from the notes which have been preserved of his Lectures[6], that he was accustomed to commence his Course by taking a general view of the distinguishing properties of common or inorganic, and organic matter. He said that

> Animal and Vegetable matters were either such Modifications of common matter as to differ from it in many instances, or derived new properties from the superaddition of a peculiar principle. Thus both Plants and Animals
4 possessed action within themselves, and a power of deriving a supply [/] of their own particular matter from other substances:—they were both capable of reproducing their kind;—and both presented two different states, the Living, and the dead.— The living, while possessed of the Living Principles; the dead when[7] deprived of it, and when they returned to common matter.[8]

(contd.)
    knew the room he was in was but part of the house but he wd not conceive that the whole house wd look bigger. W. S. Bt '.
1 whosoever] 'so' *interlined with caret.*
2 as enjoying] *in darker ink as with other insertions on page.*
3 as having] *interlined in dark ink with caret after* 'second'.
4 last,] *over erasure.*
5 *Organic.*] *replacing comma by period [ed.].*
6 Lectures] *after deletion of* 'celebrated'.
7 when] *in margin in pencil query by WC* 'then?' *and reply by Owen* 'no I think when'.
8 matter] *period inserted [ed.].*

from which, therefore; Hunter inferred , "they must have originated."

Now it is true that the ultimate elements of organized matter are precisely those which enter into the composition of Unorganized substances: But by the operation of a power, distinct from Gravitation, molecular attraction, or any of the known imponderable agents which operate on unorganized substances, these elements assume combinations of a character essentially different from those which are the result of ordinary chemical affinities.

The proximate components of Organic bodies[1] of which Sugar, Oil, and Albumen are familiar examples, cannot be formed artificially by directly combining their ultimate elements; but they can be resolved by chemical Analysis into those Elements, which are also the [/] elements of inorganic substances, and they invariably are so resolved or decomposed, when withdrawn from the influence of the power by which these Elements were first organically combined.

Modern Chemistry has detected about twenty conventional elementary substances in organized bodies but not one of these is peculiar to this Division of Nature: while of the 52 Elementary or simple undecompounded[2] substances met with in the mineral or unorganized World, the greater part never enter into Organic combinations, and some are even noxious to Life.

Hence there is a marked difference in the number of the Elements which constitute Organized and Common matter.[3]

Vegetable matter consists principally of Carbon, Oxygen and Hydrogen:— a fourth element is added in Animal matter, viz. Nitrogen.

These may be termed the Essential Elements of Organized Beings. Nitrogen is, however, present in a small proportion in some of the lower organized Plants.

The other incidental Elements in Vegetables are [4]

Phosphorus and Sulphur,	in Vegetable albumen and Gum.
Potassium,	generally
Sodium,	common in Marine Plants.
Calcium,	very generally.[5]
Alum,	rarely.
Silex & Magnium,	sparingly.
Iron and Manganium,	abundantly.
Chlorium, Iodium, & Bromium,	in Marine Plants.

The Incidental Elements in Animal Bodies, are:

1 bodies] *in darker ink over pencil with faint pencil* 'X' *interlined below and in margin pencil* '?X'.
2 undecompounded] *interlined above* 'simple'.
3 matter.] *followed by faint pencil line to indicate pause and in margin in pencil* 'Table'.
4 *The. . .are*] *underlined in pencil.*
5 generally] *on facing fol. 7v in pencil* 'mineral omni calx, omnis silex, omni ferrum ab animali'.

Sulphur	In Hair, Brain, Albumen.
Phosphorus	in Bones, Teeth, Brain.
Calcium	in Bones and Teeth.
Sodium	abundantly.
Potassium	in less proportion than in
	Plants.
Silicium and Manganium	in the Hair.
Iron	in the Blood, the black
	pigment, and Crystalline[1]
	lens
Iodium and Bromium,	in a few Marine Animals

Of these incidental Elements of Animals, Phosphorus and Calcium exist in greatest abundance. Some Chemists, as Fourcroy, and Berzelius, consider that there is an essential difference between Organized and Unorganized [/] substances in the mode of combination of the Elementary Principles.

In inorganic Nature all bodies are either simple or the result of binary combinations; either of one Element with another Element;— or of a simple body with a binary compound:—or of one binary compound with another binary compound.

Thus, Carbonate of Ammonia is a binary compound of Carbonic Acid Gas, and Ammonium.

Carbonic Acid Gas is a binary compound of Oxygen and Carbon.

Ammonium of Hydrogen and Nitrogen.[2]

An immediate combination of 3 or 4 or more substances in which all the Elements are alike united together, appears to be capable of taking place only under the influence of Animal and Vegetable Life: and the same elements which form the Volatile Salt above instanced, do, when controlled by the organizing energy, and combined in ternary and quaternary groups, constitute the greater part of all the diversified forms of proximate organized materials.

Thus, the ternary combinations of Carbon Oxygen, and Hydrogen form Vegetable Mucus, Sugar, Starch,[3] Oil. The quaternary combinations resulting from the addition of the fourth Element, Nitrogen, form Gum, Albumen, Fibrine, Animal Mucus, Casein, &c.[4]

---

1 Crystalline lens] *with pencil bracket in margin.*
2 Nitrogen] *before* 'In representing this Decomposition we produce, as it were a Dichotomy.' *with pencil deletion of line and pencilled* '? No, I don't like this expression' *in margin.*
3 Starch,] *comma inserted [ed.].*
4 &c.] *followed by vertical pencil scoring to delete following:* 'Some Chemists have attempted to reconcile the difference by supposing, that, in the *so called* (*interlined*) ternary Compounds, one of the Elements combines with a base formed of the union of the other two. Thus, Prout, observing that the hydrogen and oxygen in Sugar are exactly in the proportion to each other in which they form water, infers that these two Elements

9   The processes by which the Elements are united to form the proximate
principles of Organic bodies, are of such a peculiar nature, as to have hitherto
baffled the art of Chemistry to imitate. The parts of organized bodies may be
resolved[1] into their ultimate Elements by destructive analysis, yet the organic
agent does not change the properties of the Elements, but simply combines
them in modes which we cannot imitate.

M[r]. Abernethy used to say that he defied all the Chemists in the World to
manufacture[2] an excrement:!—Berzelius, however, assures us that *Urea* can
be produced synthetically by artificial Chemical processes:—and Wöhler has
discovered that by saturating the liquor Ammonia with Cyanogen Gas, much
Oxalic acid is formed, and may be separated from the fluid.

10   A substance bearing a remote affinity to Fat [/] has also been produced; but
these are the nearest approaches which Chemistry has yet made towards the
production of the Organic Compounds;—and we can hardly consider Urea in
that light, since, like Carbonic Acid, it is a pure Excretion.[3]

It is remarkable that Organized Bodies consist chiefly of Combustible
substances. Man, who alone has the privilege to form and maintain an artificial fire, is almost wholly dependant for fuel on the remains of Vegetables;—
and the seams of Coal which form the great source of our national wealth are
the partially decomposed and fossilized parts of Plants.[4]

Their combustible basis remains by virtue of the following state of their
Chemical composition. They contain Oxygen, Hydrogen, and Carbon in such
proportions that the Oxygen is never sufficient to convert the intire hydrogen
into Water and the whole of the carbon into carbonic acid.

The elementary combinations characteristic of Organized bodies, are only
maintained during Life. The elements, or binary compounds which operate
around organic bodies, appear, indeed, even during life, in extreme cases[5]

(contd.)
     are really so associated with each other; and that the Organic product, Sugar, is not a
     ternary compound of Oxygen, Hydrogen, and Carbon, but a binary compound of Water
     and Carbon, or a Hydrate of Carbon. [¶] Scharlan, on the other hand, assumes that the
     radical or base of Sugar, and Starch, and Alcohol, is a binary compound of Carbon and
     Hydrogen; and that these proximate vegetable principles result from a binary
     combination of the bi-carbonate of Hydrogen with Oxygen. But these attempts at a more
     simple [l *fol. 8*] explanation of Organized Combinations are only partially applicable:
     and we are compelled to admit, that though the Elements which form unorganized bodies
     are more numerous, the combinations into which they enter, are less complex than those
     which they form in the constitution of Living beings.'.
1   The. . .resolved] *inserted in dark ink over erasure.*
2   manufacture] *interlined above erasure of* 'form'*with caret and on facing fol. 10*[v] *in
     pencil in WC hand* '*Manufacture* a surreverence!' *with RO reply* 'Manufacture is the
     word[.] 'form' and 'produce' may apply to the natural production—'.
3   Excretion.] *on facing fol. 11 in RO pencil* 'at all events I wd not say calcined'.
4   Plants.] *period inserted [ed.].*
5   cases] *pencilled* 'X' *in margin.*

to have the power of disturbing or destroying[1] the equilibrium of the elements which are in organic combination, as in the burning and the freezing of the parts of a living body. In the first act, the parts are killed and irreparably decomposed: in the case of freezing, as it is only the fluids of the part which are thrown into a state of inorganic Crystallization; these fluids[2] if killed by the act, are removed by the restoration of circulation upon thawing the part, and, as fresh living fluids are supplied, the solids revive and the part recovers:—as the Experiments of Hunter on the Ears of Rabbits, &c. demonstrate.

With these apparent exceptions, all external inorganic decomposing forces are resisted[3], and overcome during life. When a vital organism no longer retains the power of resisting the tendency of its elements to enter into binary combinations, it then reverts[4], and is resolved into inorganic substances.

In ordinary experience the change is gradual and at first we doubtfully recognize death by the cessation of the vital movements; or in a highly organized species by the gradual dissipation of its caloric. [/] But the Naturalist who has enjoyed the opportunity of witnessing the energy with which the vital phenomena are manifested in certain Tropical Marine Animals, fully appreciates the force by which the surrounding active decomposing powers were resisted. At one moment he sees the brilliant Glaucus gaily inflecting its numerous painted arms, moving along the surface of the waters, seizing and devouring its prey;—in a moment its movements have ceased; it is dead;—its colours immediately[5] fade away, its form is lost, and the decomposing gelatinous body yields, and is torn by[6] the slightest touch;—defeating the most prompt endeavour to preserve it entire.[7]

The resolution of the organized body into inorganic compounds, and the dispersion of its elementary gases to form new combinations, occasion[8] the phenomena of fermentation and putrefaction; and the decomposing processes are offensive, in proportion to the quantity of Nitrogen which entered into the composition of the dead organized body.

All experience shows that the combinations of unorganized bodies depend on elective affinity [/] which is a property of the constituent elements, whilst the combining and conserving power in organized bodies is evidently not a property of the elementary parts. All the known imponderable matters, as light, heat; and electricity, influence both the composition and decomposition of organized and unorganized bodies: but nothing authorizes us to regard any

12

13

---

1 or destroying] _pencil underline of_ 'or destroy'.
2 fluids] _interlined with caret._
3 resisted, and] _interlined with caret after_ 'are'.
4 reverts] _before ink deletion of_ 'back'.
5 immediately] _interlined with caret after_ 'colours'.
6 is torn by] _inserted over erasure._
7 endeavour. . .entire.] 'endeavour' _inserted over erasure and_ 'its preservation'_deleted with_ 'preserve. . .entire.' _then inserted._
8 occasion] _pencil deletion of_ 's' _from_ 'occasions'.

one of these agencies as the ultimate efficient Cause of the activity and conservative power of the living organized matter.

By[1] means of Heat, many wonderful things may be accomplished, but heat will not act of itself. The powers of Electricity are still more wonderful than the powers of heat; but electricity we know to be governed in its mode of action by certain laws; and we know that neither Electricity nor Magnetism give any sign of Intelligence.

In the same manner, Life, as we are acquainted with it, cannot exist without motion, but motion may exist without Life. Life and Motion consequently are not synonymous terms, nor can we conceive the existence of motion without a [/] mover. The organizing principle is therefore something different from, and superadded to the common agencies of matter, over which to a certain extent it exercises a control.

Organized bodies not only differ from unorganized, in the according to Berzelius[2] ternary and quaternary nature of the[3] combination of the Elements which form their proximate constituents, but they present a very striking and peculiar character, in the laws by which the combining agent operates in the arrangement of those constituents, so as to form the individual animal or plant;—or in other words the organized whole.

We perceive design[4] and intelligence in the modes in which the fibrine, albumen, gelatine, &c, are modified and arranged, so as to form distinct organs or parts; and we see that all these parts are subordinate to the scope and exigences of the whole, and that the whole is, to a certain extent reciprocally dependent on the integrity of its parts:—A brain, a heart, or[5] a stomach have no independent existence; they have been formed with reference to the whole organized body;—remove any one of [/] these[6], and the body becomes a dead mass; every part of which, as it were, withdraws into itself.

So also when a part is removed from the whole, it generally perishes. Unorganized bodies on the contrary may be subdivided indefinitely, without any of the parts thereby losing the Chemical properties which characterized the whole.

An organized body is not, however,[7] as Kant defines it, absolutely indivisible. In the more simple species of Plants and animals, the organs essential to life instead of being concentrated in definite spaces and under peculiar forms, for the purpose of vigorous and consecutaneous[8] action, are diffused in a greater or less degree throughout the whole. Hence, a part of the body containing in a certain proportion the essential organs of Life, may in these

---

1 By] *deleting pencil double quote mark heading line [ed.]*.
2 according to Berzelius] *interlined with caret before* 'ternary'.
3 nature of the] *interlined with caret after* 'quaternary'.
4 design] *after pencil deletion of* 'a'.
5 or] *interlined with caret.*
6 of these] *on page break with* 'of these' *below line and* 'these' *repeated top of fol. 14.*
7 however] *replacing for* 'hower' *[ed.]*.
8 and consecutaneous] *interlined with caret after* 'vigorous'.

species[1] be separated, and yet retain its powers and characters as an organized whole.

Thus a branch of a Tree when detached and transplanted, may send out roots, and develop its buds into reproductive flowers. The Experiments of 16 Trembley and Rœsel have shown us to what an [/] unexpected extent the homogeneous gelatinous body of the Polype may be subdivided, and each subdivision live, and grow, and propagate. And in the lower articulate animals, where each joint contains the essential organs of the whole as in the Naids, one or more may be separated and may live. It is in animals of this simple character that the phænomena of spontaneous separation or fissiparous generation are particularly[2] manifested: and it must be borne in mind, that with reference to the divisibility of an organized whole,[3] the separated parts when they survive, tend uniformly to rise from the subordinate character, and assume the form and nature of individuals or wholes equal to that from which they were originally disjoined.

In general terms, therefore, we may say that organized bodies are distinguished from unorganized by their indivisibility:—by their incapacity to suffer division with a continuance of their characteristic properties in the separated portions, and hence it is only in Organic Nature that individuals can 17 be said to exist. Some [/] however, from observing that Crystals are only divisible in certain directions, and that the parts which result from this cleavage, as it is technically termed, are often different from the whole, regard these substances as Individuals, observing that their properties, as Crystals, are lost by indefinite division. But if we consider Crystals as Individuals in this sense, there is this great difference between them and organized bodies,—that the component Molecules are similar throughout the whole Crystal; whilst the particles which make up the Organized individual are different both as to structure and properties. And, again, in those unorganized bodies which do consist of an aggregate or commixture of heterogeneous parts, the adaptation of the different parts to the Scope of the whole (a relation which characterizes Organized bodies,) is wanting.

The Matter of organized substances is either in a solid, a fluid, or a gaseous state;—it is never found in two of these conditions, each having a 18 subserviency to the existence of the other [/] in the same being. Organized beings, on the other hand, consist invariably of reciprocally dependent[4] solid and fluid parts: thus[5] whilst the vegetable has its woody fibre, it has its sap also; and animals with their firmer bones, their muscles and cellular tissue, have likewise blood circulating through their bodies, and various fluids de-

---

1 in. . .species] *interlined with caret after* 'may'.
2 are particularly] *interlined above deletion of* 'is'; *and in margin a pencil* 'X'.
3 with. . .whole,] *interlined with caret after* 'that'.
4 reciprocally dependent] *interlined with caret.*
5 thus] *interlined above* 'whilst'.

posited within their tissues, which are just as essential to their constitution and continuance, as the containing parts themselves.

It were indifferent whether we took away the solids, (were such a thing possible,) or the fluids of a vegetable or animal;—in either case, the Individual must perish, or in a few rare exceptions, as the Rotifer & Vibrio, revert to a passive condition of life[1].

From this admixture of Solids and Fluids in organized bodies, results the variety of consistence which they present. Pliancy and softness are their general characteristics; and it is interesting to observe that these qualities are gradually diminished as the organism approaches the term of its vital existence, and as it is about to merge into that inorganic kingdom in which Solidity or hardness is the term of Perfection.

19   The only circumstance, perhaps, in the comparison of Organized and Unorganized bodies, in which they are found to correspond with one another, is the mode in which conformity or symmetry of parts is produced in both. Crystals, e.g. have symmetrical and unsymmetrical surfaces, angles, points. Animals have also symmetrical and unsymmetrical parts, which may be compared[2] to those of crystals.

The original of every Animal is a circular flattened disc, called the Cicatricula of the Egg; or the Germ-disc, which prior to the expulsion of the Ovum from the Ovary appears in the form of a vesicle:—but from this simple figure the Embryos of various Species of Organic beings quickly deviate and ultimately present a diversity of Forms.—

We distinguish, e.g. a radiated symmetrical type in the Star-fish or Medusa, where a Series of corresponding parts are arranged round a common centre, with a want of symmetry between the upper and under surfaces.

We perceive a ramified symmetrical type in many Plants and Zoophytes, 20   where the leaves and [/] Polypes are repeated symmetrically on the branches of the Tree, or Polypary.

Again, we have a *consecutive symmetry* presented in the repetition of like parts, from before backwards, as in Worms; which, however, have their upper and lower surfaces unsymmetrical.

And, lastly, we have a *bi-lateral symmetry* in the repetition of like parts on each side of a mesial plane as in the higher animals and Man:—with a want of symmetry in the parts which succeed one another, as well as in those on the dorsal and ventral surfaces.

In some animals, the *consecutive* is combined with the *bilateral* symmetry, as in the Centipede; and we perceive a trace of the same combination in the Vertebræ of the higher animals.

---

1 or. . .life] *inserted with* 'as. . .life' *interlined.*
2 which. . .compared] *interlined to replace deletion of* 'and these *likewise { *interlined with caret after* 'these'} present similar manifold modifications analogous'; *in margin in pencil*'??X which may be compared to'.

The symmetry or dis-symmetry of Crystals differ[1] from those of Organized Bodies, in being expressed by flat surfaces and straight lines.—It is only in the Crystallized state that inorganic[2] can be compared with Organic bodies, with respect to *form*. In a state of Nature the masses of common mat-
21   ter generally present [/] an indeterminate shape, and when reduced artificially to their pure constituent parts, the determinate figures are few and simple:— the cube, the hexahedron, the rhomb, the prism, are the elementary forms of the Inorganic World. Plane surfaces, and straight lines, uniting at angles that measure certain determinate numbers of degrees, are the accidents which give to Crystals their characteristic and individual shape.

With Organized bodies it is far otherwise:— The law of co-relation or of adaptation to an end, by which all the processes of the development of an organism are regulated necessarily impresses upon it a determinate and characteristic form; while the harmony subsisting between the form of the individual,—the functions it is destined to fulfil;—in short, the conditions under which it is to exist; causes endless diversity of specific shapes in the Organized kingdoms of Nature. Of the myriads of Animals and Vegetables which people and adorn the Earth no two Species are or can be[3] exactly similar.

But their forms, however they may otherwise differ,[4] are constant in this respect, that instead of being defined and circumscribed by straight lines, angles, and plane surfaces, they always present a more or less rounded or [/]
22   wavy contour; and this general form is not only the characteristic of the whole, but of each of its component parts:—the ultimate molecules possessing a distinct form into which an organized body can be resolved, are globular.

With the doubtful exception of some of the Earths which enter into the composition of the hard parts of Animals and Vegetables, none of the substances which form the living organized Being, exist in so pure a state as to be capable of assuming a regularly crystallized figure[5]. Some organized substances, however, although they do not form crystals in the living plant or animal, can yet, by artificial processes, after death, be so far separated from extraneous matter as to be obtained in a state of purity, and thus be made to assume the crystallized form:—e.g. Sugar, the pure Fats, Spermaceti, Urea:—while the greater part of organized solids and fluids cannot be made to crystallize under any circumstances.

---

1   differ] *on facing fol. 20ᵛ without insertion point* 'want of symmetry'.
2   inorganic] *after deletion of* 'the'.
3   or can be] *interlined with caret after* 'are'; *on facing fol. 22ᵛ in WC hand.* 'Else they would be *one* or *the same* Species??' *with RO response underlined* 'nullum simile est idem'.
4   differ,] *comma inserted [ed.].*
5   figure] *with pencil* 'X' *after* 'figure' *and WC note on facing fol. 23ᵛ* 'X Qu. The Enamel of the Teeth an exception?' *with RO reply* 'it is one of the earths—'.

With respect to the definite and symmetrical forms we have been consider-
23 ing, we can as little explain or [/] understand the causes which give rise to
them in crystals as we can foresee or comprehend the nature of those forces
which lay the axis of consecutive or bilateral symmetry in the germ of
Organized bodies; or which determine the differences of the dorsal and ventral surfaces in the same.

We admire, in the operation of the organizing energies the pre-established
harmony between the structures which are built up, and the purposes they are
afterwards to fulfil, in subserviency to the well-being of the whole:—But we
look in vain, either in the composition or mode of formation of Crystals for
any such manifestation of the subordinancy of a part[1] to the activity of the
whole.

A Crystal presents no diversity of elementary constitution in its different
parts: it is homogeneous; hence it is as complete in its parts as it is in its
mass:—The minutest spark of carbonate of lime has all the properties of a
crystal of this substance were it as large as a mountain. When a crystal is
formed in the midst of a fluid, the particles composing it unite in conformity
24 with the ordinary law [/] of cohesion and affinity, and each stage of the process of Crystallization is independent of that which preceded it. It grows by
superficial deposition upon the first formed part;—while the growth of an organized[2] body takes place by a simultaneous secretion throughout the whole
of its substance, and the several parts succeed each other according to a relation of the strictest causality. The radicle which bursts from the fecundated
seed of a Plant determines the growth of the Stem, which in its turn plays the
same part with reference to the leaves and flowers. The parts that appear first
are the cause of the appearance of those that follow.

The reference to the wants of the whole, which characterizes the processes
of Organization is not only manifested in the construction of intire organs, as
the leaf or flower,—the eye and ear,—but also in the disposition of the simplest elementary parts:—Thus a fibrous structure is given to muscle, because
a shortening of the part is to be produced by a contraction[3] of fibres: and the
fibres of a muscle are laid down precisely in that direction which, under
25 certain relations is best adapted to effect the destined motion of the part to
which they are attached. The same law is evident[4] in the physical condition of
the Nervous System, for without a division of the nerves into a certain
amount of non-communicating filaments, local innervation or local sensation
were impossible.

---

1   for. . .part] *with marginal query* '?X' *repeated for two lines.*
2   Crystallization. . .organized] *WC pencil note on facing fol. 25ᵛ:* 'Qu. Is this orthodox.
    Do not bones grow by deposit stratum-super-stratum—Madder Experiments &c. &c.',
    with RO pencil reply:' 'How do those experiments contravene the dicta advanced?'.
3   a contraction] *interlined above deletion of* 'an angular puckering'.
4   evident] *written over erasure and on facing fol. 26v WC query* '*observable*? *evident*?'
    with RO reply 'perhaps'.

There is no essential difference in the law of the Subordination of the parts
to the well-being of the whole, as it is manifested in Plants than as it is more
strikingly manifested in animals, but[1] the organs of Plants[2] are less dissimilar
and less numerous, though indefinitely subdivided and multiplied. Instead
also, of being hidden in the interior, they are dispersed over the surface: the
processes of interchange of gaseous elements with the external atmosphere
are not confined to a single point, but are effected by an endless multiplication
of superficial parts. Hence a great proportion of Botanical Terminology is devoted to the expression of the manifold diversities of form presented in the
leaves, leaflets[3] and leaf-stalk, composing the respiratory foliage of Plants.

26   It[4] is interesting to observe in the [/] lower animals a similar multiplication
of the same organ, yet the analogy of this condition to the aggregated cubes
or prisms which constitute the Crystal is very remote;—for the multiplied
spiculæ of the skeleton of the Sponge, the many stomachs of the Infusoria,
the hundred ovaries of the Tape-worm or the thousand mouths of the
Millepore,[5] are subservient to, and have the source of their existence in, the
individual organized whole.

Again, as the parts of an Organized body have co-relation to each other,
and to the well-being of the whole, so also the functions which maintain the
activity and existence of an organized body are reciprocally dependent upon
one-another.

Respiration maintains the activity of the Heart; the heart at each stroke,
transmits the arterialized blood to the brain, and the brain under this influence
animates every other organ, and is again subservient to respiration. Any ir-
reparable injury of one of these Mainsprings of the Machine or any considerable interruption in the activity of the Lungs, heart, or brain,[6] occasions[7]
Death:—whence they have been called the *atria mortis*.

I witnessed not long ago at the house of our President a beautiful and instructive experiment, as simple as it was conclusive in illustration of this law.
Sir Astley Cooper arrested, in a rabbit the flow of arterial blood to the brain
by compressing simultaneously the vertebral and carotid arteries; the wind[8]
pipe being left perfectly unobstructed. In a few seconds the animal ceased to
breathe—when released from the grasp it fell and lay apparently lifeless on

---

1   than. . .but] *RO insert from facing fol. 26ᵛ with insert point marked by caret after*
     'Plants' *and deleting repeated* 'but' *[ed.].*
2   the. . . Plants] 'the' *altered from* 'their' *and rest interlined with caret after* 'organs'.
3   leaflets] *terminal* 's' *inserted in pencil.*
4   It is] *inserting paragraph break for one-inch space inserted in last line of page [ed.]*
2   the Infusoria. . .Tape-worm] *RO insertion from facing 26v with* 'the thousand Millepore'
     interposed with pencil line from after 'Infusoria' *and deletion of* 'or' *after* 'Millepore'.
     Pencil insertion of 'or' *inserted after* 'worm'; 'of the Tape-worm' *following* 'the'*on fol.*
     26 also deleted.
6   brain,] *faint pencil caret follows.*
7   occasions] *pencil caret follows.*
8   The wind] *over* 'In a'.

the ground. The respiratory actions were then artificially produced by alternate pressure on the ribs—in a short time circulation was reestablished,[1] the brain received its accustomed stimulus, and sensation began to be[2] manifested. In half an hour the rabbit had completely recovered and ate. Sir Astley had determined by previous experiments that the arrest of respiration was in this case quite independent of any pressure upon the pneumogastric or sympathetic nerves—But[3] you will find the whole of this important series of experiments in the 3^d. No. of the Guy's Hospital Reports—[4]

27   Some have supposed that Life was only the consequence of the harmony or consentaneous working[5] of these prime wheels of the Machine, but we are compelled to refer this harmony to a cause which uses it as a means whereby to operate through the whole, to a cause which is independent of the individual parts, since it exists[6] before[7] these parts[8], and consequently before[9] their harmonious co-operation.

Observation of the phænomena of the development of the embryo, irresistibly leads to the idea of the existence of an organizing principle manifesting itself by properties distinct from those which characterizes any other known imponderable substance[10]. The *Sphere* of the operations of this agent is at first limited to the germ-disc of the Ovum;—for all the other parts of the egg relate only to the nourishment of the germ.

The germ is the whole organism potentially; the different parts *actually* exist by its development.

We recognize the *power* to produce the different parts of ye[11] intire organism, as residing in, and operating from the germ, when we trace their[12] *actual*

28 formation [/] in the incubating egg. The germ-disc, or membrane expands and spreads over the yolk; and the organs of the animal originate from the separation and folding of its[13] different layers:—first, the elements of the nervous System; then the digestive membranes, and then the vascular System.

From these elements the details of the different systems are successively produced. The first trace of the central part of the nervous system is not to be

---

1   reestablished,] *comma inserted [ed.].*
2   began. . .be] *interlined over deleted* 'was'.
3   But] *after deletion of* 'Y'.
4   I witnessed. . .Reports—] *entire passage on facing fol. 26^v in same ink as previous insertions. No insertion point marked.*
5   consentaneous working] *interlined in ink over* 'right interplay'.
6   exists] *deleting* 'pre-'.
7   before] *interlined above deletion of* 'to'.
8   parts] *before deletion of* 'to these parts'.
9   before] *interlined above deletion of* 'to'.
10 substance] *in left margin pencil* '?X'.
11 ye] 'e' *interlined.*
12 their] *over* 'the'.
13 its] *interlined above deletion of* 'the'.

regarded as the Brain, or as the Spinal Chord alone, but as the *potential whole* of the nervous System. In like manner, the different parts of the heart are developed apparently out of one simple homogeneous membranous tube:—and the first rudiment of the alimentary canal is not to be regarded as intestine merely, but as including potentially the whole digestive System; Since we observe that all the accessory glands,—the Liver, the pancreas, the Salivary glands, are formed by an extended vegetation, as it were, of this, at first, simple tube.

29   It can no longer be doubted that the germ is [/] not the mere miniature of the later organs, pre-existing, already formed, and mechanically[1] expanded by the generative processes, as Bonnet and Haller believed;—but that the germ is amorphous matter, vivified by an organizing principle, which, operating upon the surrounding amorphous nutrient substance, arranges and forms it into the organs by whose harmonious action, Life is afterwards to be maintained.

It does not require high microscopic powers to perceive the first rudiments of the different organs; their first appearance is sufficiently conspicuous, they may be even said to be moderately large, but simple; and we can see plainly that the subsequent complication results from a[2] transformation of the simple rudiments.

These observations are, at the present day, no longer theoretical suppositions, but positive facts;—nothing is clearer than the development of glands, from the digestive canal; and the development of the digestive canal from the reciprocally receding folds of the germinal membrane.

30   The Life of the Chick depends on the Circulation [/] of its blood over a membrane which is in contact with the porous shell;—like the vegetable, the interchange of Gases with the external air, takes place on a widely extended surface:—If the pores of the shell be closed by a coat[3] of wax or other impermeable substance[4], the embryo within is suffocated and dies:—but the harmonious action of the heart,—the allantoid vessels,—the air,—are not the cause of Life.

We can observe the period at which the respiratory allantoid bag makes its first appearance, and prior to this period, the organizing energy has laid the basis of all the important organs, by the harmony of whose future operations Life is afterwards to be maintained.

The Genius and Perseverance of Hunter elicited proofs of the existence of the same energy in the ovum even before the formation of the rudiment of a single organ[5].

---

1   mechanically] *interlined above erasure with caret.*
2   a] *interlined with caret.*
3   coat] *in ink over pencil* 'coating' *with WC query in margin* 'Wax?'.
4   or. . .substance] *interlined in pencil.*
5   organ] *followed by two inch line in pencil and* '(Pause)'; *on facing fol. 31ᵛ in RO hand in pencil:* 'Haller speaking of the cause stimulating the muscle to action, says, it is not the *Soul*—but a law derived immediately from God—First Lines—p. 237—'.

The organizing energy operates, as we have said, according to laws of Intelligence and Design; but we must not fall into the error of assigning to it the attributes of a conscious Soul. It does not reside in a single Organ; but
31   manifests itself in the formation [/] of the acephalous fœtus, and in its growth and Life up to the time of Birth. It changes the already existing nervous system, together with all the other organs in the Metamorphosis of the Larva of the Insect;—so that some ganglions disappear, and others unite. It energizes in the corporeal mass of the Tadpole, so as to produce the shortening & absorption of the Tail, and the development of the nerves of the extremities.

Lastly, this unconscious power operates under the control of Laws of Intelligence in the manifestation of the Instinct, by which, as Cuvier has well observed, the inferior animals are governed as by innate Ideas or a Dream:— But the power which occasions the Dream can only be the organizing agent, operating according to the laws of intelligence:—an agent which is present in the Germ prior to all Organs, and manifesting itself therein in the growth of the parts unconnected with any Organ.

Consciousness on the other hand[1] developes no organic product, but
32   gives origin to Ideas alone:—it is a later product [/] of development, and is connected with an organ on whose integrity it depends. But in the development of the Brainless monster,[2]—the organizing energy operates, as far as it goes, according to laws of co-relation.[3]

In plants, again, both[4] consciousness and the nervous System are alike absent, while the organizing agent is present, and operates in the disposition of the proximate vegetable principles *with reference to the interplay*[5] of the necessary Organs, and *according to the destined form* and *actions*[6] of the Species.

We conclude, therefore, that the organizing agent does not operate consciously and designedly of itself;—that it is not, as the followers of Stahl supposed, the reasonable Soul; but a power *working*[7] unconsciously, like other imponderable agents; but working according to determinate laws, which manifest in the highest degree the wisdom and design of the Law-giver;—the Great First Cause.

By the products occasioned by the action of the agent upon the matter, the Organism maintains its specific character unchanged. Out of a casually computed number of many thousand Plants and Animals there is no real transition
33   from one Species or Genus to another:—Every family of Plants [/] and Animals, every Genus, nay every Species, is subject to certain physical

---

1 on. . .hand] *interlined with caret after* 'Consciousness'.
2 Brainless monster,] *on facing fol. 33*[v].
3 co-relation.] *followed by erasure of* 'in the development' *with erasure continued to beginning of next paragraph.*
4 In. . .both] *inserted over erasure.*
5 with . . .interplay] *underlined in pencil.*
6 according. . .actions] *underlined in pencil.*
7 working] *underlined in pencil.*

conditions necessary for their existence upon the earth; to the influences of Soil;—of appropriate food;—of a certain range of temperature; according to which their Geographical Distribution is regulated.

In this vast and various multitude of creatures, through this orderly arrangement of Classes, Families, Genera and Species, the phænomena of Life and its attendant benefits and beauties[1] are distributed over the whole earth; and each Animal reciprocally enjoys the surrounding external world by Sensation and reaction.

Each species of Organism is self-existent from the period of its Creation; the Individuals have a transient existence, and some perish: but the Species long remains; and its duration was once thought to be necessarily coeval with that of the existing Sphere of its actions:—but the Species must disappear with the extermination of the reproductive Individuals, for the Genus has no power to reproduce the Species, nor the Family the Genus.

34   The history of the revolutions of the crust of the Earth teaches us that many species have become extinct;—some belonging to existing Genera, and others to Genera which are no longer represented by living beings. The study of Geology shows us, moreover, that the beings whose organized remains are discovered, have not all existed contemporaneously. In their succession they manifest a gradual approach to the forms of the beings at present in existence;—and in the Strata of the later formations, remains of animals referrible to living species occur:—but the different organized forms which have succeeded each other do not display regularly progressive stages of complication, or perfection of Structure.

Plants and animals exhibiting different degrees of complication of Structure have co-existed at different periods; their existence, and fertility, and well[2]-being seems, as now, to have been regulated by the conditions of the then external World; and a change in those conditions disturbing the harmony of the relations necessary for the [/] well-being of those existing plants and Animals, has caused their extermination; while new species appear on the stage, endowed with powers and forms adapted to the new conditions of the external world, but not necessarily superior in their Organization to the extinct Species which they have replaced.

Thus, the higher organized Sharks and Sauroid fishes which prevailed in the antient carboniferous and secondary formations disappear, and are replaced by other & lower[3] forms in the tertiary Strata. And the functions performed by the Chambered Cephalopods in the antient seas in which the secondary strata were deposited, are now principally assigned to trachelipods, which display a much lower Type of Molluscous organization.

35

---

1   and beauties] *interlined with caret after* 'benefits'.
2   well] *deleting initial dash before* 'well'.
3   other lower] *inserted over erasure with erased interlined* 'other and minor' *above line. On facing fol. 36v pencil query in W. S. Bennett hand* 'are not the forms as far as they go as perfect?'.

A comparison of the skeleton of the Ibis of the present day with that of an individual consigned by Egyptian Superstition to the cerements of a Mummy two thousand or perhaps four thousand[1], Years ago gives no evidence[2] that the legs of the bird[3] have acquired any appreciable addition of length by the practice of wading continued through the successive generations which have
36   preserved the [/] Species through that extended[4] period. And, in short[5], there is not a single fact which demonstrates, or even renders probable, that the changes of form which we witness in the succession of the organized inhabitants of the earth were produced by progressive development, or transmutation of Specific Forms.

     The individuals of each species have a characteristic durability of Life;— the operation of the Organizing energy in them is limited. To suppose a power of prolonging the vital actions in an individual beyond the specific period, is to suppose that the organizing agent has the power to develope new organs, or so[6] to modify the old[7] as must cause a transmutation of the Species:—but this supposition cannot be maintained.

     The combination of the organizing agent with organic matter, which is essential to life, can be prolonged beyond the Specific period only whilst the agent remains passive. This passive condition is manifested in the Torpid Animal; and Hunter seems to have commenced his Experiments on Organic
37   Heat with the view of determining how long Life [/] might be maintained in the higher animals by reducing them to a torpid state, by the abstraction of the Stimulus of Heat.

     This passive condition of the organism, is not, however, Life. Hunter knew well that the Specific Extent of Time assigned to the individual for the *enjoyment of the vital actions*[8] could not by this means be extended: but only that the period of its enjoyment would be changed:—that Life, instead of being continuous would be enjoyed at two different periods remote from one another, in proportion to the length of Time during which the vital actions were interrupted by Torpidity. Now, the possibility of effecting this interruption in the vital actions without destroying the condition of the Organism necessary for this resumption; and the length of Time during which the passive

---

1   or. . .thousand] *RO interlined in dark ink with caret after* 'thousand'
2   Mummy. . .evidence] *note on facing fol. 36ᵛ without evident insert point* 'It is difficult to fix the period which might depend in a great measure on the hieroglyphics on mode of wrapping &c &c or any other indication as to whether it was an ancient Egyptian or a Greek Egyptian mummy. I should therefore prefer 'a moderate chronology.' '. *In same pencil as preceding note, presumably by W. S. Bennett.*
3   of the bird] *RO interlined with caret after* 'legs'.
4   extended] *inserted above deletion of* 'long' *and in margin a pencilled* 'XX'.
5   short] *with WC insertion of* 'fact?' *above the line crossed out.*
6   so] *interlined with caret after* 'or'.
7   old] *before deletion of* 'to such an extent'.
8   enjoyment. . .actions] *faintly underlined in pencil.*

state may be maintained, is inversely[1] as the complexity of the Organization, and is consequently greatest in the Seed and[2] Egg, which are the simplest conditions of living organized beings.

Hunter calls them *living*, to express the State of Self-conservation, and of susceptibility of impressions [/] exciting to action:—but Life properly can only be said to commence when the organic energy is called into operation by the stimulus of Caloric, atmospheric air, and other conditions necessary to its activity.

The seeds of Plants, and the Eggs of Animals continue in the State of Germs only so long as no Operations of interchange with the external World are allowed to take place. They remain in a state of capacity for development, but the organizing agent is passive so long as the Ovum is withdrawn from the influence of Air and heat.

It is thus that the Ova of Insects which are transported from Tropical Climates, along with exotic plants, occasionally retain the organizing energy, and this begins to operate and develope the embryo, under the influence of the Warmth of our Hot-houses.

The seeds of Phanerogamous Plants have been known to retain the Organizing energy for 20 Years, under water; and for 100 years when kept under ground, and removed from the influence of the Atmosphere. The bulb of an Onion found in the hand of an Egyptian Mummy was planted [/] and[3] germinated and grew. It had retained[4] the passive powers of Life from[5] 2000 to four thousand[6] Years.

It is essential to this singular condition of Organized bodies that they should be preserved from the influences of the powers of external Nature:—subject to these, there is no repose for the Germ; it must either progress or retrograde;—it must develope or putrefy.

Within certain limits, the fully developed organism may fall back to the passive vital state, or become torpid when the stimuli of vital actions are removed. This state resembles that of the unincubated Ovum: but in general the external vital stimuli are essential to the maintenance of Life in the developed and active state of an Organism, and Death is the consequence of their supression.

The external conditions essential to the support of life, are Heat, Air, Water, and Food:—these produce the incessant Changes in the molecular parts of the Organized body; and they combine with the organized body to

---

1   is inversely] *interlined above deletion of* 'appears to be'.
2   and] *interlined above deletion of* 'or'.
3   and] *followed by pencil deletion of* '(is said to have)' *with* 'X'*pencilled in margin and pencil lead line from* 'grew' *to query WC on facing fol. 40*ᵛ. 'Is this satisfactorily proved?'.
4   retained] *after deletion of* 'probably'.
5   from] *interlined in pencil.*
6   to . . .thousand] *interlined in pencil with caret before* 'Years'.

40   supply the place of the decomposed [/] and excreted particles. They are therefore termed the vital stimuli; but must be distinguished from many other accidental irritants which are not essential to life. A mechanical stimulus, e.g.
which modifies the condition of a sensible membrane occasions, it is true, a
vital phænomenon, viz. Sensation:—but it[1] neither sustains nor strengthens
the Organic Energies.

The vital stimuli effect the molecular changes in the Solids by means of the
Fluids, especially the Blood;—which being duly modified and vitalized by
their influence, stimulates all the[2] Organs,— deposits therein the new materials,— and is chiefly concerned in the excretion of the old.

The Blood preserving its fluidity while circulating in the living body is not
only alive itself, but is the support of Life in every other part; & as[3] this fluid
loses its own vitality in supporting that of other parts; it is necessary that it
should be moved and circulated to be again saturated, as it were, with living
powers. Whence it is essential that there be a reaction of the Solids upon the
41   Blood: for in [/] the language of Hunter, the body dies without the motion of
the blood upon it; and the Blood dies without the motion of the body upon it.
And this leads to the consideration of another important vital stimulus: the
imponderable Agent, which operates through the nerves, and which, besides
being the regulator of the motions of the Body, also influences ye molecular[4]
changes in the organism[5].

This property of all organized bodies to suffer certain molecular changes
essential to life by means of the so called Vital Stimuli, has been termed by
Hunter the "susceptibility of impressions which lead to action" and by others,
"Incitability and Irritability".[6]

The organizing energy can effect nothing without the co-operation of the
external vital stimuli. They form the grand external impulse for the working
of the whole Machine. The combinations and decompositions occasioned[7] by
the reciprocal actions of these stimuli and the Organic particles are as essential
to the manifestations of the vital phænomena, as the actions going on between
Oxygen and Combustible matter are essential to the production of the
phænomena of Combustion.

42   The phænomena of Life have not unaptly been compared by Richerand
and other modern Physiologists to those of Combustion and Flame.
Hippocrates, and Harvey influenced by this analogy of the vital operations to
the consumption of bodies by fire, identified as we have seen[8] the vital principle with that Element, and termed it the 'calidum innatum.'

1   it] *over erased* 'is'.
2   the] *over* 'their'.
3   ; & as] *inserted in dark ink.*
4   also. . .molecular] *inserted over erasure with pencilled* '?X' *in margin.*
5   organism] *inserted over erasure of* 'different organs'.
6   Irritability".] *period inserted [ed.].*
7   occasioned] *inserted over erasure.*
8   as. . .seen] *interlined in dark ink with caret after deletion of* 'the' *following* 'identified'.

From these general observations on Life, and the stimuli essential to its maintenance, we pass to a brief special consideration of the relative importance, and specific modes of operation of the several vital Stimuli.

With respect to Plants, light is an indispensable vital stimulus: it is less essential to animals:—almost the whole tribe of Entozoa grow, live, and propagate in the recesses of Animal bodies, where light can never penetrate. The gemmules of some Zoophytes seem to be repelled by light,—avoid its influence, and move and fix themselves in the obscurest recesses of rocky places.

Other low organized Animalcules, on the contrary, court the influence of the luminous rays:—thus certain green Infusoria [Cercaria] may be observed
43   to move and collect towards that side [/] of the vessel containing them, which is exposed to the light:—and the higher organized animals are observed in general to become scrofulous and rachitic when removed from the influence of light.

An indispensible condition to animal life is the intussusception not merely of new matter, but of already organized matter; while plants receive nourishment from binary compounds, as Water and Carbonic Acid. Nutriment, however, is not merely a stimulus to vital actions, but is itself convertible into a living substance:—it is a stimulus which not only vivifies, but can itself be vivified.

Water either enters into Organic Composition as such, or contributes its elements to form a different ternary or quaternary compound. It is also necessary to the manifestation of life in its free state. It is the oil to the wheels of the machine: unless the parts of the Animal are moistened with water, action and life cannot go on.

In some of the lower organized Animals, the deprivation of Water produces the same suspension of the Vital Actions without death, as the diminu-
44   tion of heat in the higher Class.

Let a drop of water containing the parasitic Vibrio, or the active Wheel-Animalcule, evaporate;—their motions cease, apparently they die, and as they become desiccated they are no longer distinguishable from the surrounding inorganic particles. Thousands of such dried and torpid, yet not dead, animalcules may[1] float unobserved in the Air, or be perceived only as the flickering motes in the Sun-beam:—Re-moisten them, and the vital actions are resumed with undiminished activity. Spallanzani and Bauer astonish us by their statements of the extended periods of this Torpidity from Dryness, after which they have witnessed Animalcules to revive.

Atmospheric Air is so essential a condition to the vital phænomena, that the higher animals can scarce exist a minute without breathing[2]. The intus-su-

---

1   may] *over deletion of* 'might' *with* 'ay' *interlined in pencil.*
2   Atmospheric. . .breathing] *WC pencil note on facing fol. 45ᵛ* 'The Tritica *from smut of wheat [interlined with caret] I saw restored in a few minutes, by applying a drop of water on the glass on which they had been dried for Seven Years. W. C.' and RO reply*

ception of food may be pretermitted for a long time: the Amphibia will endure a month's fasting with impunity; but when deprived of Oxygen even for a very short period they perish.

45   Caloric, lastly, enters into the composition, and plays a conspicuous part in the vital actions of organized beings: it is chiefly important to those which have the weakest powers of generating heat, but it is essential to all, whether Plants or Animals.

In this condition we may perceive an analogy between the organizing processes, and the binary combinations of chemical affinities: in both cases a determinate temperature or range of temperature is necessary for the process; and a certain quantum of Caloric is absorbed during the formation of the new compound.

Whether Electricity be[1] essential to the manifestation of Life, is as yet wholly unknown:—we incline to the belief only from a probable analogy.

The dependance of living Organic beings on the vital stimuli, varies both as to degree and kind:—the experiments of Hunter and[2] Edwards show that new-born Animals have most need of external warmth, but sustain a deprivation[3] of air longer than adults.

46   The duration of Incitability in the absence of external [/] vital[4] stimuli is generally in an inverse ratio to the complexity of the Organization.

Mollusks, Insects, and Reptiles, may live for months without food or drink[5], while Man under similar circumstances will perish in a week[6].

Some Insects, and the larvæ of the Œstrus or Horse-Bott in particular, will survive a long time though immersed in irrespirable Gases. Mollusca have survived after being 24 hours under the exhausted receiver of an Air-Pump. Reptiles will live in distilled Water from 10 to 20 hours. Frogs, of which the lungs have been extirpated, have been known to survive the operation 30 hours.[7]

Complication of the organized Structure increases the reciprocal dependance of the different Organs:—hence, the more simple animals survive injuries longer than the more complex. It is familiar to all how long time the Amphibious Reptiles manifest vital actions after the severest mutilations, and

---

(contd.)

    in pencil '(The Sperm whale—1 hour, 1 hour 10 minutes 1 hour 20 minutes)'.

1 be] *in pencil over ink.*

2 Hunter and] *interlined with caret before* 'Edwards'.

3 deprivation] *with terminal* 'tion' *in dark ink with pencil* 'tion' *above.*

4 external vital] *on page break with* 'external vital' *below* 'absence of' *and* 'vital' *repeated on top of fol. 46.*

5 or drink] *interlined with caret.*

6 week] *circled in pencil and pencil note in W. S. Bennett hand* 'Not always—case of the woman that was buried under snow I forget how long but more a week—she heard the bells ring for church—' *with RO pencilled reply* 'She sucked the snow/'.

7 hours.] *period inserted [ed.].*

the enduring irritability of their muscles and nerves in relation to the simplicity of their structure is far from being a new doctrine.

47   The muscular irritability of young animals is more lasting than in the adult[1]: and the same holds good with respect to sensation. Legallois found that an Animal one day old exhibited signs of feeling, or rather an automatic excitability fifteen minutes after the excision of the heart;—while another of the same species, 30 days old, retained the same power under similar circumstances, for 2 1/2 minutes only.

The observation of continuance of the circulation after an animal has been pithed, in the cold-blooded and warm-blood animals, and in the different periods of life of the same Species, equally illustrates the same law, that the higher the parts of a whole are developed, the more the whole is dependent upon this integrity.

We come now to consider the last peculiarity of Organized Beings:—their perishability.

The masses of common matter have an indeterminate and indefinite existence, and when they cease to be, it is from disintegration and dissipation 48 of [/] their particles in consequence of destructive processes operating from without. The existence of Organized Beings, on the contrary, is terminated by circumstances inherent in the individual; and the period of their duration is determinate and definite. But here we must distinguish between the extinction of Life in the Individual, and the destruction of the organized Material.

Organized matter is maintained, and in some conditions appears to increase on the surface of the Earth from age to age: the individuals alone die, and add their quota of dead organized materials to the general stock;—Life itself, is however maintained with a semblance of interminability by transference from one individual to another until the species is wholly exterminated, as Experience, derived from the retrospective History of the Earth teaches us to have frequently happened.

Something similar to this[2] transference of life from one generation to another we also perceive to take place in the individuals themselves, in which 49 the organizing energy glides as in a stream from the [/] old and effete to the new and vivifying particles of the body. It is thus that in some[3] parasitic climbing plants, as the old root or stem dies, life is transferred to the new sprouts in which the organizing energy continues to operate, and the individual seems gifted with endless existence[4]:—

1   the adult] *inserted over erasure.*
2   Something. . .this] *over erasure.*
3   some] *pencil interlined above* 'some' *in pen.*
4   existence] *before deletion of* 'this peculiarity has not escaped the Poet, who thus addresses the Ivy:' *followed by 4 lines of space to insert quotation left blank and on facing fol 50ᵛ note by WC in pencil* 'This is a Poetical fiction. If you divide the parent stem of the ivy where it rises from the ground, its little claspers or roots which attach it to the tree will not continue the life of the plant, but it will die all the way up. This I have seen more than once: and is done by woodmen to prevent trees from being

The religious Brahmin sees the emblem of Immortality in the Banyan, whose branches extending from the parent trunk, bend down, take root, and self subsist; so that while death is busy in one part of the Individual, new life is vigorously operating in another[1].

50   Why[2] the Individual should perish, and its place and office in the œconomy of Nature be filled by the development of another being, is perhaps the most difficult problem in the whole range of general Physiology. We can appreciate the beneficence of the arrangement in the multiplication of individual enjoyment, and we are able to trace the connexion of the phænomena, but beyond that we cannot penetrate.[3]

It is insufficient as well as unsatisfactory to reply that the inorganic influences gradually wear out individual life; for then we ought to witness an impairment and progressive debility of the powers of the[4] organizing agent from the commencement of individual existence:—but we know that the organizing energy exists in such plenitude of power at the period of full maturity, that it multiplies itself in the formation of Germs:—It must therefore be a wholly different and deeper cause which occasions the death of Individuals.

We[5] may affirm that the increasing frailty of organized beings in Age is produced[6] both *mechanically*[7] by the aggregation of dense particles in the areolar tissue which causes a progressively increasing resistance to the circu-

---

(contd.)

    hide-bound and destroyed. The Tree will soon rid itself of the dead Ivy.' *Insertion followed by pencil interlined note on 50ᵛ in W. S. Bennett hand* 'I can bear witness to that, as Tony Lumpkin says,— the blow of a bill hook, through the ivy stem so as to completely divide it kills all the ivy above it and proceeding from it. I have done it myself—B.' *followed by WC continuation* '{¶}In the gallery there are three preparations of leaves which send off young shoots from their indentations, and when the leaves fall they support the young plants by their succulency until they take root. There is also a preparation of a Fern whose leaves bend down, & take root when they reach the Earth. And so do the branches of the Ivy— and so does the Rose, and many other plants, as is well known to gardeners, who purposely bend them down and confine them with a forked stick, covering that part with Earth; and the bent up extremity in one year becomes a new plant with roots which strike out from the underground buds, and are then separated from the parent bush by the knife, and transplanted. They call it *laying*, in Cornwall.'.

1   another] *before vertical pencil deletion of* 'The newly rooted branch lives and fructifies as a distinct individual; *and {deleted in pencil} thus is the Torture invented by the Roman Tyrant *is as it were[interlined with caret in pencil} exemplified in Nature,— the dead body tied to the living and that succession of Life and Death which is interrupted in most organized Species, is here continuous.—'[I fol. 49]; in margin of fol. 49 in W. S. Bennett hand in pencil* 'I would leave this out'.

2   Why] *with pencil interlining of* 'Pause' *at top of page.*

3   penetrate] *with erased comment in WC hand in pencil and brackets* '[a laugh!]'.

4   powers of the] *interlined with caret before* 'organizing'.

5   We] *inserting paragraph break for one-inch space in text [ed.].*

6   produced] *erased seven-line WC note on top of facing fol. 52v:* ' "I {two illeg. words} you, good man: who do you continue. . ." said Alexander Pope to a' {illeg. from here}.

lation of the fluids; and *chemically*[1] by the accumulation of decomposed organic matter, the ordinary affinities of which at length rise superior to those which are peculiar to the living energies.

But why should the effects of this conflict, exemplified in the gradual yielding of the vital powers be witnessed only at the latter stages of life? In maturity &[2] age, the organizing energy is subdivided and expended in various offices, and the whole exists by the operation of individual parts like a machine, which is maintained by the action of its different wheels.

In the germ the organizing energy exists in its state of greatest concentration:— the developing power is at a maximum[3]. But when the organism is carried beyond youth, we no longer perceive a simple condition of the whole with energy undivided, but a complex whole with energies variously distributed. This distribution or [/] division of the organizing energy of the whole is accompanied with a corresponding impairment of incitability. Development therefore being at an end, and the general external stimuli of life ceasing at length[4] to be responded to, Death ensues, and the continuance of Life in the Species depends on the previous generation of a germ in which the undivided energy developes a new Organism.[5]

The decomposition of the organized particles of the individual, and their excretion, is more readily understood than the death of the whole. The conversion of the alimentary substance into food, necessitates the separation of matter which contains an excess of useless elements.

Plants, which convert carbonic acid and water by ternary combinations into vegetable matter give out oxygen. In Animals, the excretions, which may be truly so called, from being totally useless in the œconomy, are carbonic acid and urea.

Plants which absorb their aliment by minute pores, throw out their *excreta* imperceptibly, in a [/] molecular state of subdivision; while Animals, which take in food in larger masses by a conspicuous aperture, expel the excrement in a corresponding[6] state of aggregation.

Much of the intestinal excretion is, however, composed of the indigested parts of the food, and those parts which are derived from the Organism itself;

52

53

(contd.)
7   *mechanically*] *underlined in pencil.*
1   *chemically*] *underlined in pencil.*
2   In. . .&] *after interlined* 'Pas' [sic] *in pencil; followed by pencil* 'maturity &' *interlined in pencil.*
3   power. . . maximum.] *with pencil line underneath drawn from margin to inserted* 'Pause' *or* 'Pas' *interlined after* 'maximum.'.
4   at length] *interlined in pencil with caret after* 'ceasing'.
5   Organism] *with horizontal pencil two-inch line across page and pencilled* 'Pause' *in margin.*
6   corresponding] *inserted over erasure* .

for[1] the intestinal mucus and the bile, cannot be regarded as pure *excretes*, since they are destined to serve certain special purposes.

Carbonic acid and Urea, besides being wholly eliminated from the Organism, are totally useless and noxious in the living œconomy. It is true that the urine is modified according to the nature of the food; and occasionally, like the intestinal excretion, carries out of the System useless parts of the aliment before they have become organized. But when[2] food has not[3] been taken as in Reptiles which have fasted for months, the urine is excreted without any alteration of its essential conditions.

54   In the pupa of Insects, which during their [/] transformation take in no aliment, the secretion of the *Vasa Malpighiana* goes on; and we know by the experiments of Wurzer and Chevreul that these secerning tubes contain Urea. So also in the Embryo of the Vertebrate animals, the corpora Wolffiana excrete Urea before the kidneys have taken on their functions.

Doctor Davy detected Urea in notable quantity in the fluid surrounding the Embryo in the uterine cavities of the Ovo-viviparous Squatina, or Monkfish. It exists abundantly in the allantoid bag of the Chick in ovo. It is present in the liquor amnii of the Dog before the 5th. week of pregnancy; and has been detected in the Human liquor amnii at the full period.

Still[4] more decisive evidence of the *essential* connexion of the urinous excretion with the organizing processes, and its independance of foreign alimentary matter, is derived from the Analysis of the contents of the Bladder of Fœtuses which have been still-born with an impervious urethra.

55   With respect to the excretion and interchange of [/] elements which take place in the respiratory actions, it is not intirely explained by the hypothesis that by this means the Element Azote which is still wanting in the formation of animal matter is obtained, or that the overplus of nutriment is thus in part excreted: because we perceive that the respiratory powers are not more feeble: nay are generally most vigorous in the animals which subsist on already animalized material:—and we know that Reptiles which take in no nourishment for months continue to consume Oxygen, and expire carbonic acid.

Carbonic acid and Urea may, then, be regarded as the constant and essential *excreta* of the Animal Organization, cast out of the Organism by the vital processes, independent of the food which may be taken in. The intestinal excrement of the higher animals serves to support the 'shard-born Beetle,' and

---

1   for] *with erased interlined word above.*
2   when] *followed by ink deletion of* 'no'.
3   not] *interlined with caret after* 'has'; *and on facing fol. 54v in WC pencil:* 'No food cannot be taken—when food has not been taken.—which is best?—'.
4   Still] *in margin pencilled* 'X?' *and WC pencil note on facing fol. 55v:* 'Might not the material be derived from the alimentary matter taken in by the mother of the fœtus, and from whom the fœtus derived its nourishment and material for all its Secretion?'.

hundreds of other coprophagous Insects: but the two essential *excreta*, carbonic acid and urea, are unfit for the nourishment of any other Animals.[1]

56   Animal Life therefore[2] is essentially connected with the never-ceasing decomposition of organized matter of which Carbonic acid and Urea are the principal results. The intestinal excretion, even in the higher animals, may be long interrupted without Death following as a necessary result; but the suppression of the Urinary excretion is soon succeeded by dangerous symptoms: and the arrest of the excretion of carbonic acid is almost instantly fatal.

Thus we have seen that Animals and vegetables not only differ from common matter in their active properties as succinctly stated by Hunter but also in their general physical qualities and constitution. The form of the organized Being is determinate, and its outline rounded or undulating; its size is limited; its duration is temporary.

57   The[3] elements combine under a [/] different law to form its proximate components:—the whole body is an assemblage of heterogeneous parts, of solids and fluids, arranged so as to form a variety of fibrous and cellular tissues, and aggregates of organs and parts differing from one another in their form, structure, and functions; but all nevertheless mutually dependant one upon another, and concurring to a common end:—the preservation of the individual.[4]

This preservation is the result of an internal activity, of incessant changes and renovations of[5] the matter entering into the composition of the Organized Being;[6] existence may be summed up in[7] the following series of actions;—incipience by a *genesis* or creation; temporary endurance as an[8] individual by *nutrition* and *excretion*[9];—prolonged continuance, as a *species*, by *generation*; modification during the term of existence or *age* ; and end by Death.— To[10] these specific acts or phænomena must be added the peculiar inherent power which living beings possess of resisting and overcoming the general

58   physico-chemical laws which govern [/] the rest of the Universe[11].

---

1 Animals.] *followed by deletion by vertical scoring of paragraph with light pencil* 'note' *in margin by first word* 'Carbonic acid is a binary compound eliminated [l *fol.* 55] in the decomposition of animal matter, and Urea approaches very near to it, if it be not actually in a state of binary composition:—at least, the compound produced by Wöhler from the Cyanate of Ammonia is very similar to it.'.

2 Animal. . .therefore] 'Animal' *inserted in pencil before* 'Life' *and pencil transposition of* 'is therefore'.

3 The] *inserting paragraph break for one inch blank space [ed.].*

4 individual.] *period inserted [ed.].*

5 of] *deletion of repeated* 'of' *in ink.*

6 Being ;] *semi-colon inserted to replace comma to accord with deletion of* 'whose' *following [ed.].*

7 summed up in] *interlined with caret after* 'be'.

8 an] *interlined with caret after* 'as'.

9 *excretion*] *over erasure.*

10 To] *inserted over* 'to'; *with pencilled* 'O/' *in margin.*

11 Universe] *followed by pencil* 'Bon' *and ink line across page.*

$\mathcal{N}otes$
to
$\mathcal{L}ecture$ $\mathcal{F}ive$

1-12 Owen's opening discussion closely resembles Hunter's in the introduction to his "Surgical Lectures" in *Hunter's Works* 1: 211–20.

2-6 Source of quote untraced. This is similar to the definition given by Hunter in ibid., p. 212. Palmer's edition of the Surgical Lectures was derived from the lecture notes of Nathaniel Rumsey taken from Hunter's 1786 and 1787 courses. Owen is probably using another version of lecture notes in the College possession. Hunter's originals were among the manuscripts destroyed by Everard Home. See also the version of Hunter's lectures published in twenty-two parts in *London Medical and Surgical Journal* N. S. 1(1837): 96–9 and following.

3-8 Owen is alluding to the recognition of "organic" as distinct from animal and plant existence. On this distinction see Johannes Müller's discussion in his *Handbuch der Physiologie des Menschen für vorlesungen* vol. 1, which Owen is utilizing in the first edition of 1834 (Coblenz: Hölscher, 1833–4). This specific edition is cited by Owen in his notes to *Hunter's Works* 4: 310 note, and a copy of this is in the RCS library. All quotations in the notes from the *Handbuch* will be to this edition. On the complexity surrounding the various editions of Müller's work see E. B. Clarke and L. C. Jacyna, *Nineteenth Century Origins of Neuroscientific Concepts* (Berkeley: UC Press, 1987), pp. 470–1. William Baly's translation of Müller (*Elements of Physiology* vol. 1 [London: Taylor and Walton, 1837-8]) was first seen by Owen in September of 1837 (Letter of Owen to Baly, 30 September, 1837,

Royal College of Physicians Archives).

4-5 The source of this quote has not been located. It is similar to a passage from Hunter's first surgical lecture in *Hunter's Works* 1: 214–5.

4-13 "Die Erfahrung zeigt also, dass bei den unorganischen Körpern die Verbindung von der Wahlverwandtschaft und den Kräften der verbundenen Stoffe abhängt, dass in den organischen Körpern dagegen die bindende und erhaltende Gewalt nicht bloss die Eigenschaften der Stoffe selbst sind, sondern noch etwas anderes, welches der chemischen Wahlverwandtschaft nicht allein das Gleichgewicht hält, sondern auch nach den Gesetzen eigener Wirksamkeit organische Combinationen verursacht. Von den imponderabeln Materien haben Licht, Wärme, Electricität, auf die Verbindungen und Trennungen der Stoffe in den organischen Körpern ebenso Einfluss, wie auf die Verbindungen und Trennungen in den unorganischen Körpern; aber nichts berechtigt uns, eines dieser Agentien ohne weiteres als letzte Ursache der Wirksamkeit in der belebten organischen Materie anzusehen." Müller, *Handbuch*, 1st ed., p. 4. Owen's discussion throughout the remaining lecture draws very heavily on the discussions in Müller's "Prolegomena", pp. 1–63, itself drawing frequently on Berzelius's *Lehrbuch der Chemie*, 5 vols, trans. F. Wöhler, vol. 4: *Lehrbuch der Thierchemie* (Dresden: Arnold, 1831). Owen is very often only making a direct translation or a close paraphrase of the Müller text, and he generally follows Müller's sequence of topics. See also my "Darwin, Vital Matter, and the

Transformism of Species," *Journal of the History of Biology* **19** (1986): 404–19.

6–9 A similar list is given in Müller, p. 1.

6–17 Ibid.

7–15 Owen's comments should be compared to ibid., pp. 3–4.

8–4 See ibid., p. 2.

9–8 The reference is likely to Abernethy's lectures at St Bartholomew's which Owen attended on several occasions.

9–11 The reference is to Friederich Wöhler's famous synthesis of urea. See Müller, p. 37 for discussion.

10–13 Ibid., p. 5.

11–3 Ibid., p. 3

11–8 See J. Hunter, *On the Blood*, reprinted in *Hunter's Works* 3:108.

12–9 Owen is possibly referring to some of Hunter's observations on invertebrates made on the isle of Belleisle when he was stationed there between 1761–3.

13–5 Müller, *Handbuch*, p. 4.

15–11 "Die organischen Körper unterscheiden sich nicht bloss von den unorganischen durch die Art ihrer Zummansetzung aus Elementen, sondern die beständige Thätigkeit, welche in der lebenden organischen Materie wirkt, schafft auch in den Gesetzen eines vernünftigen Plans mit Zweckmässigkeit, indem die Theile zum Zwecke eines Ganzen angeordnet werden, und diess ist gerade, was den Organismus auszeichnet. Kant sagt: die Ursache der Art der Existenz bei jedem Theile eines lebenden Körpers ist im Ganzen enthalten, während bei todten Massen jeder Theil sie in sich selbst trägt." (Müller, *Handbuch*, p. 18). Müller, and Owen following him, are denying Kant's claim that the cause of organic life is to be found in the entire whole organism. He agrees instead with Johann Christian Reil that the causes of life can reside in each of the analytic parts. See J.C. Reil, "Von Der Lebenskraft," *Reils Archiv für Physiologie* 1 (1796), esp. para. 6, pp. 40–4.

16–11 This closely paraphrases Müller, pp. 19–20.

17–4 Ibid., p. 18. Müller cites Fredrick Mohs, *Grundriss der Mineralogie*, 2 vols. (Dresden: Arnold, 1822–4), vol 1, "Vorrede," p. 6.

19–11 Müller, p. 24: "Der Keim ist das Ganze, *potentia*, bei der Entwickelung des Keimes enstehen die integrirenden Theile des Ganzen *actu*.. . . .Alle theile des Eies sind bis auf die Keimscheibe, *blastoderma*, nur zur Nahrung des Keimes bestimmt; die ganze Kraft des Eies ruht nur in der Keimscheibe. . . ." This section of the lecture also appears in the original draft and cannot represent an indirect knowledge of these ideas via Martin Barry's paper which Owen probably only encountered in April of 1837. See final portion of Lecture Four for the comments added at this time.

21–15 Compare to Müller, ibid. p. 20. Müller cites the first volume of E.H. Weber's edition of Friedrich Hildebrandt's *Handbuch der Anatomie des Menschen* 4th ed., 4 vols. (Stuttgart: Wolters, 1833–4), on the relations of crystals and organisms.

23–11 Compare to Müller, p. 20.

26–17 "Jede Verletzung einer dieser Haupttriebfedern in dem Mechanismus des organischen Körpers, jede grössere Verletzung der Lungen, des Herzens, des Gehirnes kann die Ursache des Todes werden, daher man sie die *atria mortis* genannt hat." Ibid, p. 23.

26–31 Astley Cooper, "Some Experiments and Observations on Tying the Carotid and Vertebral Arteries, and the Pneumo-Gastric, Phrenic, & Sympathetic Nerves," *Guy's Hospital Reports* **1** (no. 3, 1836): 457–75.

28–1 See quote above, 19–11.

29–3 "Es kann jetzt nicht mehr bezweifelt werden, dass der Keim nicht die blosse Miniatur der spätern Organe ist, wie Bonnet et Haller glaubten. . . ." Müller, p. 24.

29–16 Owen is drawing not only on the writings of Hunter and the reports of Müller here, but also on his own microscopic observations which he was conducting at this same time in preparation of the fourth volume of the *DIC* for publication.

30–15   From this point to 35–5 Owen is giving either a very close paraphrase or an exact translation of Müller, pp. 25–6. Müller's argument, somewhat obscured in Owen's discussion, involves an attack on Georg Ernst Stahl's identification of life with consciousness.

34–10   Compare 33–11 to 34–10 with following from Müller "...jede Familie der Pflanzen, der Thiere, jede Gattung, jede Art ist an gewisse physische Bedingungen ihrer Existenz auf der Erde, an eine gewisse Temperatur und bestimmte physisch-geographische Verhältnisse gebunden, für welche sie gleichsam erschaffen. In dieser unendlichen Mannigfaltigkeit der Geschöpfe, in dieser Gesetzmässigkeit der natürlichen Klassen, Familien, Gattungen und Arten, äussert sich eine das Leben auf der ganzen Erde bedingende gemeinsame Schöpfungskraft. Aber alle diese Arten des Organismus, alle diese Thiere, die gleichsam eben so viele Arten die umgebende Welt mit Empfindung und Reaction zu geniessen sind, sind von dem Zeitpunkte ihrer Schöpfung selbständig; die Art vergeht mit der Ausrottung der productiven Individuen, die Gattung ist nicht mehr fähig, die Art zu erzeugen, die Familie nich fähig, die Gattung herzustellen. Thierarten sind im Verlauf der Erdgeschichte durch Revolutionen der Erdrinde untergegangen und in den Trümmern vergraben, sie gehören theils ausgestorbenen theils noch lebenden Gattungen an. "Müller, *Handbuch*, p. 25.

35–11   Owen's major study on the anatomy of the Chambered Nautilis, *Memoir on the Pearly Nautilus* (Nautilus Pompilius, *Linn.*) (London: Royal College of Surgeons, 1832) prepared him to speak authoritatively on this group.

36–1   Owen is referring to the results of the studies on animals and plants recovered by the French from Egyptian tombs during Napoleon's expedition. The zoological results are discussed in Jean-Baptiste Lamarck, Georges Cuvier, and Bernard de Lacépède, "Rapport des professeurs du Muséum, sur les collections d'histoire naturelle rapportées d'Egypte par E. Geoffroy," *Annales du Muséum d'Histoire Naturelle* 1 (1802), pp. 234–41. See on this Toby Appel, *The Cuvier-Geoffroy Debate*, chp. 4.

36–11   Owen's endorsement of Müller's claim on this point implies that species are prevented from transmutation by an internal conservation of life force. Hence against Lamarck's transformism, which assumed the unlimited power of vitalizing materials to complexify organization and life, the expenditure of the vital force in the development of organization restrains the individual to changes within the species limits. See discussion of this issue in my "Darwin, Vital Matter...," pp. 405 ff.

37–3   J. Hunter, "Experiments and Observations on Animals, with Respect to the Power of Producing Heat," *Philosophical Transactions of the Royal Society* 65 (1775), as reprinted in *Hunter's Works* 4:131–2 and ff. Owen has synthesized together the 1775 paper and the later "On the Heat, &c. of Animals and Vegetables, (*Philosophical Transactions* 68 [1777]) in this reprint.

38–1   Hunter, "Lectures on Surgery," *Hunter's Works* 1:223–4.

39–7   This passage displays Owen's thesis that the primordial germ displays a dynamic activity of either decay or complexification. This should be compared to Darwin's comment in his "B" Notebook (B 26). See also my "Introductory Essay" above, pp. 46 ff.

40–1   See Müller, p. 27.

41–2   Hunter, *On the Blood, Works* 3:112.

41–10   See Hunter, "Lectures on Surgery, Ibid., 1:223.

42–2   "Richerand hat daher die Aeusserungen des Lebens nicht uneben mit den Erscheinungen der Verbrennung und der Flamme verglichen." (Müller, p. 29) No specific reference to Richerand is given.

42–5   See Hippocrates, in *Opera* 1: 112; and Harvey, *On Generation* ex. 71 (p. 385 Whitteridge ed.). Owen cites these two references in his subsequent *Essays and Observations* (1861); p. 113n.

43–4 Owen seems to be referring to Robert Grant's studies on this issue. See R. E. Grant, "On the Influence of Light on the Motions of Infusoria," *Edinburgh Journal of Science* **10** (1829), 346–9.

44–10 These observations were summarized close to the time of Owen's delivery of this lecture by Allen Thomson in his article "Generation" in R. B. Todd's *Cyclopedia of Anatomy and Physiology* vol. 2 (London, 1836–9) (issued originally in serial form), p. 431. Owen was fully conversant with this work and published articles in it. See also Müller, p. 32.

45–11 "Ob auch Electricität zur Entwickelung des Lebens nothwendig ist, ist uns noch ganz unklar." (Müller, p. 31)

45–14 Müller, (p. 32) cites Edwards, *De l'influence des agens physiques sur la vie* (Paris, 1824; no page number) on this same point.

46–10 This material is directly taken from Müller, p. 32, citing Edwards, *De l'influence des agens physiques....*

47–6 Müller, (p. 32) cites Julien Jean Legallois, *Experiences sur le principe de la vie* (Paris: Hautel, 1812), p. 33 on this.

47–11 See Müller, p. 33.

48–13 From this point to 52–6 is essentially a translation of Müller, pp. 33–4, adding some additional material from Müller, p. 25.

49–5 Müller (p. 33) on this example quotes from Johann H. F. von Autenreith, *Handbuch der empirischen Menschlichen Physiologie* 3 vols. (Tübingen: Heerbrandt, 1801–2),1:112.

54–3 These three authors are cited on the same point in Müller, pp. 36–7.

54–10 Reference uncertain. Possibly John Davy, "On the Urinary Organs and Secretions of Some of the Amphibia," *Philosophical Transactions of the Royal Society* (1818): 303–7; or idem. "An Account of Some Experiments and Observations on the Torpedo (Raia T. Linn.) *Philosophical Transactions* (1832), 259–78.

Lecture Six
Analysis of the Manuscript

This manuscript is located at the BMNH (O. C. 38). On the folded cover page made of a stiffer paper it has in Owen's hand "Lecture ~~VI~~ IV. Vegetables & Animals/May 13th." The title page, in Clift's hand, reads "Lecture IV. Plants and Animals:—Distinguishing Characters of." The body of the lecture itself is also titled in Owen's hand "Lecture 4th." Although clearly intended to be the fourth lecture, the splitting of the first two lectures made it the sixth lecture delivered on 13 May. The first eight folia of this manuscript are on an unmarked paper otherwise identical to the "J Bune 1832" paper used for the bulk of the previous lectures. The remainder of the lecture is then written on folded sheets of a slightly heavier "J Green & Son 1837" paper measuring 32 x 19.7 cm in half sheet. This is the same paper used for the insertion at the end of final Lecture Four discussed previously. This indicates a final copy date after the final version of Lecture Five was completed.

The holograph of this manuscript is at the RCS (Owen MS 67. b. 12). Some indications that it may originally have been intended for an earlier point in the series are found on the cover page, where Owen has written "2" and "Organisms/ Organization/Organic Actions/Vegetables and Animals." The draft is written primarily on the same "J Green & Son 1836" paper used for the draft of Lecture Five, but sections are also on an older "J Tassell 1834" paper, and some insertions have been made on "J Whatman 1836." There is some evidence that Owen was running low on writing paper near the beginning of 1837. This deficiency enables sections of this draft to be dated more specifically. A new insertion in the opening section is written on a draft agenda of the Zoological Society meeting for 8 November 1836, and a loose sheet outlining the lecture is on the reverse of a printed notice inviting Owen to the meeting of the Zoological Society Garden Council, dated 10 December 1836. Internal evidence all suggests a winter 1836–early 1837 date of composition of the first draft.

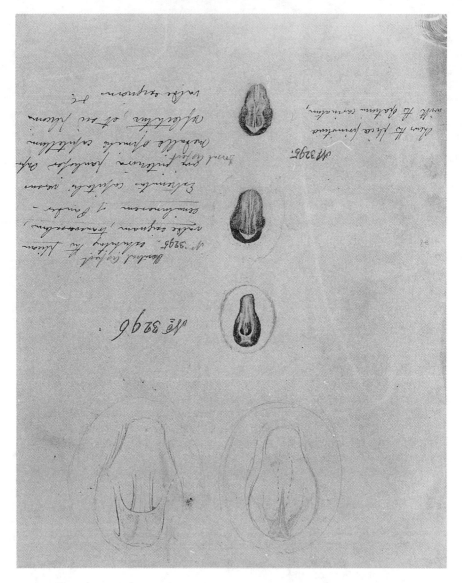

Fig. 13. Owen's Pencil Drawing Of The Development Of The Chicken (Ca. 1840), Based On Hunterian Preparations 3295 And 3296. (*Courtesy President and Council, Royal College of Surgeons of England*)

Lecture Six
13 May 1837

1   M^r. President

Having considered in the preceding Lecture the Characters which distinguish the Organized from the Unorganized divisions of material objects, and having in that comparison pointed out in how many respects Animals and Plants resemble each other; I proceed next to endeavour to establish the distinguishing characters and peculiarities of these grand divisions of Organized Nature.

It is much more easy, however, to show in what respects animals and vegetables resemble each other, than to determine their real and essential differences. "Nothing", says Cuvier, "seems so easy to define as an animal; every one conceives it to be a Being endowed with sensation and voluntary motion: but when we proceed to determine if a living object under observation does or does not belong to the animal kingdom the application of the above definition often becomes extremely difficult and uncertain."

2   It has been asserted, indeed, that no Naturalist as yet has proposed characters which can be considered either as truly applicable to all known animals, or of so precise a nature as to distinguish them clearly from vegetables.

Hunter, in bringing the powers of his deeply meditative understanding to bear on this difficult question, took a wide survey of the general dependencies of animals upon Plants, and of both upon common or inorganic matter; and thus arrived at the perception of the least exceptionable, and most important difference, between the two great Classes of Organized Nature,[1] as you will find in the introduction to the Catalogue of the Physiological Series[2]. He says

> The operations of animals and vegetables are attended with waste of their component parts. This waste is recruited by a supply of common matter, but common matter cannot be *immediately* converted into animal sub-

---

1   Nature,] *comma inserted to replace period to integrate with insertion at this point [ed.].*
2   as. . .Series] *RO pencil note on facing fol. 3^vwithout insert point marked.*

3

stance: the decay of animal substance cannot be supplied from common matter until it has, by certain changes been altered into animal or vegetable matter. Vegetables can [/] immediately convert common matter into their own substance, and be supplied from it; but animals cannot.

Since the time of M^r. Hunter, a justly celebrated foreign Botanist has reproduced the same maxim:[1] "It is the Office" says Mirbel "of Plants alone to transform inorganic matter into organized living bodies;—whereas Animals feed only on Organized matter."[2]

The principal food of Plants consists[3] undoubtedly of[4] the two inorganic binary compounds, Carbonic acid and water; but they are also nourished by animal and vegetable matter in a state of partial decomposition. It has been supposed that Plants could be supported by Carbonic acid and Water alone,

4

but the Experiments [/] of Hassenfratz, Saussure, Giobert, and Link have demonstrated that under these conditions Vegetables thrive badly, their growth is arrested, and they rarely bloom or fructify. With the assistance however of more congenial nutriment derived from dissolved, but not wholly decomposed Organic Compounds, Plants are able to effect the important purpose of converting binary Compounds into organized matter. Without this power, both the Vegetable and Animal Worlds would of necessity be annihilated, and successive Creations[5] of Animals and Vegetables would be required to maintain the present System of things.

For, in the vital operations of Animals a great quantity of organized matter is constantly decomposed and is rendered, at all events, useless as nutriment for other animals: but by means of Plants it is again changed into organic combinations on which the Herbivorous tribes of Vertebrate and Invertebrate animals subsist. Then again, by means of combustion and other decomposing processes an incalculable quantity of vegetable matter is continually[6] resolved

5

into binary compounds, or the ultimate [/] elements of matter; and thus the nutriment both of Animals and Plants would be constantly decreasing, if Plants were not endowed with the power of forming new ternary organized compounds out of common matter.[7]

---

1 maxim:] *over erasure with pencil colon and in margin in WC hand* '(maxim?)'.
2 matter] *followed by deletion of paragraph* 'The objections to this distinction are founded on the circumstance that certain Annelides, or red-blooded worms swallow sand and earth; and some coleopterous insects seem to feed on inorganic matter:— but in all these cases, the earth is in a state of minute subdivision and blended intimately with decomposing animal and vegetable particles.' *and on facing fol. 4^v in W. S. Bennett hand* 'They do— cow-shard is inorganic, no, it is the partly digested remains of organized plants, mixed with animal secretions equally organic—'.
3 consists] *interlined with caret.*
4 of] *interlined with caret.*
5 Creations] *after erasure of* 'ages of' *with* 'creations'*interlined above and in margin pencil* '?'.
6 continually] *over erasure.*
7 matter.] *followed by ink deletion of* ', in binary combinations' *with period after* 'matter'

Hence the whole of originally created organized matter does not continue to circulate, as it were, in the vegetable and animal worlds, passing entire from one being to another; but the perpetual dissolution of organized bodies presupposes the recomposition of organized matter from the elements, or from[1] binary compounds.

To Plants must be ascribed not only the power of producing new organized matter, through the influence of light and heat; but also that of multiplying the organizing principle from some unknown sources of the external world. By the constant decomposition of organized matter in the living machine, the vital actions are at length arrested;—the agent disappears:—or we may say, that the organizing principle is resolved into its general physical sources, whence it would seem to be reproduced by Plants.

6   We[2] have already spoken, in the preceding Lecture of the difference in the elementary composition of Animals and Vegetables. If we were to confine ourselves to the examination of the more complicated organizations in the two Kingdoms, we might suppose the distinction founded upon the presence of Azote as a Chemical constituent of Animal bodies to be absolute; but like

7   most of [/] the distinctions between plants and animals, this fails in its application to the simpler species:—for in the lowest organized Cryptogamic Plants, Nitrogen is distinctly present. Subject however[3] to this exception it may be recieved as a general proposition that Carbon predominates[4] chiefly in the more solid parts of Vegetables; while, abundance of[5] Nitrogen characterizes[6] those of animals.

The earth by which the cellular tissue of Animals is hardened to form their system of support and protection is generally Lime: the hard parts of vegetables, besides lime, yield also alumine and silica. Many of the Infusoria, however have silicified Skeletons.

(contd.)
    inserted. [ed.].
1 or from] *over erasure with pencil* 'or from' *as erased interline.*
2 We] *follows deletion by vertical pencil scoring of* '{¶}If we are indisposed to admit the multiplication of the organizing power out of unknown sources of the external world by means of Plants, then we must adopt the Theory of Haller and Bonnet, that the seemingly endless multiplication of the Organizing energy by growth and propagation is wholly by an evolution of pre-existing Germs:— or we must admit the inconceivable notion that the division of the organizing power consequent on propagation does not diminish the intensity of the same. The fact, however will always remain, that in the Death of Plants and Animals the organizing principle {*interlined in pencil above deletion of* 'energy'} is dissipated ineffectively, or merges into its general Physical Sources.'.
3 however] *interlined in ink with caret in insertion.*
4 Subject. . .predominates] *RO insertion from facing fol. 8ᵛ written over pencil* 'Subject to these exceptions, we may believe it as a general proposition, that' *to replace deletion of* 'Carbon, however, predominates' *with insertion point marked by lead line & caret after* 'present'.
5 abundance of] *interlined above deletion of* 'in general'.
6 characterizes] *interlined in ink above deletion of* ' exists in great proportion in'.

The elementary combinations which form what are termed the proximate components of organized bodies are more numerous but more simple in Plants than in Animals. In the former they are ternary, in the latter quaternary compounds:—in both they are divisible into acids and oxydes:—but in Plants there is a third order of substances of which we discover no traces among animals;—these are the vegetable salifiable bases.

8   With respect to the *acids*, it is to be observed that there are few which are common to both the Vegetable and Animal Kingdoms; and whilst their number is small among animals, it is very great among Vegetables.

Of the organic *oxydes*, some, as albumen, ozmazome, and sugar, are common to both Animals and Vegetables; but they occur in very different proportions in each: Sugar, which is so abundant among[1] Plants is scarcely to be detected among Animals, save under certain morbid conditions; and ozmazome which is so universally distributed among Animals, has only hitherto been discovered[2] in a few fungi.

Of the ternary compounds of Carbon, hydrogen, and Oxygen, such as Starch, Gum, Sugar, the resins, woody fibre, fixed oils, volatile oils, Camphor, extractive matter; all of which enter so largely into the constitution of Vegetables,—but a very few can be discovered among Animals: such as the Sugar of Milk, and of Urine; the resin of the bile and of urine, the elaine[3]

9   and [/] stearine of the fat; the volatile oily principle of Castoreum, and the Camphor of Cantharides.

The quaternary organic compounds of carbon, hydrogen, oxygen, and nitrogen, which form most of the proximate principles of Animal bodies, are, on the contrary, very rare among Vegetables. The most common of these are albumen, gelatine, fibrine, animal mucus, and ozmazome; the less common are the matter of Saliva, caseous matter, urea, and the pigmentary substance of the Eye.

Still, vegetables are not without several of these quaternary compounds, such as albumen, and ozmazome; and they even possess others which are peculiar to themselves, such as gluten, the matter of the pollen of flowers, indigo, and several extractive colouring principles: to which must be added the whole exclusive class of salifiable bases, quinia, cinchonia, veratria, strychnia, morphia, &c. &c. which are compounds of Carbon, united in a[4] large proportion, with a little oxygen, hydrogen, and nitrogen.

Our knowledge of the characteristics of Plants and Animals in reference to

10   their elementary parts [/] and proximate principles is derived from the labours of the Chemist, and these investigations properly belong to his department of Science:—yet we cannot perceive the attempt which Hunter made to display the simple components of organized bodies, at the commencement of his

---

1 among] *over erasure.*
2 discovered] *over erasure.*
3 the elaine] *over erasure.*
4 a] *interlined with caret after* 'united' *and transposed for grammatical sense [ed.].*

great collection, without a conviction that the Chemist in this field would have been to him of all men the most interesting. For, what is Chemistry but a species of dissection, or unravelling of the elementary components which elude the knife and microscope of the Anatomist.

It would have been an omission to have past over without notice the vast accession of Physiological knowledge which the labours of the chemists have contributed. I have endeavoured to give a succinct allusion to the important results which they have worked out, and I proceed[1] next to review a series of facts principally established since the time of M^r. Hunter by means of Microscopic observation.

In the comparison of Animals and Vegetables with reference to their primary tissues, it may be stated as a general proposition, that Vegetables are
11 more simple in this respect. This difference cannot, indeed, be appreciated in the lowest forms of the two Kingdoms, but becomes sufficiently conspicuous as we ascend to the higher organized species.

The Cryptogamic plants consist of a homogeneous aggregate of oblong or rounded cells filled with fluids or a granular substance[2]; and, like the fresh-water Polype, present no trace of proper tissue. As we ascend to the Phænogamia we find two elementary tissues, the tubular and cellular, separately developed and the whole body of the plant is invested with a distinct integument, or bark.

The cellular tissue of vegetables, however, differs from that of animals, inasmuch as the cells are closed, entire, have[3] firm parietes,[4] are separated by intervals, and are connected together in a reticulate form. The tubular tissue of vegetables presents two distinct classes of vessels, the spiral or respiratory, and the nutrient:—but both are distinguished from the vessels of Animals in having few or[5] no ramifications. Closed[6] cells and unbranched vessels are the
12 positive characters of the vegetable, [/] and compose its whole internal substance. By their various combination and arrangement they form the root, the stem, the leaves, the flowers and the fruit:—and it is wonderful how nearly all vascular plants, however different in their outward appearance, resemble one another in their intimate structure.[7]

---

1   It. . .proceed] *RO ink insertion from facing fol. 11^V to replace deletion by vertical pencil scoring of paragraph on fol. 10* '{¶}I must have felt it to be an omission of my duties to have passed over the vast accession to Physiological knowledge which has {*deleting interlined 'benefitted'*} flowed from the labours of the Chemist without briefly but succinctly alluding to the important results which have flowed from them. I proceed'.
2   substance] *followed by caret and ink deletion of interlined* 'the globuline of Turpin—'.
3   have] *interlined in pencil above deletion of* 'with'.
4   parietes,] *before pencil deletion of* 'and'.
5   few or] *inserted in ink before line with pencil interlined* 'few or' *above and pencilled* 'X ?' *in margin.*
6   Closed] *interlined above deletion of* 'Solid'.
7   structure.] *on facing fol. 13^V in RO pencil without insertion point indicated* 'This is seen by glancing over the pages of the works which illustrate the microscopic texture of

Globules or fibres[1] and not cells are the elementary parts of most of the Animal tissues, and the vessels of animals are always branched;—the component tissues are also more numerous in animals than in plants:—besides the cellular and vascular there are the nervous and muscular tissues; and in a subordinate category may be ranked the tendinous or[2] fibrous, the osseous, cartilaginous, and horny tissues, which are principally modifications of the cellular tissue, and are less uniformly present in the composition of animal bodies.[3]

Such are the principal differences between Animals and Plants as relates to their intimate[4] Structure. Hunter founds his distinction of Vegetables and

13   Animals on a comparison of one of their Vital [/] Manifestations;—viz. their powers of assimilation:—let us proceed briefly to compare them with reference to the other vital functions. Development, growth, incitability, propagation, death[5], are the general phænomena of all organized bodies, and the consequence of organization. Animal bodies alone are characterized by the possession of other properties; which are therefore called animal in contradiction to the general or *Organic*. These[6] are the powers of feeling and of voluntary moving either the whole or a part of the body.[7] Motion alone is not the characteristic of an animal:—some animals indeed[8] are as destitute of the power of locomotion as plants:[9]—neither can we deny motion to plants, since their whole life is attended with imperceptible motion,—the motion of the Cambium and Sap.

Growing plants also incline to the light; their roots extend towards the most congenial soil;—the climbing plants entwine their tendrils in definite directions around the bodies which serve as their support. The stamens, or

14   male organs of Generation at the period of fertilization, bend in [/] regular succession[10] towards the female part or Style: or the Style inclines towards, and courts each stamen in succession.[11]

(contd.)
different vegetables & parts of vegetables'.
1   or fibres] *interlined with caret.*
2   or] *over erased* 'and'.
3   bodies.] *with RO pencil note on facing fol. 13ᵛ without insert point indicated* 'As Hunter here shows us'.
4   intimate] *interlined in ink with caret after* 'their'.
5   death] *interlined above deletion of* 'perishability' *with faint pencil line to nearly illegible note on facing fol. 14ᵛ*] 'expressed, but rather Jeremy Benthamish'.
6   These] *over erasure.*
7   body.] *heavy double ink ink slash lines as if to mark pause. Period added [ed.].*
8   indeed] *interlined in pen over pencil with caret.*
9   plants:] *on facing fol. 14ᵛ in pencil in W. S. Bennett hand* 'True as to *locomotion*—but what animals have no motion? to not mix words, I would insert "indeed".'.
10   succesion] *with pencilled additions on facing fol 15ᵛ, no insert points marked:* 'Luxifrage—Lily—Expound Diagrams [;] Stamens[,] Hedysarum[,] Oscillatoria[,] *Calendula fluvialis* [,] close Dark in rain [,] *Mirabilis.*'.
11   succession] *with pencil slash line to indicate either pause or insertion.*

In the leaves of the *Hedysarum* we perceive during[1] the day time, a constant motion of the lesser folioli which resembles that of the respiratory apparatus of the higher animals in being alternate although here probably the analogy ceases. In some of the lower organized plants, as the *Oscillatoriæ*, the entire[2] plant is in constant motion like the pendulum of a Clock. The flowers of *Calendula pluvialis* close in the absence of Light, or when rain comes on:—

Those[3] of the *Mirabilis* [4] expand in the evening, and close on the approach of the morning Sun.

Lastly, some plants, as the *Dionœa Muscipula* and *Mimosa* manifest conspicuous motions consequent on external stimulus;—and the same movements[5] take place in the Sensitive plants whether excited by Mechanical, galvanical, or chemical stimuli, or by changes of Temperature and degree of illumination[6].

Hunter, who was among the first to consider Plants and Animals as members of one great Family in [/] Nature studied with particular attention and much detail the phænomena of these motions.

He divided these motions into three Classes.

The first kind of motion was that of the particles of the fluids and solids in the operations of growth, repair[7], excretion, secretion, &c.

The second kind of motion has immediate reference like the first, to the internal operations of the individual, such as growth, excretion, &c. but is a motion not of[8] particles but of whole parts, in pursuit of external influence; "arising from the stimulus of want," to use M^r. Hunter's expression.

The examples which he gives of this kind of motion are the Movements of the Tendrils and Claspers of Cirriferous Plants, and the more conspicuous motions of the leaflets of the *Hedysarum*.

The third kind of motion is exclusively produced by external stimulus, and is also a motion of whole parts. In some cases it seems to have reference to the internal operations of the individual, as in the case of the opening and closing of the leaves and flowers of certain plants, from the absence or presence of Light, which may affect the respiratory or excretory functions of the leaves. In other cases the motion has no immediate reference to the internal operations of the individual, but is simply the consequence of the Stimulus;—as in the approximation of the leaflets of the *Dionœa Muscipula* which

15

16

---

1 during] *over erasure.*
2 entire] *over pencil* 'entire'.
3 Those] *preceded by em dash as if to indicate new paragraph.*
4 *Mirabilis*] *after deletion of* '*Convolvulus* '.
5 movements] *on facing fol. 15^v with no insert point* 'Preparations'.
6 illumination] *WC note on facing fol 15^v* 'See Dr Daubeny's paper in the last part of Philos Trans on this subject. Some well-imagined experiments'.
7 repair] *follows erasure of* 'and'.
8 not of] *after deletion of repeated* 'not of'.

appears to be designed to confine the stimulating body:—and in the bending downwards of the leaves of the *Mimosa*, as it were,[1] to avoid or recede from the object which irritates.

Hunter's first kind or Class of Motions is common to both Plants and Animals, and corresponds with the 'Insensible organic contractility' of Bichat. The effects resulting from it are much greater in Plants than in Animals, and therefore we must suppose it to exist in a much higher degree in the more simple division of Organic Nature.

Thus, a Vine by virtue of this property of its Tissues, will grow upwards of twenty feet in a single summer;—while a Whale probably does not increase to the same extent in as many Years.

17   Hales found that a small Vine was capable of sustaining, and even of raising a column of Sap 43 feet high, [Vol. I. p. 112.114. Expt. 36.] while the forces moving the nutrient fluids of a Horse could only support a column of blood 8 feet 9 inches high.[2]

When, also, we consider that the sap of the tallest tree must be sustained and raised from the root to the most distant branches, it must appear that the propelling power in such Vegetables far exceeds that in any animals:—and, indeed, it is such as the texture and construction of a vegetable only can support.

With respect to the motions of the nutrient fluids in Vegetables it must be observed that it differs from that of the blood in Animals in being wholly dependent on influences which are external to the individual, especially upon the actions of Light and Heat. It is always a more simple Motion than the circulation of the Blood:—it is always unaccompanied by, and is independent of a special internal propelling Organ or Heart.

18   Some Physiologists have attempted to account for the motions of the vegetable juices without having recourse to any inherent vital properties of the surrounding Tissue. Thus, Dutrochet, observing that water contained in an upright slender glass-cylinder began to manifest an ascending and descending rotatory movement when heat was applied to one side of the vessel, sought to elucidate the motions of the Sap in Vegetables and its rising at the returning warmth of Summer on the same principle.

But the currents of the Water here obviously depend on the ascent of the warmer particles with which the rotation commences, and all the particles on the heated side of the Cylinder rise, while those next the colder side descend. But later[3] and carefully repeated observations on the motions of the Sap have demonstrated that this fluid moves in different directions in different vessels, which nevertheless are under the same circumstances as regards Heat. This phænomenon may be seen in fine Sections of the leaf-stalks of the Fig, and

19   still more satisfactorily in uninjured leaves where the different streams of Sap

---

1   as it were,] *interlined above deleted* 'in order'.
2   Vol. . . .high.] *faint vertical pencil score through this. Deletion uncertain.*
3   later] *over erasure.*

may be observed.

Schultz who has successfully studied the motion of Sap in higher organized plants, regards it as a perfect circulation, ascending in one set of Vessels descending in another; and these communicating by transverse vessels.

In the lower organized plants the motion[1] of the nutrient fluids affords at the present day a well-known and favourite phænomenon to the Microscopic Observer, and Instruments are constructed for the express purpose of observing it.

In the *Chara*, (a simple aquatic plant) the granular[2] sap rotates in closed elongated spaces or cells, obviously from a cause distinct from the partial application of Caloric; yet at the same time independent of any contraction of[3] the parietes of the cell containing the Fluid.

This interesting phænomenon of Organic motion or rotatory motion of the Sap in cells, or the closed cavities of the joints of Plants, is termed a *Cyclosis*.

20   A similar cyclosis of the Sap has been observed [/] (by Meyen) in the cells of *Valisneria spiralis*, and in the hairs of the radicles of *Hydrocharis morsus Ranæ*. It seems to depend on a faculty of attraction and repulsion in the parietes of the containing cells; and is closely analogous to the motion of fluids along the surface of the membranes of many Animals which is occasioned by the action of microscopic vibratile cilia:—yet no cilia have been detected in the cells of the Chara.

In the higher organized vegetables ordinary capillary attraction undoubtedly performs an[4] important part in the propulsion and support of the columns of Sap. And since light has a manifest influence on the growth of plants, we must suppose that this external stimulus attracts, or otherwise aids in the movement of the nutrient juices.

In Animals, on the contrary the motions of the nutrient juices are nearly independent of external impulse, the principal circulating force is the central contactile organ, the Heart. The actions of this Muscle depend, indeed in

21   some degree[5] on the changes of the Blood in Respiration by the [/] influence of the external atmosphere:—but this relative dependency, and the support afforded to the containing vessels by the surrounding atmospheric pressure are the only external influences to which the circulation in animals is subject. And the phenomena of Hybernation show that the dependency of the Heart's

---

1   Schultz. . .motion] *with WC note in pencil on facing fol. 20*[v] 'Your old friend M[r]. Solly could give you the latest discoveries on this subject, in consequence of various plants examined by him and M[r]. Havell at the rooms of the Society of Arts Adelphi— I saw the objects beautifully distinct. {¶} In the *Chara* , &c &c the circulation seen beautifully.'.

2   granular] *over erasure.*

3   contraction of] *over erasure.*

4   performs an] *over erasure with erased pencil interline of* 'performs'.

5   in. . .degree] *interlined with caret after* 'indeed'.

action in respiration is not an essential one.[1] In some of the lower animals, as in the Sertulariæ and Plumulariæ, the motion of the nutrient fluids resembles the Cyclosis of the Sap in the lower vegetables.

The more conspicuous motions of Plants, such as are exhibited in the tendrils of climbers, the leaflets of the *Hedysarum*, of the *Dionœa*, and *Mimosa* are more easily and satisfactorily distinguishable from those of Animals; although the actions of the Fly-Trap and Sensitive plant being induced by external Stimulus present the nearest apparent similarity to the muscular motion in Animals.

With respect to the climbing of Plants, some Botanists, as Palm, have considered this action to be sufficiently explained by the circumstance that these plants describe circles with the extremities of their twigs, and, by this mode

22   of Growth, attain near [/] objects, and twine round them:—Some, as the Hop, always turning to the left;—others, as the Convolvulus as constantly to the right:—while the Tendrils of the Vine have the faculty of twisting in either direction.[2]

But there are some phænomena which cannot be explained on this simple[3] mechanical principle; and the knowledge of which probably induced M^r. Hunter to refer his second Class of Vegetable motions to an inherent irritability, or susceptibility to peculiar stimulus. I allude to the climbing of the *Cuscuta*, the tendrils of which, it is well known, will only entwine themselves around living plants; showing indisputably that the inclination of their extremities towards one side is not a[4] necessarily inherent law of their growth but an action;—and an action which is the consequence of a particular stimulus, or organic attraction.

Nevertheless, it is to be observed that this is an action which would never take place without the external Stimulus; and that it has no analogy whatever

23   to a voluntary motion. The rising and [/] falling of the leaflets of the Hedysarum in like manner are the necessary consequences of the stimulus of light, and are only suspended during its absence.

The motions of the arms of the Polype, (which Mr. Hunter in his Croonian Lecture terms a *random action*, and considers to be of the same kind as those of Plants,)[5] differ from all the motions we have been considering in this essential particular, that they originate from an internal impulse,[6]—while the movements of the leaves of the Mimosa are invariably a passive response to the application of Stimuli.

---

1   And. . .one.] *RO pen insertion from facing fol. 22^v written over pencil* 'and the phenomenon of hybernation show that the dependency of the heart's action in Respiration is not an essential one.' *with pencil line to insert point.*

2   direction.] *period inserted [ed.].*

3   simple] *over erasure.*

4   a] *interlined with caret after* 'not'.

5   (which. . .Plants,)] *pencil parentheses.*

6   impulse,] *before pencil deletion of* 'or are voluntary;'.

There is also an essential difference in the nature of the motion itself in the Polype, and the Sensitive plant. If we touch the feeler of a Polype, it recedes from the irritant by a true contraction of the part within itself: which contraction appears to result from the injury experienced by that part of the nervous System which is disseminated through the feeler touched. In the case of the Sensitive plant there is nothing like this contraction of the part touched, but only [/] an articular plication of the neighbouring part,[1] without any of the dimensions of the irritated leaf being altered.

24

As[2] the nonspontaneous motions of the Sensitive Plants[3] are extensive and comparatively energetic, so the power[4] on which they immediately depend results from the organization of particular parts of the plant, and is, as it were, concentrated in such parts.[5]

Nevertheless, the motion of the part is not produced by the contraction of a fibre drawing the moveable towards the fixed part, but by the distension of cells which push the moveable from the fixed part.[6]

The most simple mechanism employed in these movements of Vegetables is that which effects[7] the opening and closing of leaves and flowers;—and it is most easily studied in the latter. The flower of the *Mirabilis longiflora* expands its infundibuliform corolla in the evening, and closes[8] it at day-break. This flower is formed by the confluence of five petals, each of which have a [/] mesial nervure. The five nervures[9] which support the membranous substance of the corolla are the sole agents, both of the movements of expansion, or the awakening, and of the closing or sleep of this nocturnal flower.

25

But, it will be asked, how can the same nervure draw the petal towards the centre by curving inwards, and push it outwards by an excentric curvature?—The anatomy of the nervure, and a simple experiment, explains this apparent anomaly.—the external part of the nervure is composed of a cellular tissue permeated by the Sap; the cells being disposed in a longitudinal series, and decreasing in size from within outwards, so that, in the turgescent state of the cells, the tissue which they form is curved with the concavity outwards; and it is this tissue of the nervures which produces the

---

1   part,] *in pencil in W. S. Bennett hand on facing fol. 25ᵛ* 'and where the stimulus is strong, depression of the leaf stalk'.
2   As] *in margin pencil* 'X'.
3   Plants] *over erasure with pencilled* 'X?' *in margin and in RO pencil on facing fol. 25ᵛ* 'Expound—Diagram[,] Point out parts'.
4   power] *with erased interlineation of* 'such' *above*.
5   parts] *in RO pencil on facing fol. 25ᵛ with horizontal pencil line for pause* 'Intumescence—Texture—series of Cells—sap—Oxygen.'.
6   part.] *in RO pencil on facing fol. 25ᵛ* 'pair of bent springs—opposed—keep each other straight—Experiment—*Lindsay {with faint deletion}'* .
7   effects] *over erasure*.
8   closes] *over erasure*.
9   nervures] *over erasure*.

expansion of the flower.

The internal part of the nervure is of a more complicated structure: it is occupied by an extremely delicate fibrous[1] and globular tissue situated between
26   a stratum of tracheal vessels [/] which separate it from the sap-cells[2]; and a
stratum of superficial cells occupying the most internal part of the nervure: if
the respiratory and fibrous part of the nervure be separated from the external
cellular portion, the latter curves outwards, while the fibrous part curves
inwards. This invariably happens, and the two opposite curvatures are
constantly maintained. Hence the cellular, or nutritious tissue of the nervure
is the agent of expansion; while the respiratory and fibrous tissue produces
the closing of the flower.

But, how do these parts operate;—what changes are effected in them to
produce these motions;—and on[3] what cause do they depend?—

Dutrochet has determined this question as regards the expansion of the
flower, by a very simple experiment:—he took the nervure of one of the
petals of the *Mirabilis* when the flower was closed and immersed it in water,
upon which it became strongly curved outwards, taking the form
corresponding to the petals of the awakened or expanded flower:—He then
27   placed the same [/] nervure in Syrup, upon which the nervure curved itself in
the opposite direction. The explanation of these simple experiments is equally
simple and intelligible:—The cells of the nervure which were comparatively
empty before the first immersion, became distended with water by
endosmosis or imbibition, and produced the outward curvature from their
mechanical arrangement. When placed in the Syrup, the water of the cells
escaped by exomosis into the denser exterior fluid, and the opposite curvature
was produced.

It is not so easy to explain the mode of action of the fibrous and respiratory
tissue which composes the inner side of the nervure, although we know that
it is this structure which effects the curvature of the petals towards the centre,
and closes the flower. It is not by the simple elasticity of the gaseous fluids
contained in the cells or spiral tracheæ[4] operating when the Sap-cells of the
opposite side of the nervure are empty; because if the nervure is retained in
28   water for six hours, it begins to curve in the opposite direction [/] to that
which it took on the distention of the cells upon its first immersion. This
change of direction, produced by the action of the fibrous and respiratory
part of the nervure overcoming the force of the Cellular part, takes place
independently of any change in the degree of Light to which the part is
exposed; but one condition is essential to the action, viz. that the fluid in
which the part is immersed should contain *Oxygen*:—In deoxygenated water,
no action of the respiratory and fibrous part of the nervure takes place

---

1   fibrous] *over erasure.*
2   sap-cells] *over erasure.*
3   on] *over erasure.*
4   tracheæ] *over erasure.*

however long it is immersed, and the nervure remains permanently bent outwards by virtue of the distention of the longitudinally disposed cells by endosmosis or imbibition of the surrounding fluid.

Dutrochet concludes, therefore, that the fibres which are placed between the stratum of spiral tracheæ and the stratum of air-cells, and which curve the nervure by extention and not by contraction, act under the influence of the surrounding Oxygen:—And here, for the present, rests[1] the explanation of the[2] mechanical changes in the parts on [/] which the singular phænomenon of vegetable motion depends.[3] It is obvious that the distention of the cells of the expanding mechanism of the flower must, in the natural state, arise from a different and higher cause than mechanical imbibition; the fluid must be impelled into the cells by an action. *Internal action*[4] is still more certainly connected with the force which closes the flower, yet both are widely different from muscular contraction.

In the Plants whose leaves exhibit the phænomena of alternate collapse and expansion, the moving powers are situated in enlargements at the base of the leaf-stalks, and stalklets. The irritable intumescence is considerable in the stalklets supporting the leaflets of the common French-Bean:—but the properties of this part have[5] been most studied in the Sensitive plant, where they are most conspicuous. The essential structure is the same as in the nervure of the petal of the flower of the *Mirabilis* or Convolvolus[6], but it is only in this simple condition that it can be [/] successfully studied, and the opposing powers separately experimented upon.

As the moving force in each case acts by extension, like a curved spring which has been temporarily straightened by an antagonist force, we can readily understand why a partial[7] incision of the enlargement of the leaf-stalk in the mimosa should be followed by its incurvation towards the wounded side. This phænomenon was first discovered experimentally by Lindsay; but the fact was first publickly made known, with a series of original experiments, by Dutrochet.

As the fibrous structure of the intumescence is that which acts in the Sensitive, as in other plants, in the sleep or collapse of the leaves, and as the incurvation is in the same direction when induced by external irritation, we infer that it is the fibrous and respiratory structures which act when the plant is touched; and that the vegetable irritability is the property of a fibre which

---

1   rests] *interlined with caret after* 'present'.
2   the] *interlined with caret.*
3   depends.] *before erasure of two inches of line.*
4   Internal action] *faint pencil underline.*
5   have] *corrected in pencil from original* 'has'.
6   Convolvolus] *sic. This is spelled correctly* ('convolvulus') *in the RCS original draft, fol. 54.*
7   partial] *over erasure.*

curves itself by oxygenation: Hence a weak and sickly Sensitive Plant is not[1]
31   susceptible of Stimulation, and [/] the exhaustion of the oxygen in the
respiratory part of the intumescence may be the cause why the motions of the
leaves of the Sensitive plant become gradually more feeble, and finally cease,
upon reiterated stimulation, as shown in the Experiments of Hunter[2] which
you will find in the Catalogue of these specimens—[3]

With respect to the part of the intumescence most susceptible of the
impressions, Dutrochet has conjectured that it is the globules which surround
the fibres. M[r].[4] Mayo, has observed that the under part of the tubes is the
most sensible to touch or pressure. Hunter, and others, have determined by
Experiment that an impression upon a part of the plant is propagated to other
parts, and excites them to action. Yet with all these Analogies to the
phænomena of the instinctive motions of animals, there remains this great and
important difference, that the motions of the parts of Vegetables always
follow, and are dependent upon, a physical Stimulus being applied to the
irritable part; while in Animals they arise out of an internal determination from
32   parts not moving, as the Nerves, to the moving powers.—[/] And in this
comparison we must be careful not to confound irritability with sensation.—

Plants are irritable, but not sensible. The muscles of an animal, out of the
body contract from irritability, not from sensibility. If sensation existed in
plants, there would be a manifestation of consciousness; but we see only the
Automatic repetition of the self-same action under every kind of stimulus:—
even the Vegetable which from the high development of its irritability has, in
common language obtained the erroneous name of *Sensitive*, responds to
every kind of stimulus, whether Mechanical, Galvanical, or Chemical, by the
same unvarying inflection of its leaves and leaflets.

The remarkable properties of the *Mimosa sensitiva*, *Hedysarum gyrans*,
and *Dionœa Muscipula*, are, then, referrible to irritability,—or excitability;—
to a property which the muscular fibre of Animals possesses, and by which
that fibre contracts from the application of stimuli[5] independent of the
operation of the nervous energy. It is the superaddition of the nervous system
33   by which an [/] Animal can excite contractions of its irritable parts
independently of external Stimuli, which forms the grand differential
character which distinguishes it from, and raises it above the Vegetable.

M[r]. Hunter conscious of this distinction, and adopting the Hallerian view[6]
of the independence of the irritable quality of the Muscular fibre of the

---

1   not] *interlined over erasure with* 'not susceptible' *also written by RO on facing fol. 31*[v].
2   Hunter] *deleting period to continue with pencil addition.*
3   which. . .specimens—] *RO pencil addition continued on to facing fol. 32*[v] *with deletion
    of* '*introdu {double deleted } Catalogue of these specimens'.
4   Mr.] *in pencil.*
5   stimuli] *after pencil deletion of* 'many' *and before deletion of* 'distinct from and'.
6   and. . .view] *RO interlined in pencil with restoration of deleted* 'of' *following for
    grammatical sense [ed.].*

nervous system,—has placed his preparations illustrative of the nature and properties of Muscles immediately after the examples of those plants in which irritability is most conspicuous:—and philosophically defers the consideration of the nervous system until he has first illustrated all those functions relating to individual preservation which are common both to Plants & Animals; as Incitability and Motion from Stimulus, Nutrition, Respiration, Excretion[1].

34   Amongst the important motions in the animal system which go on independently of the nervous system, and in consequence of external Stimulus, those of the innumerable microscopic cilia which beset many membranes in the highest as well as the lowest classes of Animals, are not the least extraordinary. I have observed the oscillatory motions of the cilia on a detached portion of the Gill of a Bivalve Mollusk to continue without ceasing for more than three days, until they[2] were detached by the putrefactive process; yet the substitution of fresh-water instead of sea-water in the vessels containing other portions of the same Gill, produced an instantaneous paralysis of the vibratile cilia and all motion ceased.

    I regard, therefore such[3] ciliary movements as are not under the control of the will, (for in some cases, where the cilia are the organs of locomotion, these motions[4] can be voluntarily arrested) but the non-spontaneous ciliary motions I regard as[5] of the same essential nature as the constant vibrations of the leaflets of the Hedysarum,[6] to arise from an inherent irritability, excited by a certain external stimulus;[7] but to be, in both cases, independent of a nervous System.

35   The presence of light is essential to the vibrations of the leaflets of the Hedysarum;—a surrounding medium of sea-water seems equally indispens-able to the vibrations of the cilia in the edible mussel[8]:—without these exter-nal stimuli they cease:—the animal can neither arrest, excite, or accelerate the vibrations of these cilia; but their actions continue in a separated portion of the membrane endowed with them as long as the stimulus is supplied, and the necessary physical conditions of the irritable part are present.

---

1   Excretion] *followed by deletion of paragraph* '{¶}Cuvier has been said, but erroneously, to have adopted the Hunterian Arrangement of the Organs of Animals, in his "Leçons d'Anatomie comparée"; but he differs from Hunter at the outset, by proceeding from a consideration of the Bones and Muscles to describe that peculiar system which in Animals principally {l *fol. 33*} determines the contraction of the irritable fibre.' *followed by horizontal pencil line across page for pause.*

2   I have. . .until they] *note on facing fol. 35ᵛ in W. S. Bennett hand* 'In going from plants to animals shd not something be said of those organisms of the debateable land Corallines &c'.

3   such] *RO insertion over erasure with second* 'such' *interlined above in pencil.*

4   motions] *interlined in insertion.*

5   as are . .as] *RO insertion from facing fol 36ᵛ with insertion point marked by line and caret after* 'movements' *before deletion of* 'to be'.

6   Hedysarum,] '='*mark inserted above comma in pencil.*

7   stimulus;] *with pencil* '='*mark inserted.*

8   mussel] *over erasure.*

If this be a correct view of the nature of the non-spontaneous[1] ciliary motions, it is obvious that they are no test of animality. When we see the embryo or gemmule of a Sponge moving through the water by the action of the superficial cilia, we witness the effect of a high degree of that irritability on which the inflection or vibration of the leaves of Plants depends, but nothing more. There is not the slightest sign of Spontaneity in the motions of the gemmule: it goes blindly, affected only by the stimulus of light which it seems to recede from, like the petals of the *Mirabilis*, or other nocturnal

36 flowers, which close at the approach of [/] day. Nay, the Embryos of some indubilitable Plants[2] manifest the same non-spontaneous[2] locomotions, although the locomotive organs in these have hitherto escaped detection. This phænomenon has been frequently observed in the Algæ;—and has been particularly described by Trentepohl in the embryos of the *Conferva*[3] *dilatata*;—and by Treviranus in those of the *Conferva limosa*. The Zoocarps of Bory St. Vincent, also pour out self-moving embryos or germ-grains, which after a time, fix, and are developed into jointed filaments like the parent:—but these are not, therefore, animals[4].

Ehrenberg has recently affirmed, in an interesting Memoir in Poggendorf's

37 Annals [Poggendorf's Annals. 1832],[5] that he can distinguish [/] the locomotive Embryo of the Algæ from the simplest Monad,[6] as easily as he can distinguish a Tree from an Eagle.

The *nerves* are the organs on which spontaneous motion and sensation depend; and the possession of these chiefly distinguishes the Animal from the Vegetable: the nervous System has therefore been termed the essence of an Animal. All the other Systems in the body are subject, in their plan of Arrangement, to the modifications of the nervous system, upon which, therefore, the primary divisions in the natural classification of the Animal kingdom are founded.

It has been objected that a nervous system has not been detected in all animals;—but as in every Species in which nerves have been indubitably[7] detected, sensation has been found to depend exclusively upon them, we are hence led to assume that all animals in which Sensation is observable, must

---

1   non-spontaneous] *interlined by RO in ink with caret.*
2   nonspontaneous] *modified in pencil by interline of* 'non' *from* 'unspontaneous'.
3   *Conferva*] *deleting pencil dash on line break after* 'Conferva'.
4   animals] *followed by pencil deletion by vertical scoring of paragraph* '{¶} Nothing like Animality or Spontaneity, as I have said before, can be detected in the moving microscopic germs, whether of the Polype, the Sponge, or the marine Algæ, which have just been mentioned:— they never stop to take food, or vary their course, to avoid or attain a definite object; but proceed in an unvarying course 'till irritability is exhausted, like the Infusoria.'.
5   [Poggendorf's. . .1832]] *inserted in text in brackets to replace original footnote reference by Owen [ed.].*
6   from. . .Monad] *inserted above line with caret.*
7   indubitably] *with erasure of pencil interline of* 'indoubtedly' *above.*

have it depending on a nervous system.

Recent and more accurate researches on this system have demonstrated its existence in many of the Animals in which nerves were (a few years ago) denied to exist; and where the nervous matter is not aggregated into the filamentary form, there is much reason to believe, with Hunter, that it is present, but diffused through the homogeneous substance of the body, as in the *Hydra* or fresh-water Polype:—and the extreme degree to which this and correspondingly simple *Acrite* [1] animals are divisible, may depend in a great degree on the dissemination, or[2] the absence of a centralized aggregation, of the nervous particles.

There is not any of the great divisions or classes of the Radiata of Cuvier, in which nervous filaments have not been detected in the higher organized Species, to[3] which we progressively ascend from the most simple forms, without any violent interruption. Among the Entozoa the nervous substance is dispersed in the Hydatid and Tape-Worm: but it has been detected in the filamentary form in the *Distoma hepaticum* by Mehlis; and[4] in the *Linguatula* (by myself) where it is even aggregated in the ganglionic form, near the head, as shown in this preparation. In[5] the great *Ascaris* and *Strongylus* the nervous filaments are easily demonstrated midway between the great longitudinal Vessels. Ehrenberg has described the nerves [/] in the Medusa; D[r]. Grant has delineated them in the Beröe;—Cuvier and Tiedemann have long ago demonstrated the nervous filaments in the Star-fish and[6] other Echinodermata. Lastly, by the labours of Ehrenberg, the Infusoria have been proved to possess all the essential structures, as they had long been known to manifest the grand characteristics of a true animal.[7]

In the simplest Infusoria, Ehrenberg has found a mouth and a complicated stomach;—in others a mouth with teeth, an intestine and anus. In the *Vorticella* the digestive bag is situated within the parietes of the Cup;—and in the Rotifers, or wheel-animalcules, Ehrenberg has demonstrated the mouth and teeth; the female organs, muscles, ligaments, a vascular system, nervous filaments, and ganglions and rudimental organs of vision. These ocelli, or coloured eye-specks are of great importance in regard to *determining*[8] the existence of the[9] Nervous System of the lowest animals. In the *Nereis* in which the nervous system is palpable, the pigment of the eye-specks covers a bulbous enlargement of a nerve, like a cup:—[/] We may therefore safely infer that the same sensitive substratum exists in the *Planaria* and all the

---

1 *Acrite*] *pencil written over pen with pencil* 'X?' *in margin.*
2 , or] *inserted in pencil over* 'on'.
3 to] *preceded by open parenthesis without close parenthesis elsewhere in text.*
4 ; and] *faint as if intended erasure.*
5 In] *inserted over erasure with erased pencil* 'In' *interlined.*
6 and] *with pencil cross-out. Retained for grammatical sense [ed.].*
7 animal.] *with pencil line to indicate pause following.*
8 *determining*] *pencil underline.*
9 existence of the] *RO interlined in pen with caret and in margin pencilled* 'X?'.

lower animals which have ocelli, although the arrangement of the nervous matter in the filamentary form may not be present or discernible in the rest of the Body.[1]

When we thus enter into a close examination of those organisms which are placed, as it were, on the lowest step of the Animal and Vegetable Series, and carefully compare the motive and irritable phænomena manifested by them, and which have been supposed to destroy the distinction between the two organized kingdoms, we shall find good grounds for attributing even to the lowest class of Animals the grand characteristics of Sensation and Voluntary Motion.

Many aggregated marine animals have a branched and plant-like form; and are fixed, as it were by a root, to the bottom:—but the individual vital endowments of each polype, and its voluntary movement on its common stem, exhibit a multiplied animal organization, not the organization[2] of a Vegetable.

41　　　　If the Sponges and other marine organized [/] productions are more doubtful with respect to their vegetable and animal nature, the absence of all voluntary motion, either of the whole or of individual parts, points out, notwithstanding their chemical constitution, their close relationship with the[3] Vegetable Kingdom.[4]

The attributes[5] of Sensation and Voluntary Motion modify, as might be expected, all the other Functions which Animals possess in common with Plants:[6]

For example, as regards nutrition, Vegetables which are fixed to the Soil absorb immediately by their roots the nutritive parts of the surrounding

---

1　Body.] *followed by pencil bracket and partial underline of sentence as if to indicate pause.*

2　not the organization] *with* 'not' *crossed out in pencil and* 'instead of' *interlined above and then erased and* 'the organization' *in different ink than rest of line.*

3　points...the] *interlined in same ink as* 'the organization' *on fol. 40 above deletion of* '*I apprehend decides the question; and they{deleted in pencil} *must be ranked with the marine{deleted in ink}'.

4　Vegetable Kingdom] *deleting terminal* 's' *on* 'Vegetables' *with* 'Kingdom' *inserted. Period added [ed.]. On facing fol. 42ᵛ WC query* 'Qu What will Mʳ. Stokes and Mʳ. Broderip say respecting the Vegetable nature of Sponge?' *with reply by RO in pencil* "I don't know what Mʳ. Stokes will say— but after much consideration of the subject I find that I know little or nothing satisfactory about the nature of Sponge. The inclination of my opinion is to consider it a kind of link (one of many) between animals & plants. The animals smell when burnt, *& {interlined with caret.} an apparent voluntary motion in the absorption & ejection of water from certain foramina seem to claim for it a modification at least of animal life—'.

5　attributes] 's' *inserted.*

6　Plants:] *before ink deletion of* 'and these modifications combined with the two peculiar qualities constitute more especially the Animal nature.' *In W. S. Bennett hand pencil interline of* 'what? the attributes?' *above both* 'modifications' *and* 'qualities' *with both words underlined in same pencil and in margin in same pencil* 'I don't quite understand this.'.

fluid;—the roots are indefinitely subdivided; they penetrate the smallest interspaces, and seek, as it were, at a distance the nourishment of the plant to which they belong. Their action is tranquil but unintermitting; being only interrupted when dryness has deprived them of the juices which are necessary for them.

Animals, on the contrary, which are rarely stationary, but which have the power of moving, not [/] only parts of the body, but the intire body from place to place, require the means of transporting the provision necessary for their subsistence:—Accordingly, they have been gifted with[1] an internal cavity appropriated for the reception of the nutriment, upon[2] the parietes of which open the pores of the absorbent vessels, which have been justly compared to internal roots.

An internal cavity is requisite for Animals on another ground: their food must first be digested.

Plants are supported by water containing carbonic acid, or the already dissolved organized material of the soil:—But nutrition in Animals does not immediately commence by the absorption of such fluids as the soil or the Atmosphere furnish, but their food consists of substances already in organic combination, which must be prepared and submitted to instruments for dividing and comminuting it, as well as[3] to the the action of solvent fluids:—Thus Digestion or the preparatory assimilation of the Food is intirely peculiar to Animals.

They alone are endowed with organs of Sensation which guide them in the choice of aliment:—they alone possess labial and other prehensile organs for seizing the food—[4]teeth and jaws for comminuting and destroying its vitality if living;—muscular actions by which it is swallowed, and a reservoir for its reception endowed with chemical and vital powers for dissolving and assimilating such parts as are proper for nourishment and which are selected and taken into the System through the purely vital sensibilities of the absorbent internal surface:—while such parts as are unfit for nourishment, are expelled.

Whether the nutrient fluids be[5] absorbed from an internal surface as in animals, or be received by an absorbent external surface, as in plants, they[6] require to undergo certain changes by exposure to the influence of the atmosphere, before they are[7] finally adapted to the repair and growth of the body.

In the more perfect vegetables, the leaves are the instruments of

42

43

---

1   been…with] *interlined with caret above erasure.*
2   upon] *after erasure of one word.*
3   , as well as] *inserted in different ink, with erased pencil interline of* 'as well as' *above.*
4   food—] *dash over deletion of semi-colon.*
5   be] *interlined in different ink with caret.*
6   , they] *comma inserted with* 'they' *over erasure and erased pencil* 'they' *interlined.*
7   they are] *inserted over erasure of* 'it is'.

respiration;—[1]in the Aphyllous or leafless plants the respiration takes place over the whole surface, as in many of the more simple Animals:—

44   But a process of such essential importance to Life generally, and so[2] especially related to muscular motion, has very soon[3] its express apparatus in the Animal kingdom:—and before we leave even the Radiate Classes, we find, as in the Holothuria, an appropriate organ, which in a small space exposes a great extent of vascular surface to the Atmosphere.

As[4] we proceed to survey the ascending Scale of Animal Life, we find the respiratory apparatus becoming[5] gradually more concentrated in its form, and more energetic in its action; but varying extremely in its structure, according to the circumstances under which the different tribes of animals exist.

A more important difference between Plants and Animals is manifest in the products than in the instruments of respiration.

In Plants, the assimilating processes principally consist in the decomposition of the binary Compounds Water and Carbonic Acid, and their conversion into the ternary combinations of Carbon, Oxygen, and Hydrogen which form the Proximate Constituents of Plants. Now, during this Conver-

45   sion, there results an [/] overplus of Oxygen, which is separated and set at liberty by the superficial respiratory organs, the leaves; which at the same time absorb carbonic acid from the Atmosphere. These processes go on with greatest activity in the Sunshine; but in the Shade, during Night, and in unhealthy and withering Plants, a portion of Oxygen is absorbed from the Atmosphere and Carbonic acid is liberated; but always in less proportion than they take in during the Day; or in health.

Breathing, therefore, seems in Plants to be a simple correction of the assimilative processes: but it is essential to their life and health. By respiration they remove from the Air a part of the carbonic acid exhaled by Animals, and maintain a sufficiency of Oxygen.

Animals subsist on already organized matter, and their substance contains, besides, Carbon, Oxygen, and Hydrogen; also Nitrogen, which in[6] many Plants is[7] intirely wanting.[8]

46   We have already seen, that for the supply of organized matter to repair the loss which is constantly [/] experienced in the Animal kingdom, Animals are dependent on Plants; and these reciprocally derive from the respiratory product of Animals a proportion of the Carbon which is essential to their

---

1 ;—] *over erasure.*
2 so] *interlined in different ink with caret.*
3 has very soon] *interlined in different ink above deletion of* 'is not long left without'.
4 As] *after pencil deletion of* 'And'*with* 'as' *altered to* 'As'.
5 becoming] *interlined in same ink as previous inserts.*
6 in] *interlined in same ink as previous interlines with caret.*
7 is] *inserted in ink in left margin.*
8 wanting] 'ing' *inserted in same ink as previous inserts over* 's' *pencil line to WC pencil comment on facing fol. 46ᵛ* ' "which is intirely wanting in many Plants": or "which, in many plants is intirely wanting." as "Plants may want water" &c. &c.'.

support.

The changes effected in the atmospheric air by the respiratory apparatus of Animals is of the same essential nature under all conditions whether in the Night or Day, in health or in debility:—A proportion of Oxygen universally diminishes, and the quantity of Carbonic acid gas is increased.

The respiratory act among Animals can be accelerated or retarded by the will of the individual. Animals are informed of the necessity of respiring by the feeling of a want,—an uneasiness;—just as they are admonished by the necessity of taking aliment by the painful sensations denominated Hunger and Thirst.

The essential business of respiration in the two grand classes of Organized beings is different, and forms an important distinction between them:—The excretion of *Oxygen* is the grand object of the respiration in Plants;—that of
47   *Carbon* the end [/] for which relations are established between Animals and the surrounding Atmosphere.

Another marked difference between the respiration of Plants and Animals is the involuntariness of the act in the one, and its voluntariness in the other:—its occurrence with unconsciousness in the one, and consciousness in the other.[1]

Vegetables, like Animals have their peculiar excretions as well as internal secretions.

Besides the Oxygen already spoken of, they exhale moisture which is principally water, abundantly from their leaves; or from the surface of their stems when they have no leaves. By means of the leaves also, plants throw out those substances which they may have absorbed by their roots, but which are calculated to injure them.

Various apparently glandular organs may be readily demonstrated in the leaves, or flowers, or other parts of vegetables, which excrete a variety of acrid, glutinous, saccharine, balsamic, and odiferous substances.
48   Lastly[2], [/] the Stamens secrete the Male fecundating matter, and the pistil the fluid by which they are moistened.

Among the internal secretions, the æriform fluids are not the least curious, of which those that distend the bladders of the floating fuci are remarkable examples, and present a close analogy to the air-vesicles of the Hydrostatic Acalephæ. The other secretions of Vegetables are as we have said[3] of infinite variety.[4] They are all stored up in cells contained in different parts of each individual plant, and seem rather to stand in the relation of subserviency to other and higher organized beings, than to fulfil any important purpose in the

---

1   other.] *pencil pause line inserted before paragraph break with WC pencil comment on facing 48ᵛ* 'Epilepsy–Coma–Fainting–Swoon–&c breathing without consciousness.'.
2   Lastly] *on page break with long space to indicate paragraph division.*
3   as…said] *interlined in different ink with caret.*
4   variety.] *restoring period after pencil cross out of* 'gummy, resinous, oleaginous, balsamic, camphoric, &c.' *to replace* ';—' *following* 'variety' *[ed.].*

œconomy of the Vegetable secreting them.

In Animals on the contrary, most of the secretions are prepared for the especial purposes of the individual;—as the Synovia of the Joints, the limpid fluids which bedew the cellular and serous membranes, and fill various cavi-
49   ties of the [/] body, the chambers of the eye and ear particularly;—those that moisten and defend the surfaces of the mucous membranes, as the tears, and mucus: those that are subservient to digestion, as the Saliva, gastric juice, pancreatic and biliary secretions; those that lubricate and prevent the surfaces exposed to the air[1] from drying; as the sebaceous and oleaginous fluids of the Skin, and cerumen of the ears:—those that are laid up as reservoirs of nutriment, or defenses from the Cold, as the fat, marrow &c. Those that are the vehicles for the worn-out parts of the body, as the Urine and Perspiration.[2]
50   Nor is the List exhausted; for numerous [/] Species of animals have peculiar fluids, secreted for especial and peculiar cases;—as the venomous fluids of the fangs of Serpents, and of the[3] stings of various Insects; the inky fluid of the Cuttlefish; the fœtid secretions of the anal glands of numerous quadrupeds; the glutinous filaments which form the silken coccoon of the larvæ of Lepidoptera, and the web of the Spider; the wax with which bees build their Cells, &c.

Secretion is therefore, in Animals a much more extensive and important function than among Vegetables:—and the apparatus by which the various products are eliminated, is far more complicated among the former than the latter.

In vegetables we do not find any apparatus set apart for the excretion of matters derived from a change in the constituent particles of the organs once formed:—but among Animals the apparatus by which this depuration of the System is accomplished is one of the most important of all, for the preservation of the Individual.
51   Secretion in Vegetables is a function much [/] more under the influence of external circumstances than it is among animals; it is also more subject to periodical changes among the former than among the latter:—and whilst the function is mostly called into activity by the stimulus of Light, Heat, &c. in the one, it is influenced by the[4] internal and peculiar stimuli transmitted through the medium of the Nervous System in the other.

---

1 to the air] *interlined with caret.*
2 Perspiration.] *restoring period to replace* ':—' *after* 'Perspiration', *followed by pencil deletion by scoring of:* 'those that minister to the reproduction of the Species, as the fluids of the female germ or ovum; the spermatic and prostatic fluids of the Male:— and finally the yolk which is prepared in the ovary of the Bird, and appended to the germ to become the first nourishment of the incubated chick;— and the milk which is poured out by the Mammal for the first aliment of the newly born being.' *with RO pencil note opposite* 'Mammal' *on facing fol. 50*[v] 'Mammiferous animal'.
3 of the] *inserted in pencil with caret.*
4 is influenced …the] *inserted over erasure.*

When therefore it is said that the nerves are essential to Secretion, the expression has reference to their being, in Animals, the exclusive conductors of the energy which effects the separation of the elements of each secretion from the blood. It may be and probably is the same energy which is the immediate cause of the formation of the different secretions in vegetables but it[1] operates here through different conductors.[2]

From the general review which has been taken of the physical construction and vital phænomena of Organized Beings, and from[3] a comparison of the modifications which they present in the Animal and Vegetable kingdoms we cannot fail to perceive how many important characters they possess in common: &[4] how grand a series of Instruments they form, by which the great end is at last obtained, of making Matter subservient to the manifestation of Mind[5].

---

1 it] *with interline below of* "it so'.
2 When...conductors.] *RO pen insertion over erased pencil of same on fol 52ᵛ to replace pencil deletion of paragraph* '{¶} The idea, however, that the Nerves are essential to secretion, entertained by some Physiologists, arises out of too limited a view of that *general organic [interlined by RO in pencil above deletion of* 'important'} 'Function'.
3 from] *inserted in ink in margin.*
4 &] *inserted in different ink over dash.*
5 making...mind.] *RO insertion at bottom of page with deletion of* 'making' *on page break and* 'making Matter subservient to the Manifestation of Mind' *repeated on fol 52. Pencil* 'Bon' *on fol. 52 followed by double ink line to mark termination of lecture.*

Notes
to
Lecture Six

1–15 Cuvier reference uncertain. See W. S. MacLeay's comments on the plant-animal distinction in his *Horae Entomologicae, or Essays on the Annulose Animals* vol. 1 (London: Bagster, 1819), 196, also unreferenced, which is Owen's probable source. A copy of MacLeay's rare work is in the RCS library.

2–4 MacLeay (ibid.) cites Lamarck as the author of this claim.

3–2 J. Hunter, "Lectures on Surgery," Lecture 2, as quoted in *DIC* I (1833), p. 2. Owen has slightly altered the punctuation from the published version.

3–7 Charles-François Mirbel (1776–1854). Owen is quoting verbatim an unreferenced paraphrase of a comment by Mirbel in Macleay, *Horae* II, 193n.

3–9 Compare to Müller, *Handbuch der Physiologie des Menschen für Vorlesungen* (Coblenz: J. Hölscher, 1833), p. 45.

4–9 See ibid., p. 38. Owen is endorsing Müller's view that the conservation of life force is the principal limitation on the development of life. This underlies Owen's opposition to transformism since there must be a strict conservation of the expenditure of organic force and degree of organization. On this see my Sloan, "Darwin, Vital Matter," pp. 407–11.

5–13 See Müller, p. 38.

7–13 Ibid, pp. 5–6. Müller is drawing on Berzelius' discussions in *Thierchemie*. See note 4–13, p. 233 above.

11–6 Owen's discussion in this section is a useful summary of microscopic anatomy as it was conceived on the verge of the introduction of the cell theory by Schleiden and Schwann in 1838.

12–5 Owen is distinguishing plants and animals on the basis of structural differences that the cell theory was to abolish.

13–2 J. Hunter, "Lectures on Surgery 1" *Hunter's Works* 1: 214.

13–7 See Müller, *Handbuch* 1: 39.

14–7 See Hunter, "Croonian Lecture on Muscular Motion, No. 1" *Philosophical Transactions of the Royal Society* (1776) in *Works* 4: 200 and Owen's footnote on the same. See also Müller, *Handbuch*, p. 40.

14–17 Hunter, "Croonian Lecture," *Works* 4: 201.

15–3 Ibid , 199-201.

15–9 Ibid., 200. Owen has only para-phrased Hunter's text in this quote.

15–12 Ibid.

16–7 See Owen's note to ibid., p. 201.

16–10 See similar claim, without reference to a specific treatise of Bichat, in Owen's note to ibid., p. 199.

17–2 Stephen Hales, *Vegetable Staticks*, in *Statical Essays* [London: Innys, 1731-33] I, p. 112, exp. 36. Owen is repeating Hunter's account of these experiments in *Hunter's Works* 4: 204.

17–15 See Müller, *Handbuch*, p. 44.

18–7 Ibid. Owen is essentially translating Müller's discussion of Dutrochet's experiment. Müller seems to be referring to Dutrochet's *Recherches anatomique et physiologique sur le structure intime des animaux et des vegetaux* (Paris: Baillière, 1824). This is cited by Müller on p. 40.

19–6 Ibid. Müller is citing C.H. Schultz's works, *Ueber den Kreislauf des Saftes in Schöllkraut* (Berlin, 1822), and *Die Natur der lebendigen Pflanze* (Berlin, 1823) in support of this claim.

19–14 Owen's allusion is to the contemporary interest of several microscopists in the motions of the fluids in both plants and lower animals. See my "Darwin, Vital Matter. . .," pp. 373–88. See also Müller, p. 44.

20–3 Ibid.

21–8 Owen was exploring independently at this time the analogy of the lower plants and animals in discussions with Arthur Farre. See Introduction, p. 58.

21–14 Compare with Owen's discussions in his notes to Hunter's "Croonian Lecture" *Works* 4: 200–01. See also Müller, *Handbuch*, p. 40.

21–15 See ibid. Müller cites Ludwig Heinrich Palm, *Pflanzen* (Stuttgart: Löflund, 1827), p. 48, in support of this claim.

22–13 See Hunter, "Croonian Lecture" *Works* 4: 200 and Owen's note to same. See also Müller, p. 40.

23–6 Hunter, *Works* 4: 200.

24–2 There . . . altered] This passage appears almost verbatim again in ibid., p. 201.

27–2 See note 18–7.

33–12 Muscular fibers in the Hunterian series followed the preparations on sap and blood in the first series ("Parts employed in Progressive Motion"). The parts of the nervous system are commenced in Series 8 and continued in Series 9. See *Synopsis of the Arrangement of the Preparations in the Gallery of the Museum, of the Royal College of Surgeons* (London: Carpenter and Son, 1818). The 1818 arrangement is summarized above, Table One, p. 13.

34–17 Owen's point is elaborated in his note to Hunter's first Croonian lecture, *Hunter's Works* 4:201. He is concerned to maintain a distinction between animals and plants by his claim that the apparent nervous actions of the sensitive plants still do not involve anything like nervous tissue, and are mechanical responses to external actions rather than internally activated.

35–13 Owen is alluding to Robert Grant's studies on sponge reproduction in his "Observations and Experiments on the

Structure and Functions of the Sponge," *Edinburgh Philosophical Journal* **13** (1825), 94–107; 343–6. This is cited in the parallel discussion by Müller, *Handbuch*, p. 41. Owen's discussion down to 37–3 is closely similar to the discussion in Müller.

36–6 Müller, ibid., cites G. R. Treviranus, *Biologie, oder Philosophie der lebenden Nature* (Göttingen, 1803), vol. 4, p.634

37–3 Müller, p. 42 makes this same reference to Ehrenberg's landmark study on organic atomism,"Ueber das Entstehen des Organischen aus ein facher sichtbarer Materie," *Poggendorfs Annalen* **24** (1832): 1-48.

38–13 Compare with the discussion in Müller, p. 42. Müller cites the claims of Otto, Mehlin and Nordmann to refute the claim that lower animals lack nervous tissue.

38–6 The Acrita is the fifth subdivision of the animal kingdom, comprising the microscopic infusoria, created by William Sharp MacLeay through his subdivision of Cuvier's Radiata in his *Horae Entomologicae* of 1819. See on this Owen, "Acrita," *Cyclopedia of Anatomy and Physiology*, ed. R. B. Todd (London, 1835) 1: 47–9.

39–1 C. Ehrenberg, "Vorläufige mittheilung einiger bisher unbekannter Structurverhältnisse bei Acalephen und Echinodermen," *Müllers Archiv für Anatomie, Physiologie und Naturschaftliche Medizin* (1834), pp. 562–80.

39–5 Müller, p. 42 on this point cites Ehrenberg's *Organization der Infusionsthierchen* (Berlin, 1830).

40–1 Müller, p. 43 cites his own study on the optical system of the marine annelid *Nereis* , "Memoire sur la structure des yeux chez les mollusques, gastéropods et quelques annélides," *Annales des sciences naturelles* **22** (1831): 5-28.

41–5 Müller, *Handbuch*, p. 41, discusses this in the context of a summary of Grant's experiments on the sponges. See note 35–13.

47–2 Compare the discussion from 44–6 with Müller, pp. 45–6.

Fig. 14. Watercolor Illustration By George Scharf, Dated March 1837, From Roof Of The Newly Remodeled College Of Surgeons. The view is to the southeast looking toward the Church of St. Clement Danes and Portugal Street. The roof of the new large main gallery is to the right, and that of the new smaller side gallery is directly ahead. The new lecture theater is on the ground floor below the window well (see plan, fig. 4, p. 46). Owen's apartment at 43 Lincoln's Inn Fields is through the small facing window to the left. The building to the far left is the former Old Duke's Theater, serving as Copeland's China Warehouse at this date. Subsequent remodelings added a new museum extending into the Copeland building in 1854 (Fig. 15). These structures were destroyed in the 10–11 May 1941 bombings. *(Courtesy President and Council, Royal College of Surgeons of England)*

Lecture Seven
Analysis of the Manuscript

The copy-text for this lecture is located at the BMNH (O. C. 38). The stiff paper covering folder is entitled in Owen's hand "Lecture 6 Lecture 7 Organization of Animals, Blood" and otherwise lacks a delivery date. The initial page of the lecture is headed "Lecture VI/ Proxmiate Components of Animals. *Blood*" in William Clift's hand. The body of the lecture is composed uniformly on the same "J Green & Son 1837" paper folded into half sheets as used in the manuscript of Lecture Six. This is the final manuscript existing in either draft or copy, and it bears a prominent "The End" in Clift's hand on the final page. Comment is warranted on one uncertainty about the dating of this manuscript. The *Life of Richard Owen* (I: 110), drawing upon Caroline Clift Owen's missing diary, mentions that the *third* lecture was delivered on 16 May, and it was the "first given entirely without notes." This comment cannot be reconciled with any of the possible datings of Lecture Three, and must be presumed to be in error. The use of the same paper employed in Lecture Six, clearly dated to 13 May makes the May 16 delivery date the most plausible option for this lecture. Although the cover page suggests that it was intended as two lectures, there is no obvious division point in the manuscript, and several pages have been deleted by vertical pencil scoring as if to shorten it for delivery. This supports the conclusion developed in the Introduction that these first seven lectures were probably then followed by an exposition of material based directly on the first volume of the *Descriptive and Illustrated Catalogue* utilizing either the printed text or the manuscript draft used in the preparation of the *DIC*.

The holograph is located in the Owen archives at the RCS (Owen MS 67.b.12), and bears on the cover page the title "Proximate Components of Animals—Blood." The draft contains no direct evidence of the date of original composition except paper watermarks. The main body of the holograph manuscript is composed on the same "J Green & Son 1836" paper used for lectures Five and Six. A new introduction, integrating this lecture draft with that of lecture Six, is written on "J. Whatman Turkeymill 1836" paper. All internal indications support the assumption of a composition date of late 1836 or early 1837. Reference to Johannes Müller's *De globularum sanguinis usu* (see note 38v–6 below) and to Henri Milne-Edwards' article on the blood in R. B. Todd's *Cyclopedia of Anatomy and Physiology* (see note 55–5 below) give more specific indications of dating of the final draft.

Fig. 15. Interior Of New East Museum, Designed By Charles Barry And Completed In 1854. This added a new gallery, approximately the same dimensions as the original 1837 main hall, to the smaller gallery depicted in Fig. 14. *(Courtesy President and Council, Royal College of Surgeons of England)*

Lecture Seven
16 May 1837

1

Mr. President,

In our[1] previous Lectures[2], the general characters of Organized Beings have been[3] reviewed, and the qualities which distinguish them from unorganized substances.

A comparison[4] was next established between the two great Classes into which the organized kingdom is resolvable; and we[5] attempted to define the almost imperceptible boundaries which distinguish Plants and Animals.[6]

2

Lastly,[7] the principles of the classification of the Animal Kingdom were sketched out,[8] and we[9] traced the main features in the progressive complication of the Animal structure[10], as exemplified in the different Grades of fixed and immutable species, up to that point when each system of Organs presents a relative perfection[11], with just proportions, and harmonious[12] co-ordination

---

1 our] *RO interline in pencil above deletion of* 'the'.
2 Lectures] *before pencil deletion of* 'which I have had the honour to *read from [below erased interlined* 'deliver'} this Chair'.
3 have been] *interlined in pencil above deletion of* 'were'.
4 comparison] *after pencil deletion of* 'general'.
5 we] *interlined in pencil above deletion of* 'I'.
6 Plants and Animals.] *interlined by RO in pen with caret before pencil deletion of* 'those beings which live and grow and propagate;— which cover the otherwise cold and barren mineral world with fertile soil; which clothe and adorn it with verdure, and enliven it by the phenomena of their diversified actions, and incessant changes of condition.' *and deleting by cross-out of original ink insertion by RO from facing fol.* 2ᵛ: 'so different when viewed as they exist at one extreme of the *organized {interlined with caret} scale, so closely assimilated *in their simpler forms of {interlined below deletion of* 'at the opposite'} organization—which alike,'. *Period inserted [ed.].*
7 Lastly,] *before deletion of* ' I sketched out'.
8 were...out] *interlined by RO in pencil.*
9 we] *interlined by RO below in pencil with caret.*
10 structure] *interlined in pencil above deletion of* 'Organization'.
11 perfection] *before pencil deletion of* 'of Structure'.
12 harmonious] *after deletion of* 'a' *and* 'harmonious' *altered by interline from* 'harmony of'.

and adjustment suitable, as to the constitution of[1] the material instrument of the exigencies and will of Rational Man.

We[2] have now to commence the consideration of the diversified products of the organizing energy:—to examine first the elementary tissues, their mechanical, chemical, and vital properties, and afterwards to describe the several organs which result from their combination, tracing the gradual complication of each organ, according to the plan upon which the Founder of the Collection has arranged his series of illustrations of this extensive subject; and connecting the consideration of the varied forms and structures which will be presented to our view, with their adaptation to the exigencies of the Individual, and at the same time as they illustrate[3] laws of development, and those general analogies which may be fairly traceable through the intire series of the[4] Animal Kingdom.[5]

3   The Hunterian Collection[6] illustrative of the Organization of Animals, commences with a series[7] of the Component parts of an Animal, under the head of "Animal Matter."—& In the arrangement of these specimens the Founder appears to have been guided by Physiological rather than Anatomical considerations;—the nervous and mucous elementary tissues being preceded by others of a more complex character.[8] We have first[9] a coagulum of Blood; then follow the textures concerned in the active and passive Organs of Motion;—next an example of a secreting Organ:—to this succeed[10] specimens of the Nervous System which peculiarly characterizes the Animal:—then follow the compound textures which connect and protect the more important Organs:—and, lastly, the supplementary substance, Fat; which Hunter rightly regarded as being rather appended to, than forming an essential part of the Animal body.

We detect, therefore, in this series, partly a Classification of Tissues, and partly a classification of Organs. Now[11], the organs of an Animal, physiologically considered, seem most naturally to resolve themselves

1st. Into those immediately concerned in the generation and regeneration of the whole machine.

---

1   to...constitution of] *interlined by RO in pencil with caret.*
2   We] *in ink over pencil .*
3   as. . . illustrate] *interlined by RO in ink over pencil erasure with caret before deletion of* 'the'.
4   the] *interlined in ink with caret.*
5   Kingdom.] *before deletion of* 'organizations.' *with pencil* 'Kingdom' *following. Period added [ed.].*
6   Collection] *interlined in ink over pencil with caret above erasure of* 'Series'.
7   series] *over erasure with erased pencil* 'Set' *interlined above.*
8   &...character.] *with* '&' *inserted in pencil and pencil circle around passage.*
9   we...first] *interlined below deletion of* 'The' *to replace deletion of* 'The Series commences with'.
10 succeed] *altered in pencil from* 'succeeds'.
11 Now] *after pen line over pencil as if to indicate insertion, but without facing insertion.*

4   2ndly. Into the immediate organs of Motion. [/] &[1] 3rdly. Into those organs which influence all the other parts of the machine, by conveying impressions to, and reactions from, a Central Organ.[2]

The first are the *Digestive*, nutritive and reproductive organs.

The second, are the *Muscles*.

The third, the *nerves*.

These are the parts essential to the performance of the functions of every Animal, and always exist; subject to different degrees of separate or distinct elimination throughout the Series. We may afterwards place, in a subordinate category, those parts, which, in the development of the organized being attain to no other essential properties than the physical or passive qualities of Solidity, Elasticity, Tenacity, &c:—and such are Bones and Shells, Cartilages and ligaments, tendons, fasciæ, connecting and investing membranes, and cuticular productions.

Amongst the organs of Nutrition and reproduction[3] might be classed[4] every part of the body, since each organ acts, as it were, the part of a gland to itself:—the muscular tissue attracts from the blood the fibrinous particles subservient to its reproduction, and assimilates[5] them to itself:—the nerve attracts

5   and retains the [/] albuminous matter. In the tissue of the Bone, the jelly and the earth are elaborated from the Blood:—in short to use[6] the language of Dr.[7] Prout, a sort of Digestion is carried on in all parts of the body, to fit for absorption and future appropriation, those matters which have been already assimilated. But the especial phenomena which characterize the organs of Nutrition are not merely Nutrition, but those processes of separation and recombination of Organic particles which ultimately effect the formation and perfection of the common nutrient fluid, the blood; and which serve to convey it to and from all parts of the System.

The perfection of the Blood is the end and aim of all the parts of the reproductive apparatus, properly so called[8] of the Individual: the digestive system elaborates and provides the basis of the fluid; the respiratory and secretory organs remove the particles which detract from[9] its nutrient properties; the vascular mechanism of propulsatory organs and branched canals distributes it to every part of the body:—Hunter, therefore, may be said to have represented this System of Organs by the simple Coagulum of Blood which he places at the beginning of his series of Constituent parts of the Animal

---

1   &] *inserted in pencil in margin.*

2   Central Organ] *with subsequent deletion of interline above of* 'the cerebro-spinal'.

3   reproduction] *before pencil deletion of* 'we'.

4   be classed] 'be' *interlined by RO in ink over pencil* 'be' *with* 'class' *altered in pencil to* 'classed'.

5   assimilates] *altered in ink from* 'assimilate'.

6   in...use] *inserted in ink with pencil* 'use' *interlined.*

7   Dr.] *interlined with caret.*

8   , properly....called] *circled in ink and inserted from next line with lead line to insert point.*

9   detract from] *interlined by RO in ink above underlined* 'are incompatible with'.

body;—& in the exposition of his Physiological Collection we are, at this Point, called upon to[1] consider the nature and properties of the fluid[2], out of
6 which all the[3] structures of the Animal [/] body are formed, sustained, and repaired.

The[4] quantity of the Blood[5] in the human body has been variously estimated, at from 8[6] to 30^{tbs}. pounds. It[7] exists in two states, the arterial and venous, in the same[8] individual, and presents modifications of still[9] greater variety and importance when traced through the different classes of the animal Kingdom[10].

In the highest and vertebrate division, it is a fluid of a bright or dark red colour, according as it is moving from or towards the respiratory organs.

In the *Mollusca* it is generally a colourless fluid; but presents a reddish tint in the genus *Planorbis*, and a blueish or greenish tint in some other rare exceptions.[11]

Among the *Articulata* we find a whole Class, and that the lowest of the Series, the *Annulata*[12] in which the blood reassumes, as it were the characteristic red colour; but there are some Species here in which the blood presents only a yellowish tint as in the Aphroditæ; while[13] in the *Clepsina*, one of the Leech tribe, (Hirundiidæ)[14] the blood is entirely destitute of Colour.[15]

In the lowest division of the invertebrate Series with one or two exceptions, the nutrient juices are colourless.

Berzelius[16] supposed that the common House-Fly[17] presented an exception

---

1 &...to] *RO insertion in ink with lead line from facing fol. 6^V to replace ink deletion of* 'and I now proceed to'.
2 fluid] *after deletion in ink of* 'all-important'.
3 all the] *over erasure.*
4 The] *after deletion of* 'The blood' *with succeeding* 'the' *altered to* 'The'.
5 the Blood] *interlined above deletion of* 'which'.
6 8] *before erased* 'tbs'.
7 It] *inserted in left margin with period and capitalization inserted for grammatical sense [ed.]*
8 same] *with pencil* 'X?' *in margin.*
9 still] *interlined in pencil without caret above* 'greater'.
10 Kingdom] *after deletion of* 'Series.'.
11 exceptions.] *RO comment on facing 7^V* 'and is red, according to Horne in the *Teredo nasulis* (Phil. Trans: 1806.' *No insert marked.*
12 the Annulata] *interlined in ink with caret.*
13 while] *interlined above deletion of* 'and'.
14 Hirundiidæ] *altered from* 'Hirundinidæ'.
15 entirely...Colour] *interlined above deletion of* 'altogether white.' *with WC pencil. comment on facing fol. 7^V over pencil drawing of jar and chemical flask* 'Limpid or colourless *is often considered as Synonymous with* white[.] Which is the true expression? as I have never seen white blood except once in the Human Mesenteric veins—'.
16 Berzelius] *inserted in ink by RO in margin replacing deletion of* 'Some have'.
17 House-Fly] House-l-Fly.

7   in regard[1] to the colourless nature [/] of the blood in the Class of Insects; but
the bright red fluid which may be pressed from the head of that Insect is
nothing more than[2] the pigmentum[3] or colouring matter of the large compound
eyes.[4]
Constituents[5]
     Under ordinary circumstances, the blood, while circulating, or recently
drawn from the vessels of a living Animal, appears as a homogeneous
fluid;—but, if it be examined microscopically while flowing through the
capillaries of a transparent membrane, it is seen to be composed of minute red
particles[6], suspended in a colourless and transparent fluid; or rather as M[r].
Hunter justly expresses it, "there is no appearance but that of globules moving in the vessels."—"In such a situation," he observes, "the other parts
called the coagulable lymph and serum are not distinguishable, on account of
their being transparent, or colourless, and the globules do not, strictly speaking, constitute a part of the fluid, but are only diffused through it." (On the
Blood. p. 16.)
     On a close examination,[7] however, the colourless fluid constituent of the
Blood is indicated by an optical appearance occasioned by the difference in
the rate of motion of that part of the fluid which is in the centre of the
Capillary, and that which is in contact with its inner parietes; and whose motion is consequently retarded.

8   If Blood be examined immediately after [/] extravasation, then the colourless fluid part is more plainly perceptible; and to this menstruum in which the
globules are suspended in the living body, the term *Liquor Sanguinis* is applied.
     What would seem the first and most simple act in the mechanical analysis
of Blood, vizt. the separation of the globules from the fluid in which they are
suspended is,[8] not the change which takes place[9] in the ordinary act of
Coagulation;—in this process the *Liquor Sanguinis* is decomposed; one of
its constituents, vizt. the fibrine or coagulable lymph assumes the solid form;

---

1   in regard] *interlined in ink with caret.*
2   nothing more than] *interlined with caret.*
3   pigmentum] *last three letters over erasure.*
4   eyes.] *WC comment on facing fol. 8[v] 'Is this proved?* I think there is too much red
    matter, & that it is also in the Body—*Fiat experimentum:* by cutting off the head, &
    then crush the body—when flies make their appearance which will soon happen. Also
    the Earth-worm.—'.
5   Constituents] *inserted in margin following horizontal ink line across three inches of the
    manuscript.*
6   particles] *interlined above deletion of* 'globules'.
7   On...examination,] *RO pencil comment on facing fol. 8[v]* 'On closely watching this
    circula'. *No insertion point indicated.*
8   fluid...is] *inserting RO note in ink on facing fol. 9[v] replacing* 'liquor sanguinis' is'.
    No insertion point indicated.
9   not...place] *interlined by RO in ink to replace deletion of* 'nevertheless, somewhat
    difficult to effect; for we do not obtain the intire fluid'.

and, as its particles approximate and blend together, they entangle the red globules, and withdraw them from the serous constituent, which now alone retains its fluidity.

In[1] order to separate the globules from the *liquor sanguinis* we must take the blood of some animal, as a Frog, or Newt, among the Batrachian Reptiles, in which the red globules are of too large a size to pass through white filtering paper. By pouring the Blood at the moment of extravasation upon the filter, a portion of the *Liquor Sanguinis* may be obtained distinct from the globules, and the colourless fluid may then be observed to separate into its constituent parts, the fibrine forming a firm mass, as usual, in the act of coagulation; but presenting a semitransparent opaline[2] colour, instead of
9   the red colour of [/] ordinary crassamentum.[3]

We have here the pure fibrine obtained after separating the *liquor sanguinis* from the Red globules—The red-globules are retained on the filter, which is shown in this preparation.[4]

The more common process of obtaining the colourless coagulum is by washing the crassamentum (or the combination of fibrine and red globules) which forms in the ordinary act of coagulation, with distilled water;[5] but it is a mistake to suppose that pure fibrine is obtained in this way; it is always more or less combined with the nuclei of the red globules which remain after the colouring matter has been dissolved and washed away.

Without having recourse to filtration we may[6] separate the pure *fibrine* from the serum and red globules by the following method:[7] the blood of a Frog should be received in a watch-glass, and the flakes of fibrine, as they successively form by coagulation, should be removed with the end of a needle, shaking off the red globules which may adhere to them. In this way the whole of the fibrine may be separated, and the red globules are left entire and floating in the Serum, and ready for numerous subsequent[8] microscopical and chemical experiments.

A ruder and more compendious way of separating the fibrine from the other constituents of the blood, is to beat up newly extravasated blood with a wisp or stick; when the fibrine, as it coagulates, adheres to the stick, and the
10   whole may thus be successively removed; leaving the blood [/] apparently unchanged, but consisting of the red globules and the serum alone.

By these various modes of mechanical analysis we obtain a knowledge of

---

1   In] *with bracket in margin and on facing fol. 9ᵛ in RO pencil* 'Demonstrate'.
2   semitransparent opaline] *interlined with caret above deletion of* 'an *opaque white*'.
3   *crassamentum*] *with RO ink note on facing 10ᵛ* 'Preps'.
4   We. . .preparation.] *RO insertion from facing 10ᵛ with arrow marking insertion point.*
5   water;] *before deleted caret and deleting originally intended insertion from facing fol. 10ᵛ* 'as in this preparation/'.
6   Without...may] *inserted to replace erasure with* 'Without...filtration' *interlined above deletion of inserted* 'Beside' *and* 'we may' *on line.*
7   by...method:] *interlined with caret.*
8   subsequent] *before deletion of* 'inter'.

the proximate constituents of the blood, and can separately demonstrate them.

In the ordinary, or as Hunter preferred to call it, the spontaneous act of co-agulation, the serous constituent is separated from the fibrine and coloured particles.[1]

By the filter we are enabled to separate the naturally fluid constituents which form the Liquor Sanguinis from the Coloured Globules, where these, as in some of the cold-blooded Reptiles, are of comparatively large size.

By interfering with the act of Coagulation, and preventing the aggregation of the consolidated particles of the fibrine, this constituent may be obtained distinct from the globules and serum.

Under some peculiar conditions, a portion of the fibrine separates from the serum and red globules in the fluid state, and afterwards coagulates into a firm uncoloured mass.

Hunter relates that he once saw in the blood of a female, a separation be-tween the two constituents of the *fluid part* [2] of the blood, which took place almost immediately after being drawn; the serous part swimming on the top, while the fibrine remained still fluid *below*.[3]

When the coagulation of the blood takes place slowly, the fluid fibrine may be skimmed off, free from the red globules; and the part so taken *will coagu-late immediately*.[4] This is also an observation of Hunter's, and a very impor-tant one, as it throws considerable light on two questions which have since divided the opinions of Physiologists:—The first is, whether the particles of fibrine in the circulating blood are chemically dissolved, or mechanically sus-pended in the serum? The second question relates to the constituent of the blood upon which the act of coagulation depends.

This latter point will be considered when we come to speak of the coagula-tion of the blood:—I shall only here observe that the recent ingenious experi-ments of Müller with reference to this subject fully confirm the opinion of Hunter, (that coagulation is the sole act and property of the fibrine.)

With respect to the fluid condition of the fibrine the observations of Hunter which I have just quoted strongly support the view which he adopted,—that the fibrine is a distinct part of the blood, while circulating. It cannot be admit-ted that the fibrine is actually dissolved in the Serum, without assuming that its known chemical [/] properties are materially modified while in the living vessels;—and the more natural explanation is, that the fibrine is merely suspended in the mass of the blood in a state of extreme subdivision, and possessed of a transparency too perfect to admit of being distinguished under ordinary circumstances.[5]

11

12

---

1 particles.] *period inserted for grammatical purposes [ed.].*
2 *fluid part*] *double underline in pencil.*
3 below] *inserted in ink over pencil with pencil pause line following. Period inserted [ed.].*
4 *will. . .immediately*] *underlined in pencil.*
5 circumstances] *faint pencil line across 1/3 of page before next paragraph to indicate pause.*

Having now considered the general properties and composition of the blood, as discovered by mechanical analysis, the next point which demands our consideration is the nature and properties, and relative importance of the proximate constituents.

Fibrine,[1] Gluten, or Coagulating lymph, formed, in the opinion of Hunter the most important constituent[2] of the blood. This deduction he legitimately drew from the fact that it was present in the blood of Animals in which the red particles were deficient; and as he found the coagulating lymph capable of undergoing, in certain circumstances, spontaneous changes, which were necessary to the growth, continuance, and preservation of the Animal; while to the other parts,—(i.e.)[3] the Serum and red Globules,—such uses could not be assigned; Hunter regarded it, with still more reason, as the essential part of the Blood in every Animal.

The objections to these facts and inferences which arise out of the researches, and in many respects truly valuable[4] researches of M. M. Prevost and Dumas, will be considered when we come to speak of the red globules.

13

Pure Fibrine obtained free from the mixture of the nuclei of the red globules, presents, in the coagulated state, the form of a reticulate[5] mass of whitish and very elastic filaments without any trace of globular structure[6]. When dried it loses 3/4ths. of its weight, and becomes yellow, hard, and brittle, but without acquiring any transparency.

Immersed in Water it re-assumes its softness and pliability, but is not in any degree dissolved.

It has no peculiar taste or smell.

When exposed to a decomposing state of heat, it melts, swells, inflames, and leaves behind a shining coal which contains Nitrogen.

If the carbonaceous residuum be still subjected to heat, it is converted into a grayish-white half-molten substance equal to 75 per Cent of the weight of the dried fibrine.

These ashes are neither alkaline nor acid:—when dissolved in muriatic acid a trace of silex is left. There is also some phosphate of Magnesia and a distinct trace of Iron; but the principal earthy constituent is phosphate of Lime.

Coagulated Fibrine is insoluble both in cold and hot water; but after long boiling, it contracts, hardens, and is so changed in its physical properties that it falls to pieces under the least pressure.

---

1   Fibrine] *in margin pencil* 'X' *and WC pencil note on facing fol. 13*[v] 'M[r]. Hunter's general expression was *Coagulable* lymph and is evidently different from coagulating in signification;—as being *capable* of *being coagulated* though not yet coagul*ating* or coagul*ated.*'
2   constituent] *after deletion of* 'and essential'.
3   (i.e.)] *interlined in pencil above dash.*
4   truly valuable] *inserted over erasure.*
5   reticulate] *RO interlined with caret.*
6   without …structure] *RO ink insertion from facing fol. 14*[v] *with insertion point marked with caret.*

14   No gas is evolved during the process, but the [/] water has lost its transparency, and now contains in solution a new substance formed from the elements of the fibrine. This solution has no similarity to a solution of Gelatin.

Fibrine, Coagulated Albumen, Casein, and Cruorine, or the colouring
matter of the blood[1] do, none of them, give out gelatine when boiled in
Water.

Of the numerous properties[2] of Fibrine which Modern[3] Chemistry has unfolded[4] none, perhaps, are more interesting than those which relate to the influence of the neutral salts on this substance. Carbonate, sulphate & Nitrate of
Potass or Soda although not sufficiently powerful, like acetic acid, to dissolve
the fibrine after it has once been coagulated, yet retard its[5] coagulation when
mixed with recently drawn blood.

This fact gives some insight into the mode in which the cooling salts operate beneficially in Inflammatory cases where the blood is overcharged with
fibrine, and has so strong a tendency to coagulate in the vessels & on the superficies of organs.[6]

---

1   , or...blood] *RO interlined in ink with caret, with comma inserted below line.*
2   properties] *after deletion of* 'chemical'.
3   Modern] *after deletion of* 'the accurate'.
4   unfolded] *before pencil deletion of interlined* 'and on which it is not my province to
    dwell,'.
5   has...its] *all written over WC pencil note* 'All fluids of different densities when these
    dissolve heat; do they not?'.
6   Of...organs.] *RO ink insertion from facing 15ᵛ headed with pencil* 'Ø' *symbol to
    replace pencil deletion by vertical scoring of* '{¶} Fibrine is remarkable for the
    property it possesses of decomposing Super–Oxyde of Hydrogen (hydrique–
    superoxyde) by simple contact, and of converting it into water by the liberation of
    Oxygen, without itself undergoing any change. {¶}* If {*over erasure*} the quantity of
    fibrine introduced into the *suroxyde hydrique* be very great, the decomposing actions are
    so forcible as to be accompanied by an extrication of heat. {¶} If concentrated acids, the
    nitric excepted, be poured upon, it swells and becomes gelatinous and transparent; and
    an acid compound body is the result. If treated with dilute acides, the moist fibrine
    contracts, and a neutral combination with the acid is the result.{¶} The acid compound
    of fibrine and mineral acids is insoluble in water;—the neutral compound is soluble.
    {¶} Acetic acid immediately permeates fibrine, and converts it into a colourless jelly,
    which readily dissolves in warm water. When this solution is boiled a little, nitrogen
    is eliminated but [I *fol*. 14; *ink bracket around all following on fol.* 15] but no
    precipitation ensues. {¶}The neutral salts, as Sulphate and Nitrate of Potass and
    Soda have a similar tendency, but though *not* sufficiently powerful to dissolve the
    fibrine after it has once coagulated, they retard its coagulation, when mixed with
    recently-drawn blood. {¶} This is perhaps one of the most important of the
    Chemical facts ascertained relative to the fibrine, since it gives us an insight into the
    mode in which these cooling salts may operate benefically in Inflammatory cases, {*WC
    pencil note on facing 16ᵛ* ' Plaister of paris {*sic*} *sets* very quickly, when mixed with
    hot water—when compared with Cold—Mixed with Urine, or Size, it is exceedingly
    slow; and more manageable by a novice in Casting in Plaister.'} where the
    blood is overcharged with fibrine, and has *so great{interlined above* 'such'} a tendency
    to coagulate, both in the vessels, and on the superficies of Organs. {¶}Fibrine,
    like Albumen and Casein, is dissolved into a jelly by the caustic Alkalies, even when

15   The proportional quantity of Fibrine contained in the Blood of a   Mammi-
16   ferous Animal differs in the [/] arterial and venous blood. Professor Müller
took 3004 grains of blood from the carotid artery of a Goat which yielded 14
1/2 grains of fibrine. He took 1392 grains of blood from the jugular vein of
the same animal, which yielded only 5 1/2 grains of fibrine.
       These facts show the high probability that Azote, which is so abundant in
fibrine, is principally absorbed into the System by the pulmonary capillaries.
And when we examine the blood[1] of different classes of Animals with the
same view viz (i.e.) of ascertaining[2] the proportion of fibrine which it contains, that proportion will be found to stand in almost a direct ratio with the
extent and rapidity with which the circulating blood is submitted to the respiratory influence of the Atmosphere. Thus in Birds, in which the nutritive
functions and muscular energies are most active, the blood contains a greater
proportion of Fibrine than in Mammals:—while in Reptiles and Fishes the
proportion is diminished to a degree inferior to that of mammiferous venal
Blood.
       In the invertebrata, the fibrine is stated to exist in a still less proportion in
the blood, than in the Reptiles and Fishes.[3]
17   It is highly interesting to trace this important relation[4] between the formation of fibrine and the accidental or temporary augmentation of the
Respiratory function in the Human Subject, under circumstances of injury or
disease.
       If a solution of continuity of an important part is occasioned by a wound or
fracture:—or if a process of ulceration threatens to open a passage for extraneous matter into a large serous cavity, Arterial action is increased and respiration is accelerated, to a degree which[5] in the Bird is normal[6], but which in

---

(contd.)
     these are much diluted;—and is not changed into a saponaceous substance like gelatin
     under similar circumstances. {¶}Both Berzelius and Chevreuil {sic} have obtained a
     trace of fatty matter from fibrine, by digesting it in Alcohol or Æther. {¶}The
     elementary Composition of Fibrine is as follows: in 100,000 parts there are *53–360.
     of Carbon {*transposed by pencil line from original location preceeding* '19–934. of
     Nitrogen} 7–021. of Hydrogen. 19–685. of Oxygen' *with horizontal ink line
     to mark pause at this point.*
1   blood] *in pencil on facing 17ᵛ* '2784'.
2   viz…ascertaining] *pencil line to show transposition of* 'viz (i.e.)' *from after* 'of
     ascertaining'.
3   Fishes.] *before pencil deletion of paragraph* 'It would be interesting to compare the
     Blood of the Flying Insect with that of the less-perfectly breathing Articulate and other
     Invertebrata with reference to the relation which in the Vertebrata evidently subsists
     between the Amount of Respiration [|*fol.* 16] and the quantity of Fibrine in the
     Blood.'.
4   relation] *double pencil pause line in margin.*
5   which] *followed by deletion of* 'is natural' *with interlined* 'may be' *above. The* 'may
     be' *deleted for grammatical sense [ed.].*
6   is normal] *interlined in pencil.*

the Mammiferous Animal constitutes Fever:—and the result is, the production of an increased quantity of fibrine in the circulating fluid, which is afterwards poured out and organized, to repair or prevent the consequence of the injury which excited[1] the fever.

It is this superabundance of Fibrine, accompanied with, and perhaps producing some slight change in the state of combination of the constituents of the blood, which makes the red globules more rapidly separate and subside; and occasions the colourless stratum on the top of the Crassamentum, usually called the Crusta inflammatoria.

This colourless stratum is also increased by an unusual slowness of Coagulation in the fibrine which is sometimes observable in inflammatory states of the Blood.

18   Fibrine is an Animal constituent which is [/] not confined to Blood. It exists in a dissolved or suspended state in the[2] Chyle and the Lymph:—it is abundantly present in the muscular fibre, and in the contractile tissue of the Uterus; but no trace has been detected in the contractile and elastic middle coat of Arteries.

We have next to consider the second ingredient of the liquor sanguinous— viz the[3] Serum. This fluid[4] was regarded by Hunter as the second part of which the whole mass of blood was composed, and into which it spontaneously separated itself after extravasation.—He believed it to be common to the Blood of all animals; but thought, contrary to the opinion of some Physiologists of the present day, that it existed in greatest proportional quantity in those animals which have red blood. [On the Blood, p. 42].

The serum[5] is the proper fluid part of the blood, and constitutes, in Man, nearly three fourths of the weight of the whole.—[6]

19   Serum manifests alkaline properties when tested with yellow or red vegetable colours:—the quantity of water which it contains enabling it to suspend every salt which is soluble in water.

When separated during the ordinary act of coagulation, (and the firmer the clot the more serum is squeezed out) its degree of fluidity resembles that of warm olive oil. But M^r. Hunter entertained a belief, that although serum contained a large quantity of matter coagulable by heat, yet it was in a more fluid state when circulating, than when separated by[7] the Coagulation of the

---

1   excited] *altered in pencil from* 'excites'.
2   the] *over erasure of* 'Chyle'.
3   We…viz the] *interlined in pencil before original paragraph beginning* 'Serum.'
4   This fluid] *inserted in pencil with* 'fluid' *interlined over deletion of* 'The serum'.
5   The Serum] *after pencil deletion of paragraph* '{¶} Serum is a fluid of a yellowish colour, and sometimes of a greenish tinge, which depends on a peculiar colouring principle, and not on any partial solution of the red globules;—for these, as Hunter rightly observes, are suspended without being dissolved in the Serum. {¶} It has a faint saltish Taste, and a specific gravity of from 1,027 to 1,029.'.
6   whole.] *before pencil deletion of* 'The crassamentum when completely formed, but still moist, and not compressed, constituting the remaining fourth.'.
7   by] *before ink underline and subsequent pencil deletion of* 'extravasation, and'.

Fibrine.

Hunter dwells more than once on what may be termed[1] the *passivity* of the Serum as regards the act of coagulation, and contrasts it in this respect with the fibrine, whose higher endowments of Vitality the whole of his elaborate Chapters on the Blood were principally devoted to establish.

Nevertheless it is in the serum of the blood that the material exists which goes almost exclusively to sustain and renovate the highest system of Organs in the animal body, vizt. the Brain and Nerves.

If serum be exposed to a heat of 160°. to 165°. of Farenheit, it is coagulated into a white coherent mass, from which a fluid may be obtained by pressure. The coagulated part is the Albumen [/] the fluid part is termed the Serosity.

20

The albumen is the principal constituent of the Serum, and that to which it owes its most characteristic qualities.[2]

In its ordinary fluid state, albumen is intimately combined with soda;—and some Chemists attribute its coagulation to the separation of the alkali, and the decomposition of the Albuminate of Soda:—a term given to signify the condition of the albumen in the circulating blood.

I shall not here[3] dwell on the results of the numerous elaborate Analyses of the Serum, conceiving their general nature to be well known. The latest chemical discovery relates to[4] the existence, in Serum, of a white, and slightly opalescent substance, fusible at 95°. Farenheit:—not forming an Emulsion with water;—soluble in alcohol:—not saponifiable, and appearing to contain azote. To this new substance the term *seroline* has been given.[5]

Hunter, although disposed to place little weight on Chemical Analysis, as teaching the use of the Blood in the living body; yet he was one of the first who endeavoured to acquire a more extended and just knowledge of the properties of the blood, by observing the changes which its component parts, and especially the Serum, underwent during the operation of Chemical reagents:—and a considerable Section of His "Treatise on the Blood" is devoted to this [/] analytical inquiry.[6]

21

He compared the Serum of the Blood with the fluid of dropsical cavities;—the Amniotic fluid—and other analogous secretions;—and showed that it differed chiefly in the greater quantity of the coagulable matter, or albumen, which it contained.

The care and accuracy which characterize[7] all his investigations, and his

---

1 what...termed] *interlined in ink over pencil with caret.*
2 qualities] *with RO pencil note* 'Prepn' *on facing fol. 21^v.*
3 here] *interlined in ink over pencil with caret.*
4 relates to] *interlined in ink over pencil with caret above deletion of* 'is' *with RO pencil note on facing fol 21^v without insert point indicated:* 'whose discovery'.
5 given.] *followed by horizontal pencil line to indicate pause.*
6 inquiry] *before deletion of* '{¶} Hunter was the first discoverer of the true fluid part of the Serum, now termed the Serosity.'.
7 characterize] *altered in pencil from* 'characterizes'.

power of seizing upon minute, but essential differences, inappreciable to ordinary observers, are remarkably exemplified in his comparison of the albumen of the Blood with that of the Egg.

> The coagulable part of the Serum, he observes, seems to be in some degree the same with that in the white of an Egg, Synovia, &c; but not exactly;—for these Secretions contain, as I conceive, a quantity of the coagulable lymph united to them.

Berzelius was the first who clearly indicated the differences subsisting between the albumen of the Blood and that of an Egg; and he warns his readers of the results obtained by those Chemists (as Chevreuil, Thenard, & others[1]) who regarding the two kinds of albumen as identical; operated only upon the more easily obtained Albumen of the Egg.[2]

---

1 & others] *interlined in pencil above pencil deletion of* 'Gay-Lussac, Prevost and Dumas,'.

2 Egg] *before vertical pencil deletion of fols. 22-24 which have been pinned together with straight pin* '{¶} Whether Mr. Hunter's conjecture as to the cause of the difference which he perceived, be accurate or not, cannot at present be determined; since the chemical properties of fibrine appear to be almost {*interlined in pencil above deletion of* 'the'} same as those of Albumen; but it is rendered probable by the fact that the albumen of the Egg is coagulated sooner by heat;—{*followed by pencil deletion of* 'the process commencing at the temperature of 149. Farenheit,'} and more readily by the ponderable Chemical Agents.{¶} Thus for example, Æther and Oil of Terpentine {*sic*} coagulate the white of Egg; but not the Albumen of Serum, according to Tiedemann and Gmelin. The white of Egg is also coagulated when shaken up with æther; while, when the serum of the blood is similarly treated, the fatty matter merely rises with the æther to the surface, leaving the Serum below fluid, and otherwise unchanged. {¶} The albumen of blood, Hunter found, in his series of Chemical Experiments, to be coagulable not only by heat, but by alcohol and acids. With a mixture of alcohol of ammonia he found the coagulation of the albumen was so far produced as to turn the Serum into a milky fluid. {¶} In his experiments with Goulard's Extract, Hunter, however, did not distinguish between the coagulation of albumen, and the precipitation of mucous matter, and other changes unconnected with the subject immediately under his investigation [l *fol.* 22] and respecting which he formed his conclusions. {¶} We have already observed that the coagulation of the fluid Albumen is always accompanied by the separation of the Soda which is combined with it in the fluid state; yet it is by no means determined, or generally allowed, that this separation is the essential condition or cause of the coagulation. Berzelius candidly acknowledges his ignorance of the change which Albumen undergoes during Coagulation; and the explanations which {*with* '?∴' *inserted in margin*} have been offered in {*interlined in pencil* 'c'} place of *those {*interlined in pencil above deletion of* 'them' *with pencil underline of* 'them, which'}, which makes it depend on the separation of the Soda, are not satisfactory or conclusive. {¶}{*Inserted pencil pause mark in margin*} Dried coagulated Albumen presents an Amber-colour, and differs from dried fibrine in being transparent. When remoistened with water it softens, swells, becomes opake, and resumes its original aspect;—and what deserves particular attention, retains its solubility. The analogy, therefore, which has been suggested by Dr. Turner to subsist between the solution of Albumen and that of Silica, and the explanation based by the same able Chemist on that supposed Analogy, with respect to the cause of the coagulation of Albumen that it

25   The Serosity, or[1] that part of the Serum which remains fluid after the albumen has been coagulated by heat, contains a little muriate of Soda; a little free Alkali, and about 1/50th. of its weight of Animal matter, which is principally a residuum of albumen, and which is immediately coagulated by Galvanism or the mineral acids.

Brande supposes that the albumen is held in solution in this case, as in the serum, by virtue of the free alkali which is present.

As, in reference to the main function of the blood, the inquiry into the relative proportion of the nutrient principles it contains, is one of immediate interest and importance, we should expect to find this subject engaging the especial attention of Hunter. He accordingly determines by Experiments and Observations, which if not characterized by the minute accuracy of the Chemistry of the present day, are sufficiently conclusive for the purpose; that the composition of the Serum varies in the same species at different periods of life; and varies still more in Animals of different Species.

The serum of a Man 56 years of age, who had met with a slight accident, (and was of a healthy constitution,) coagulated by heat almost wholly into a pretty firm coagulum, separating only a small portion of the fluid which is not coagulable by that means.

26   The Serum of the blood of a Man 72 years of age, and[1] of a healthy consti-

---

(contd.)

depends, vizt. on the feebleness of its affinity for water, and the slight causes which consequently decompose the compound, as in the case of Solutions of Silica, is invalidated by this essential difference,—that the Silica remains insoluble after being once [I *fol.* 23] precipitated, while*albumen {*after pencil deletion of* 'the'} retains its solubility. {¶} {*pencil pause mark in margin*} Coagulated Albumen differs from coagulated fibrine in not exerting any influence on the decomposition of the hydric-superoxyd. This difference, like the one previously adduced, may, however, be explained without supposing that there exists any essential difference in the Chemical constitution of the two bodies. In the latter instance it depends on a difference in the mode of aggregation of their component particles:—for it is a remarkable, though well-known fact, that round and smooth bodies when immersed in a liquid charged with Gas, do not disengage it,*while {*before erased deletion of* 'though'} the same gas is rapidly driven to the surface when rough and sharply angular substances are immersed in the fluid.—{¶} Berzelius, indeed, states, that albumen in a state of Coagulation possesses so completely all the chemical properties of fibrine, that so far as modern Analysis has gone, they must be regarded as the same substance, in a Chemical Sense; differing only in some slight, and at present inappreciable modification of the arrangement of their constituent particles. {¶} Destructive Analysis of the Combustible parts of Albumen, produces

	Gay Lussac.	Michaelis
		Arterial blood.
Nitrogen.	15.705.——	15.562.
Carbon.	52.883.——	53.009.
Hydrogen.	7.540.——	6.993.
Oxygen.	23.872.——	24.436.

1   and] *interlined with caret.*

tution, hardly coagulated by heat; became only a little thicker, and formed a small coagulum adhering to the bottom of the vessel. On adding about $3/4$ths. of Water to the blood of the man aged 56, and heating it as above, it coagulated much in the same way as the Serum of the Old Man of 72. ["Blood, p.".]

The observations upon which Hunter founded his opinions as to the relative quantity of nutrient albumen in the Blood of other Animals, and at different ages, and various circumstances as regards exercise, rest &c[1] are peculiarly characteristic of his ever-active and original mind.[2]

Many a Medical Philosopher before Hunter, had fully appreciated the different culinary qualities of the Table-meats afforded by our various domesticated and wild animals;—but no one except Hunter had dreamt of deducing the quantity of nutrient albumen in the Blood in animals of different ages, and under different circumstances as regards exercise or rest, &c. by observing the [/] quantity of the uncoagulable serosity or gravy which[3] different roasted meats afforded. ("On the Blood," p. 50.)

The first Physiologists of the present day are fully aware of the importance of this inquiry of which Hunter laid the basis in these simple observations;—and Müller in Germany, Berzelius in Denmark, Prout in England, and Edwards in France have all contributed their meed of Praise to the minute, accurate, and extensive investigations of M. M. Prevost and Dumas[4], by which the proportions of albumen and of the other nutrient constituents of the Blood have been determined in so great a variety of Animals, and in the same species under these[5] different circumstances, and Individual Temperaments, which render the results so obtained[6] practically applicable in the consideration of the Diseases of Mankind. In this diagram you may observe the principal results of these investigations.[7]

Blood-Globules.[8] The red part of the Blood Hunter chose to consider last, although it had been as he justly stated, more the object of attention than the other two, vizt. the coagulating lumph and serum, he believed it to be the least important constituent; observing, that it was not a universal ingredient in the

27

---

1   and...&c] *interlined in ink over pencil with caret.*
2   mind.] *before pencil deletion of* '{¶} It is the privilege of true Genius to*deduce {*interlined in pencil above underlined* 'found'} *the* most unexpected discoveries*from {*interlined in pencil above* 'on'} the most commonplace and every-day occurrences. {¶} We may have all listened to the vibratory noise occasioned by putting a finger into the Ear; but it required the genius of a Wollaston to deduce from that simple fact, one of the most recondite laws of Muscular motion.'.
3   which] *before deletion of* 'the'.
4   Dumas,] *before deletion of* 'and of M. Lecanu'.
5   these] *modified in pencil from* 'the'.
6   so obtained] *interlined in ink over pencil with caret.*
7   In...investigations.] *RO ink insertion from facing* 28ᵛ, *with deletion of second* 'The' *before* 'principal' *with deletion of* 'are embodied in this Diagram' *after* 'investigations.'.
8   Blood-Globules] *follows a double space in the manuscript to indicate a pause.*

blood of Animals, like the coagulating lymph and the serum; neither was it to be found in every part of those animals which have it in the general mass of their blood.

28   Lastly, Hunter, in seeking to determine the [/] relative importance of the different constituents of the blood, by the truly philosophical, but most difficult investigation of their respective periods of formation in the development of the fœtus, made the interesting discovery that the vessels of the embryo of a red-blooded Animal circulated, in the first instance, colourless blood like those of the invertebrate Animals.[1]

> The red globules, he observes, are formed later in life than the other two constituents; for, we see while the chick is in the Egg, the heart beating, and it then contains a transparent fluid before any red globules are formed, which fluid we may suppose to be the serum and the lymph.

This discovery, which is more than once mentioned in the immortal Treatise on the Blood, has been very recently reproduced by two French Physiologists, M. M. Coste & Delpeck, and was[2] received with all the consideration which its importance justly merits by the Academy of Sciences, without its being suspected that our great Physiologist had embraced it, with all its legitimate deductions in the extended circle of his investigations.

In considering, however, the relative importance of the solid particles[3] and the fluid constituents[4] of the circulating blood, in which they are naturally suspended, we must avoid the error into which M^r. Hunter fell, of regarding
29   them[5] [/] as peculiar to the Vertebrate animals[6]: For, contrary to the opinion expressed by Hunter, Globules of a definite form do exist in the blood of most invertebrate animals. Analogous[7] colourless but more minute globules also exist in the blood[8] which first circulates through the vessels of the Embryo Chick.[9]

One cannot help perceiving that Hunter did not enter into the inquiry of the part of his subject with the same dispassionate and unbiased feelings as he viewed the others; for, while acknowledging that a good deal of information

---

1 Animals.] *RO pencil note on 29^v facing* '—See Valentyn'.
2 M.M. Coste. . .was] *RO ink note on 29^v facing over pencil* 'M. Coste & Delpeck'; *deleting* 'and' *before* 'received'.
3 particles] *interlined over deletion of* 'globules'.
4 constituents] *interlined above deletion of* 'parts'.
5 them] *interlined in pencil below deletion of* 'the colouring matter as an essential' *and above* 'the {before deletion of* 'red'} globules an essential' *on bottom of page.*
6 as...animals:] *interlined with caret in ink over pencil above deletion of* 'an essential part of those globules:'.
7 'Analogous' *and* 'but...minute'] *interlined with carets in ink over pencil before and after* 'Colourless'; 'C' *altered to lower case [ed.].*
8 blood] *after pencil and pen deletion of* ' white'.
9 Chick.] *period inserted in pencil before vertical pencil deletion of* '(according to the observations of *Coste & Delpeck {interlined above deletion of* 'Prevost and Dumas'} which I had the pleasure of hearing read to the Institute, in 1832.)'.

respecting the red globules of the blood was to be gained by magnifying glasses, he alludes to the discoveries of Malpighi and Leeuwenhoeck in somewhat disparaging terms; urges objections to microscopical observations generally[1]; and endeavours to invalidate by implication the well-connected and accurate[2] observations of Hewson.

Hunter was, however, all the while ignorant of the real form of the blood-globules:—he supposed them to be spherical, and therefore concluded that a highly magnified impression received by the eye must be imperfect; "the cen-
30   tre being too far from[3] the circumference to be [/] seen at one distance, and the circumference when seen, bringing the centre within the focus, so as to obscure it." [On the Blood, p. 73.]

This objection, however, which is a purely theoretical one based on the supposed spherical forms of the red particles,[4] does not apply to them for they[5] are not spheres, but flattened discs, and which, when viewed as they lie flat on the object glass[6] are well calculated to convey an accurate idea of their contour under much higher magnifying powers than are merely necessary to determine that point.

Hunter was equally in error in supposing that the shape of the globules must be the same in all animals;—a conclusion, however, which he certainly never arrived at from observation, but from a hypothetical, and not very intelligible notion, that "the shape must depend upon a fixed principle in the globule itself." [p. 72.]

With whatever suspicion microscopical observations *might be*; and to whatever objections the instrument *was* open, when used with very high powers, ten or even five years ago, such cannot reasonably be attached to the achromatic Engiscopes of the best Makers at the present day:—but on the contrary, there is much reason for expecting that the most important advances in physiology will be effected by means of these Instruments.
31   With such an Instrument as the one[7] now on the Table, [/] the experienced investigator of the ultimate composition of the animal fluids and tissues may safely trust to the forms & superficial characters[8] of the ultimate fibres,

---

1   generally] *interlined in pen over pencil above deletion of* 'which in certain cases are no doubt applicable and just' *with WC note on 30ᵛ facing* 'This probably was levelled at Mr Hewson who was then or lately overwhelming every body with the Blood globules, and there might have been some soreness from his robbing Dr Hunter and Mr. Hunter without proper acknowledgement.'. *This is followed in different pencil* 'I had not seen what follows when I wrote the above memorandum.'.
2   accurate] *after deletion of* 'generally'.
3   far from] *interlined in pencil above and in ink below deletion of* 'near'.
4   , which...particles,] *RO note on 31ᵛ with insertion point marked by caret.*
5   them for they] *inserted to replace deletion of* 'the red blood-globules, which' *with deletion of interlined* 'red particles' *above* 'blood-globules' *and* 'for they' *interlined above deletion of* 'which'.
6   as...glass] *interlined above deletion of* 'horizontally'.
7   one] *before pencil deletion of* 'belonging to the College, and'.
8   &...characters] *interlined with caret in ink over pencil.*

globules, or granules which are manifested to him by a magnifying power of from 200 to 600, linear dimensions. I do not mean to assert that the same freedom from error exists with regard to the *size* of the objects requiring similar high powers for their display.

The exact admeasurement of such minute objects is an operation of[1] difficulty, but one of considerable importance. Observations, such as those made by Home and Bauer on the size of the Blood-Globules, by means of the ordinary Micrometer are necessarily defective, inasmuch as the globules placed upon this instrument and the divisions drawn on its surface can never under high powers[2] be simultaneously in the focus of the object glass; and as the globules, from the mode of treatment adopted by these Observers were greatly altered from their natural form, as I shall presently explain, the amount of error was proportionally augmented.

Shick's screw-micrometer,[3] the instrument invented by Dr. Wollaston called the Eriometer, should be employed for this purpose; or the image may be projected by the *Camera lucida* on a scale, and its limits then accurately determined.

32   Whatever process, however, is adopted, the [/] dimensions of the part must be repeatedly[4] taken in each case with the greatest exactness possible, and out of a number of Observations, a mean should be established. It is thus alone that the degree of certainty of these admeasurements, and the amount of possible error may be ascertained.

It[5] is the practice of the best modern microscopic observers[6], to take some small object of constant size, and adopt it as the standard of comparison[7]:— and as it fortunately happens that the Human Blood-Globule in the adult al-

---

1   an ...of] *interlined in ink over pencil above deletion of* 'a subject of much'.
2   under...powers] *interlined with caret in ink over original pencil interline of same phrase with pencil caret after* 'be'.
3   Shick's screw-micrometer,] *inserted by RO in ink over pencil before original* 'The' *with* 'The screw micrometer' *in RO pencil on facing 32ᵛ. This is followed by WC pencil note on 32ᵛ:* ' I am not sure that Mr. Bauer had not this Instrument? I know he had one with which you looked at the object in the microscope with one eye; and with the other naked eye you looked at a scale at the side of the Instrument, & calculated accordingly—as well as the squares ruled with a diamond point on the glass on which the object was placed; (4000 in an inch.?)'.
4   repeatedly] *interlined in ink over pencil with caret.*
5   It] *follows vertical pencil deletion of paragraph* '{¶} But even with the greatest care, discrepancies occur as to the *dimensions* of many of the elementary parts of the human body, in the observations of the best Observers, who are now agreed about their *forms* *&* superficial characters{*interlined in pencil with caret*}. These discrepancies are *scarcely {on line in ink and interlined above in pencil} avoidable when the same objects are observed by Instruments of Different Constructions, and their dimensions determined by different processes.'.
6   is...observers] *interlined above deletion of* 'has recently been proposed, therefore,' *with* 'observers' *on 32ᵛ facing.*
7   comparison] *before deletion of* 'in all places'.

ways presents[1] the same dimensions, taking the majority under the focus, and is always at hand, this will doubtless become the standard for all future Micrographers.

It is also fortunate for this most useful purpose that the dimensions of the blood-globule have been the subject of more careful & repeated[2] Observations than those of any other Microscopic object, and the admeasurements

33   given by Wollaston, [/] Kater, Weber, Wagner, Lister, Prevost, Edwards, Mayo, and Müller are so little discordant that the mean of their Observations may be received with the greatest confidence.

Wollaston......................................	20 1/10,000 Line.
Kater..	1/5000[th][3]
Weber...	
Wagner..	
Lister..	
Prevost..	
Edwards...	
Mayo..	1/5000[4]
Müller[5]..	

This mean gives[6] the diameter of the Human Globule as[7] 1/4000th part of an Inch. At the same time I may observe that my own direct admeasurements accord with those which attribute the dimensions of 3000th parts of an inch—[8]

But, perfect[9] accuracy in this respect is not[10] essential for practical purposes:—for, however Observers may differ as to the dimensions of the Blood-Globule itself, if they take it as the Standard whereby to compare the ultimate fibres[11] of nerves or muscles, or the diameter of injected capillaries;

---

1   in...presents] *interlined with caret above deletion of* 'is of'.
2   careful & repeated] *interlined in ink over pencil above deletion of* 'accurate'.
3   1/5000th] *inserted in pencil.*
4   1/5000] *in pencil revised from* '1/3000'.
5   Wollaston...Müller] *with WC pencil note on facing* 34[v] 'Is there no essential difference in the *size* of the *globule* in people of different dimensions? (O'Brien and Miss Crachemi.) I should calculate there would be, hypothetically:—taking in differences in Food—&c. &c. &c among vegetable & animal feeders in India and England. &c: in the negro & Albino.'.
6   This...gives] *interlined above deletion of* 'whereby'.
7   as] *interlined above deletion of* 'may be estimated at' *with WC pencil note on* 34[v] *facing with pencil bracket* '}' 'This I think was also M[r]. Bauer's calculation 1/4000'.
8   At...inch—] *RO insertion from* 34[v] *facing to replace erased pencil interline after* 'Inch.'.
9   But, perfect] *follows erased pencil line with* 'But' *interlined above deletion of* 'Happily' *with* 'Perfect' *altered to* 'perfect'.
10   not] *after deletion of inserted* 'however,'.
11   fibres] *inserted over deletion of* 'filaments'.

it can always be determined whether any of these are larger or smaller than the globule by 2. 3. or more[1] times its diameter. This being determined, however inaccurate the observer may have been in regard to his direct ad-measurements the comparative reference to the globule will always enable

34   another[2] to judge with certainty of [/] the relative size of the part he may have[3] been observing.[4]

In examining the red particles[5] of the blood, they must not be diluted or floated in water: if this be done, they will be seen of a form quite different from that which they possess in the living body; for they lose their flatness and almost[6] instantly become globular.[7] It was perhaps from this cause that Leeuwenhoeck, Fontana,[8] and all the older observers described the Human blood-globules as spheres.

To Hewson belongs the merit of having first dissipated this error, and of having described their true form.

The old statement has, however, again been revived by Home and Bauer, but fortunately the error can be accounted for; as it is incidentally mentioned in their descriptions that the globules were diluted with water; and the altera-

35   tion[9] of form, which such treatment produces, [/] seems not to have been suspected by these Observers.[10]

A small portion of Water alters the form of the globules, almost instanta-neously from a flattened circle or ellipse to a sphere: a larger proportion de-

---

1   more] *RO interlined in pencil above* '4'.
2   another] *interlined with caret above deletion of* 'us '.
3   may have] *interlined above deletion of* 'has'.
4   observing] *before pencil deletion of paragraph with bracket around it and* 'X X X' *in margin* '{¶} I entertain no doubt but that by the aid of the excellent Instruments now manufactured, and in the hands of honest and experienced observers, by whose labours in this field the boundarys of Natural Science have been of late years so greatly enlarged; that some of the most difficult problems in Physiology will be*resolved {partial underline in pencil} by the {pencil underline} admeasurements of the plan which has just been explained. *> ' {pencil '>'}.
5   red particles] *inserted below erased pencil interline of* 'red particles'.
6   almost] *interlined with caret.*
7   globular.] *interlined above deletion of* 'round.' *with WC note on facing fol 35ᵛ:* 'Spherical? If so, how do we know that they are not spherical in the living body but become flat from want of that support which the water affords to them—? —What form would the contents of an egg assume on a flat surface deprived of the support of the shell, *in air. ?*'.
8   Leeuwenhoeck, Fontana,] *interchanged with lead line and circle around* 'Fontana'.
9   alteration] *interlined above deletion of* 'change in the' *with* 'of' *inserted before* 'form' *and before deletion of* 'of their parts'.
10 Observers] *before pencil deletion of* 'We can only regret that the value of a useful Instrument is apt to be temporarily, though unjustly, depreciated by the propagation of such errors, which are almost inseparable on its vicarious employment.' *with WC pencil note on facing fol. 36ᵛ* '—Mr Bauer had had many years experience in the use of the microscope—& is no fool.—Sir Everard could only see, at most, what Mr Bauer showed *him.*'.

stroys it by dissolving the Colouring matter or capsules.[1]

In order therefore to examine the globules microscopically, a drop of Serum, or of a solution of[2] Salt or Sugar should be first placed upon the object-glass, and then a drop of fresh blood thinly spread through this fluid. The majority of the globules observed are always of the uniform or average size characteristic of the Species;—but even in Human Blood, one may be observed, here and there, of rather larger size.

In the Embryo state, the discrepancies in this respect are more considerable. In the Embryo Dog the majority of the globules are as large as in the blood of the mature animal,[3] while other globules are seen of double the[4] dimensions. [/] In the foetus of the Goat, the globules observed by M. M. Prevost and Dumas were double the diameter of those of the parent[5], exhibiting in this respect, as in the structure of the heart and the low condition of respiration, an analogy to the condition of the Reptile.[6]

The size, as well as the form of the blood-globules, is the same both in arterial & venous blood:—In a Frog which survived extirpation of the lungs 30 hours, Professor Müller found no change in the shape of the globules.

The form of the blood globules varies in different animals; but whether they be circular or elliptical they are always flattened.[7]

In man, and in the mammalia they are round discs. In Birds and Reptiles they are oval discs.

36 (margin)

---

1   or capsules] *interlined with caret before pencil deletion of* 'Hunter supposed that the intire globule was dissolved, as is still *indeed, {interlined with caret}* maintained by Hodgkin and Lister. *{with RO pencil insertion in margin* 'parts not dissolved'*}* But Hunter had ascertained that the globules were insoluble in water when saturated with the neutral salts, as well as in Urine, or Serum. [p. 16.]' *and on facing fol. 36ᵛ WC pencil note* '—One can imagine that a spherical globule would become flat by its gravity when unsupported,—but that a naturally *flat* body should become a sphere when supported *or pressed {interlined with caret}* on all sides alike *{interlined with caret }* seems rather paradoxical, and almost too much to believe.' *Followed by WC query:* 'Qu. Would not the Salt, or Sugar act chemically on the globule; and by hardening, prevent its expansion into a Sphere, if naturally one? & having once fallen down flat,— The greater density of the Salt or Sugar would be unfavourable to its rising, or plumping up, & consequently it would remain a disc.—Is there any method of showing the edge of a disc? else the eye may be deceived as to its degree of flatness.—If they were discs would all the globules fall flat, as they appear to do:—would not some be supported on edge? particularly in such dense fluid as sugar & water?'.

2   Salt] *follows deletion of inserted* 'a neutral'.

3   animal] *inserted over erasure below erased pencil interline of* 'Animal'.

4   the] *altered from* 'These'.

5   parent] *interlined above deletion of* 'adult animal'.

6   Reptile.] *before pencil deletion of* {¶'} In the blood of a Frog, of which the globules are remarkable for their large size, the *few {inserted in pencil}* exceptions which occur *to the size of the measrments{sic interlined with caret in pencil}* are usually less than the average.' *RO pencil note on fol. 37ᵛ facing:* '{Also in Chick at first round as in Invertebrate then oval—'.

7   flattened.] *before pencil line across page and RO pencil note on 37ᵛ facing* 'Diagram— form, *relative size {above deletion of* 'Structure, & size'*}* structure—/'.

In the higher cartilaginous fishes they are oval discs as in the Batrachia.[1]

In some of[2] the bony fish as the Perch, & Stickleback[3] they approach more nearly to the rounded form than in the higher Oviparous Vertebrates.[4]

38ᵛ[5]   In all these classes the blood-globules consist of a middle colourless nucleus, which is not always of constant size, and of a red homogeneous capsule.

Ehrenberg[6] states that the nuclei of the blood-discs consist of an aggregate

---

1   Batrachia.] *WC note on facing 37ᵛ* 'I recollect that Sir Everard told me perhaps 25 or 30 years ago, when Dr Thomas Young was but a young man, that he had told him that the globules of blood in the Skate were *oval* & I think *flat*—long before I knew Herbert Mayo. but I do not know whether Dr Young published it in the Transactions or elsewhere: but I believe it was then received as doubtful.'.

2   some of] *interlined by RO with caret in ink with RO pencil insert in margin* 'Ehrenberg'.

3   as…Stickleback] *RO interline in ink with caret.*

4   Vertebrates.] *Followed by pen and pencil deletion by vertical scoring of the remainder of fol. 36, all of fol. 37, and the first portion of fol. 38ᵛ reading* '{¶} They are nearly circular in the blood of the Carp; and Rudolphi has described them as perfectly circular in the Shad, (Clupea Alosa, Linn:) but it is probable that this may have arisen from [l *fol.36*] the accidental or intentional admixture of water, by the contact of which, the elliptic flattened blood-discs, of the ovipara *like the round discs of mammalian blood {interlined in pencil with caret}* instantly swell out & assume the spherical form. It is possible that Mr. Hunter may have observed the globules of the amphibia, (p: 72.) under these circumstances, and thence was led to disbelieve the statements of Leeuwenhoeck and Hewson respecting their elliptical form in these and other Ovipara. {¶} Among the Invertebrate animals the blood-discs are much less regular in their forms, than in the Vertebrate classes. Their surface is uneven and tuberculated like that of a raspberry: their contour is extremely variable; they change their figure *from the slightest course {WC interlined in pencil above deletion of pencil underlined: 'with the greatest facility'}*, and their size is considerable. {¶} In the blood of the River Craw-fish the mean diameter of the globules is 70 1/10,000ths. of a line; several were 67, and some 72. {¶} In the Oyster, I have detected still wider differences in the size of the globules of the blood in the same drop of this Creature's blood,—some globules were 60 1/10,000; others only 64, and some no more than 40 1/10,000 of a line in diameter. {¶} The diameter of the Human blood discs we have estimated at 1/4,000 of an inch. {¶} Those of the ruminant Mammalia, at least of the Goat, still smaller. {¶} The longest diameter of the blood discs of the Frog is four times that of the human blood disc. [l *fol. 37*] The blood discs of the Salamander are somewhat larger and relatively longer than in the Frog. {¶} Those of a Lizard are 2/3ʳᵈˢ. the size of a Frog's. {¶} Those of a Bird are half the size of a Frog's, and consequently twice the size of the blood-discs of Man, and other mammalia. {in margin in RO pencil slash mark and 'Diagram'} {¶} It would be highly interesting to examine the blood discs of the Ornithorhynchus and Echidna, with reference to the affinity which they might manifest in the size and shape of these parts to the Oviparous Classes:—and from the long period of Time that common salt preserves blood in its ordinary fluid state, I am in great hopes of receiving the blood of these animals in a state fit for examining this point from my friend and correspondent Mr. George Bennett.' *This all replaced by insertion following from fol. 38v.*

5   38v] *From here to fol. 38 all inserted from 38v.*

6   Ehrenberg] *after deletion of* 'Thus'.

of equal-sized granules, which in Mammalia have 1/5000 line in diameter.
The capsule or coloured part[1], is generally wanting in the Invertebrate-classes
and in the higher Articulate classes[2] the blood-discs seem to consist of the
granular nuclei alone.

The[3] component granules are larger than[4] in the Vertebrate classes, and
hence the exterior of the blood-globules appears uneven & tuberculate.

In Insects the solid particles of the blood are remarkable for their variety of
figure.[5]

In the *Mollusca*, which in all the apparatus of organic life most resemble
the Vertebrata, the globules are of a regular form, but vary in their dimensions
and are[6] invested with a colourless[7] capsule, which is very large[8] in proportion to the nucleus.[9]

38   The blood discs are most compressed and flattened in Fishes and
Batrachia, especially in the Newts and Salamanders. In the Frog the long[10]
diameter is ten times greater than their thickness.

When viewed as suspended in insoluble menstrua, some will present their
edge to the observer; and in Mammalia, Birds, and most Reptiles, the discs in
this position appear as lines, whose diameter is uniform or unbroken, with[11]
rounded extremities.

The large blood-discs of the Newt are favourable objects for the examina-
39   tion of Form:—[/] they present no projections from their lateral surfaces but
are uniformly flat. The blood discs of the Frog, however, when viewed
edgeways manifest more or less clearly an eminence on each side;—[as in the
Diagram.][12]

In other animals the nucleus of the blood-disc is too small to make any

---

1 coloured part] *interlined with caret above deletion of* 'of the blood-discs'.
2 and...classes] *interlined with caret above deletion of* 'was so high' *with* 'higher'
   interlined below 'high'.
3 The] *after deletion of* 'Hence /'.
4 seem to. . .larger] *over WC pencil* 'See page 41 or give it M. O. I fancy omission
   here, for there appears to be an indistinct reference <' *RO pen then continues over earlier
   pencil comments* 'It's interesting to consider that in the Vertebrate animals with oval
   globules these are first round', *followed by WC reply in pencil* 'You recollect Sir
   Everard's *blood* globules and *lymph* globules of all sizes?' *All of this part of insertion
   from 38v.*
5 in...figure.] *interlined.*
6 of...are] *interlined with caret above and below line.*
7 colourless] *after deletion of* 'large'.
8 very large] *with* 'very' *interlined with caret and* 'large' *altered from* 'larger'.
9 nucleus.] *before deletion of* 'than in other animals'.
10 the long] 'the' *modified from* 'their' *with* 'long' *interlined by RO in ink.*
11 with] *RO insertion in pencil to replace deleted* 'save at the'.
12 Diagram.] *before pencil deletion of* '{¶} This appearance is of great importance in the
   determination of the long-debated question whether the globules possess or not a central
   nucleus. It depends on the greater relative size of the nucleus in the Frog, which is the
   only animal in which the nucleus exceeds the smallest diameter of the disc, and the
   presence of which consequently affects the contour of the blood disc, when viewed
   edgeways.'

projection, and in Man, where D^r: Young, D^r. Hodgkin, and Mr Lister have given good reason for believing that the sides of the disc are slightly concave, the nuclei are too small to make any projection from the surface.

But in all *these* cases their presence may be shown by their effect upon transmitted[1] light. When the blood discs are viewed with a low magnifying power, they appear like small black points:—When the power is increased to a degree corresponding with that of the Instruments used by Della Torré, and Styles they assume the appearance of a white circle with a black point in the centre; and we can readily understand how they came to be described by these Observers as being perforated in the centre, and fashioned like rings. A magnifying power of 300 linear [/] dimensions adapted to an instrument with a good aperture and clear light, intirely dissipates the optical illusion by which these earlier observers were deceived. The central point now assumes the appearance of a luminous spot, in consequence of the colourless nature of the nucleus; and by varying the position of the globule as well as the direction of the rays of light, the observer may easily convince himself that the globules are intire.

But the reality of the existence of the central nuclei can be demonstrated on evidence more satisfactory than reasonings founded however unexceptionably, on optical phenomena:—they can be mechanically separated from the external colouring capsule, and examined as distinct elements.[2]

Professor Müller applied Acetic acid to the blood-globules of a Frog while under the Microscope, and saw the dissolution of the coloured capsule, while the nucleus remained undissolved.

I have repeated this experiment, the confirmation of which is easy and satisfactory. The nuclei are of an elliptical form, corresponding to that of the blood-disc.

In the Salamander they manifest the peculiar flattened form characteristic of that Animal's blood discs.

In Man, Professor Müller states that they are round, and yellower than the surrounding transparent fluid.[3]

If the fibrine be removed from the blood of a Frog received in a watch-

---

1 transmitted] *RO interlined in ink with caret.*

2 elements.] *before pencil deletion of paragraph* '{¶}I have never been able to effect this with human blood-discs; and from their small proportional size in these, few I apprehend will be able to perceive them satisfactorily; and some able {*deleted*} observers still doubt their existence:—but comparative observations in this, as in most other doubtful points of human anatomy, lead to satisfactory and just conclusions. {¶} The accidental observation of the external projection of the nuclei in the large blood discs of the frog led to the series of experiments which proves their existence; and *will lead* to the determination of their characteristic properties, which, hitherto, have been only conjectural.'.

3 fluid.]*WC pencil note on facing 42^v* '? What surrounding transparent fluid? unless there be *three* parts—i.e. The *red {interlined with caret }* *colouring {before deletion of* 'red'} matter—the disc or white, and the nucleus or yolk; which there appears to be.'; *Followed by RO reply:* 'An apothecaries globe is red but transparent—'.

glass in the manner previously described, so as to obtain the serum and red globules or discs suspended in the fluid, the nuclei may be obtained in a purer state, for chemical investigation, by the following method, than when liberated by acetic acid. Add water to the Serum and globules in the watch-glass till it is full;—let it stand for a short time till the globules have sunk to the bottom;—then gently let down the glass and its contents into a vessel of clear water, taking care not to disturb the globules.[1] Leave it in this situation for two hours until the red stratum at the bottom of the watch glass shall have turned white:—then slowly remove the watch-glass and its contents from the tumbler of water.

The stratum now examined microscopically will be found to consist[2] of colourless oval granules,[3] one fourth the size of the blood-discs.[4]

By multiplying this Experiment, and removing [/] the watch glasses at different periods after their immersion, the stratum of globules will be found to be gradually smaller from the progressive solution of the colouring matter in the water.

By these observations, conjoined with the identity in size and shape, of the nuclei obtained by solution of the Capsule in water, and those obtained by the action of Acetic Acid,[5] a satisfactory proof is established[6] that the products of the latter experiment are not contracted and corrugated globules, as has been objected, but genuine unchanged Nuclei.

In general, the form of the nucleus corresponds to that of the intire globule:—even in the Salamander they are as flat as the globule itself; but in the frog they are proportionally broader, and some approach nearly to a circular form.

The nuclei of the blood-discs have the general chemical properties of Coagulated fibrine and albumen: they are easily dissolved in Alkalies, but are difficultly acted upon by acids. Even in acetic acid they remain unchanged for twenty-four hours, and are then suspended as a[7] brownish impalpable powder; while fibrine, under similar circumstances forms a transparent solution.

The modes in which the blood-globules are affected by different chemical re-agents are best studied in the globules of the Frog's Blood separated from the fibrine by the processes already described.

Chlorine changes the colouring matter of the blood-globules first to a

---

1   globules.] *with WC pencil note on facing* 42ᵛ 'A *tumbler* must be a most
    inconvenient part of the apparatus.—Why not a deep plate or*flat {*interlined with caret*}
    bason?—it is impossible to do it steadily & neatly in a *tumbler*, without great chance of
    capsize. "A *vessel* of clear water—'.
2   consist] *followed by ink deletion of* 'no longer of the blood discs, but'.
3   granules] *followed by ink deletion of* 'four times smaller.'.
4   one. . .discs.] *RO insertion in ink.*
5   Acid,] *followed by deletion in ink of* "we obtain'.
6   is established] *RO interlined in ink with caret.*
7   as a] *interlined in ink above deletion of* 'in'.

43   brown, and then quickly [/] turns them white, and it coagulates the albumen
in[1] the globules.

Neither Oxygen nor Carbonic acid affect in any way the form of the blood
globules.

Solution of Caustic potass does not alter their shape, but diminishes their
magnitude, and progressively and rapidly dissolves not only the capsule but
the nucleus.

Liquor Ammoniæ still more quickly dissolves the globules, and changes
them at the moment of contact from a flattened circle or ellipse to a Sphere.[2]

Alcohol does not otherwise affect the globules than by the contraction
which ensues from the coagulation of the albumen.

Strychnine and Morphium produce no change in the globules.[3]

Coagulation. The state of the circulating blood in which the red globules
are freely suspended in the *liquor sanguinis*, is only maintained during life.
Whether Coagulation, and the decomposition of the fluid consequent thereon,
i.e. the separation of the crassamentum from the Serum be in all cases an act
subsequent to, and occasioned by the death of the blood;—or whether it is to
be regarded, as Hunter believed, as a vital action of the Blood, analogous to
contraction in a muscle, is yet a *questio vexata*.

There is also a difference of opinion as to the immediate as well as the remote cause of coagulation; but we have now sufficient data to determine the
question, as to the constituent on which Hunterian coagulation depends,[4] in
44   favour of the [/] Hunterian doctrine.

Hunter believed that the spontaneous coagulation which ensues upon extravasation was peculiarly the property of the fibrine.

The Physiologists of *Geneva*, M. M. Prevost and Dumas, have since attributed it to the red globules, and consider the fibrine to result from an aggregation of the colourless nuclei of the globules.—Their proposition is thus
enuntiated—[5]The attraction which keeps the red matter fixed around the white
globules having ceased along with the motion of the fluid, these are left at
liberty to obey the force which tends to make them combine and form a
network in the meshes of which the colouring matter is included.[6]

---

1   in] *underlined in pencil with WC pencil* 'X see MS' *in margin.*
2   Sphere]*WC pencil note on facing 44v* 'Hence Its use in fainting by plumping up the
     vessels which were before collapsed.—!!!'.
3   globules.] *followed by double horizontal pencil line in margin as if to indicate significant
     pause. WC pencil note on facing 44v* 'I believe there is something more than this to be
     observed—there is an attraction of the water to the alcohol, by some chemical affinity or
     other. I have seen an hydatid, and a turgid ovum soon emptied by being put into
     alcohol.'.
4   the question. . .depends,] *RO interlined below line in ink over pencil followed by
     deletion of* 'this point' *on original line.*
5   Their. . .enunciated—] *RO pencil insertion from facing fol. 45v with insert point
     indicated by lead line.*
6   included.] *followed by deletion by vertical scoring in pencil of* '{¶} Dr. Prout has
     proposed a modification of this opinion, and ascribes a share of the act of coagulation to

Hunter, however assigned no share whatever in the act of Coagulation to the red globules:—he conceived that they were passive and became[1] mechanically[2] entangled in the coagulating lymph, and added to [/] its weight, so as to make it sink deeper in the Serum.

The fibrine of the lymph coagulates and entangles in the coagulum a portion of the globules of the lymph, but only a small portion, for the greater part remains suspended in the serum of the lymph—thus manifestly showing that the globules have no share in the formation of the fibrineless coagulum—[3]

Having accurately determined the different relative[4] specific gravities of the Serum, Fibrine, and red globules, Hunter[5] supported his position by arguments drawn from the colourless condition of the coagulum formed under circumstances in which the red globules sink, before the fibrine begins to coagulate:—and this opinion has been amply confirmed by the recent experiments of Professor Müller.

He took pure arterial blood from the *arteria ischiatica* which runs between the two masses of muscle at the back of the leg of a Frog; received it into a watch-glass, and with a needle removed the colourless Coagula as they formed.

By diluting the blood in the watch-glass with a little serum, he was able to perceive with the microscope, the formation of the coagula in the interspaces of the globules, to which the globules merely adhered; without undergoing any of those changes of structure and position conjectured by M. M. Prevost and Dumas.[6]

Müller also diluted the blood of Frogs with sugar-water, which does not dissolve the colouring matter of the globules; and succeeded in separating the large globules from the *liquor sanguinis* by passing it through fine white

---

(contd.)

the particles of fibrine: these he considers, as well as the red particles of the blood, to be in a state of extreme self-repulsion during the life of the Animal, by which self-repulsion the union of these particles is prevented, except as the economy of the animal may require, and may determine. {¶} After death, however, or in blood withdrawn from the body of a living animal, the property of self-repulsion, more especially among the fibrinous parts of the blood ceases and they readily cohere and occasion the coagulation of the blood.' *In margin pencil 'X' with quote marks in pencil at head of each line.*

1 passive and became] *RO interline in ink over pencil with caret above deletion of* 'merely'.

2 mechanically] *with WC pencil comment on facing fol. 46ᵛ* 'that they were *passive and* merely mechanically'.

3 The fibrine. . .coagulum—] *RO insertion in ink from facing fol. 47ᵛ with insert headed by* 'a') *to correspond with superscripted* 'a' *after* 'Serum.'.

4 relative] *inserted by RO in ink with RO pencil interline of* 'relative' *above* 'serum'.

5 Hunter] *interlined by RO in ink above pencil deletion of* 'be'.

6 Dumas.] *followed by deletion by vertical pencil scoring of* '{¶} When the point of a needle was introduced among the globules, they could be pulled about. In one minute a clear coagulum begins to form, in the blood, which is hardly perceptible until a needle is introduced into the watch-glass to remove it.'.

46   filter-paper. The fibrine coagulated after this separation as completely as if the
red globules had been present; but it was, of course, colourless, like the
crusta inflammatoria in rheumatic or in inflammatory blood.

Lastly, Müller retarded the coagulation of the fibrine, by adding a concentrated solution of subcarbonate of potass to recently drawn blood, and thus
obtained a colourless coagulum at the upper part of the clot, from which the
heavier globules had subsided.

Hunter went further, in a similar experiment and skimmed off the fluid coagulating[1] lymph, below which the red globules had sunk; and found that
the part so taken off, coagulated immediately. [p. 34.]

This action of fibrine, Hunter calls *spontaneous*; but it appears rather to be
the *inevitable* consequence of extravasation and rest;—it is no otherwise
spontaneous than as it does not require artificial applications to produce it,
like the Albumen.

When the conditions of the fluidity of the fibrine cease to be, its coagulation necessarily, not spontaneously, ensues.

What then are the Conditions of the fluidity of fibrine?—To this question
Hunter devoted his best attention; sensible that its solution would throw more
light on the nature of the Blood than any point of view, whether Chemical,
47   Mechanical or microscopical, under which it [/] could be considered.

After various experiments and observations he drew the conclusion, that
motion in living vessels was the chief condition of the fluidity of the blood;
yet not the essential condition. For he had seen[2] blood, extravasated into the
sac of a hydrocele, which, when let out 65 days after, then coagulated and
separated into its different parts, having previously remained fluid, but
somewhat thicker than ordinary, although motionless.[3]

He had observed also that blood which had been swallowed by a Leech,
remained fluid for a considerable time in the living digestive cavity of that
animal:—but then it might have been in constant motion whilst in that
Situation.

The conclusions that Hunter drew from the[4] continuance of fluidity of the
blood in the vessels of a torpid animal where he supposed it to be at rest,[5] are
however invalidated by the fact that the high degree of irritability of the heart
in the hybernating Bat or Hedgehog, enables it to circulate carbonized or
black-blood; and such circulation does go on, though slower, than when respiration, & sensation, and action, and digestion, are in full operation in the
same Animal Machine.

Blood when out of the living body coagulates during continued agitation or
motion, as well as when at rest.—It coagulates in the same temperature as the

---

1   coagulating] *with faint underline in pencil and on facing fol. 47ᵛ WC pencil comment*
'Coagulable lymph'.
2   seen] *with faint underline in pencil and pencil* 'X' *in margin.*
3   although motionless.] *RO pencil insertion with original period altered to comma.*
4   from the] *followed by pen deletion of repeated* 'from the'.
5   where. . .rest,] *interlined by RO in pen with caret over pencil.*

body it came from, as well as in a higher temperature, or a lower temperature.
48   —It coagulates [/] in a Vacuum as readily as in the air.

It coagulates in a vessel from which all air is excluded; and when surrounded by non-atmospheric gases.[1]

Hence we may conclude that the state of admixture of the proximate constituents of the blood, on which its fluidity depends, is maintained by the influence of the vitality of the containing parts.[2]

Hunter compared the act of coagulation to the contraction of the muscular fibre.[3] Dr. Stevens [On the Blood, p. 132] states that he has seen the fibrin of inflammatory blood contract, on the application of common salt (a chemical stimulus) (after coagulation has commenced,) with almost as much rapidity as the muscles, when the same stimulus is applied to their fibres in the living body;[4] but the sudden[5] aggregation of the atoms of the fibrine of which this supposed act of contraction consists.[6]

49   Hunter[7] founded his Opinion principally from [/] observing that those causes of Death which are accompanied by an uncoagulated state of the blood are likewise followed by an absence of the usual stiffening of the Muscles. Such are the singular phænomena which occur when an animal is killed by a stroke of Lightning, or by a strong electric shock; or a violent mental emotion:—or when death is produced by Hydrocyanic acid, or acrid vegetable poisons;—or the bite of venomous serpents. Or when an animal is hunted[8] to death; or killed by a blow on the stomach.

In all these cases, the blood has been found fluid, or grumous; and the

---

1   gases] *with erased WC pencil comment on facing 48ᵛ.*
2   parts.] *followed by deletion by vertical pencil score of* '{¶} According to the experiments of Schroeder, the blood coagulates with remarkable rapidity after violent destruction of the Brain and Spinal chord; so that in one minute after the operation, coagula are found in the great vessels. *{pencil bracket mark inserted here }* {¶}That the act *of {followed by deletion of* 'the'} coagulation of the blood is a chemical and not a vital one, is, I think, proved by the following fact: that blood, which has been kept *fluid, *{followed by light pencil deletion of:* 'and of a bright arterial colour'} for many months by means of common Salt, nevertheless coagulates after being so long removed from the living vessels, when the Salt is removed by the addition of Water.'. *On facing fol 49ᵛ erased WC pencil comment:* 'I suppose this curious fact has been satisfactorily proved, or rests on good authority. It is easily proved or disproved—'.
3   fibre.] *period inserted for grammatical sense [ed.].*
4   Dr.Stevens. . .body] *RO insertion from facing* 49ᵛ *with insertion point indicated by lead line and caret. Period after* 'body' *deleted to maintain original semicolon [ed.].*
5   sudden] *RO interline in ink with caret.*
6   of which. . .consists.] *RO interlined in ink with* 'consists' *on facing 49ᵛ over undeleted original* 'which at all resembles relaxation.'. *Followed by deletion of* '; and it never takes place under any of the stimuli which excite muscular contraction.'. *WC pencil comment on facing* 49ᵛ 'Electricity—Galvanism—Electricity—Lightning said to prevent coagn.}'.
7   Hunter] *RO insertion in margin to replace deletion of* 'He'.
8   hunted] *with WC pencil comment on 50ᵛ facing* 'Hunted , and *frightened* to death, are nearly synonymous, and consequently comes under *mental emotion*!!'.

muscles do not become rigid.

If we reason with Hunter, we must suppose that the last act of Vitality which produces the stiffening of the Muscles, also enables the fibrine to co-agulate and to contract; and that in both cases the residuum of Vitality essential to the phenomena in question is exhausted and destroyed in the violent deaths before-mentioned.

On the other hand, we must admit, that if the aggregation[1] of the particles of the fluid fibrine be necessarily an act of Life, it[2] continues alive for months, nay even years, when mixed with common salt.

50   Hunter seems also to have regarded the coagulation of the blood as the act of Nutrition or of Organical reproduction:—but I should [/] be as little disposed to imagine the reproduction of muscle to be the consequence of a simple extravasation and coagulation of the blood, as the formation of any other organized product out of the blood.

It is true that a coagulum of Blood when formed in the living body, does not necessarily undergo the changes of *dead* matter, or, unless of large size[3], irritate the surrounding parts like dead matter;—but on the contrary, after being reduced to its fibrinous constituent, it possibly[4] may become organized, like the effusions of fibrine from inflamed serous membranes.—

Of the final intention of the coagulation of the Blood, and of the many beneficial results of that property, in the arrest of Hæmmorhage; the union of parts; &c, it is not in my province to speak:—It is to be observed, however,[5] that in all these cases it is essential to the salutary operations of the Coagulum, that it be not too extensive; otherwise it irritates as a foreign body; impedes, instead of accelerates the union; and sometimes is even the cause of the continuance of an obstinate hæmorrhage.

That the Blood while circulating, is, as Harvey believed, a living fluid, the cumulative evidence furnished by Hunter and others[6] satisfactorily demonstrates;—and it is admitted as a fundamental Dictum in Physiology by the 51   ablest Professors of that Science in the present day.

Hunter justly observes, that those who have formed the idea of the blood being a passive and inanimate fluid, can have no adequate notion of the manner in which it is capable of performing its great functions in the Animal Machine. [Surgical Lectures.[7]]

> If the Blood had not the living principle, it would be in respect to the body, as an extraneous substance. . .It is not only alive itself, but is the

---

1   the aggregation] *with illegible RO pencil insertion in margin.*

2   it] *deleting preceding* 'that' *for grammatical purposes [ed.].*

3   unless. . .size] *RO interlined in ink over pencil above deleted* 'if of some small extent'.

4   possibly] *RO interlined with caret in ink.*

5   It. . .observed] *with RO interlines of* 'is to be' *over deleted* 'may only' *and* 'however' *with caret.*

6   and others] *inserted by RO in ink from facing 51*ᵛ.

7   Lectures.] *deleting single set of double quote marks after period [ed.].*

support of life in every part of the body. . . .It is an essential part of the
compound of solids and fluids which constitute the living animal body.
[Blood, p. 149.][1]

If all the Arteries of a limb be tied, the motion of that limb is annihilated,
and local death and mortification ensue.—It is the arterial blood which gives
their vital capacities to all the other parts of the System; and which, through
some unknown interchange with organized matter, loses its own life; while
dispensing[2] vitality to other parts:—therefore, in the forcible language of
Hunter "it must have motion, and that in a Circle, in order to be again saturated with living powers."

52   Since the blood circulates through the whole body in the space of one minute, the same parts [/] must in that brief period lose and regain their vitalizing
capacities.[3]

In the torpid Mammalia, the organic functions are maintained, though fee-
bly, by a slow circulation of venous blood.

In those unfortunate cases of malformed heart where the blood throughout
the System is in a venous condition, the digestive and secretory functions go
on;—and it is the animal functions which are chiefly affected. Life, however
is often maintained until the full development of the sexual system;—but as
this important change requires a corresponding perfection in the rest of the
frame, the imperfect condition of the vascular machinery is then generally[4]
found to be incompatible with the longer continuance of vitality.

Hunter having proved that the blood supported the life of the Solids,
next sought to determine in which of its proximate constituents the vitalizing
and nutrient properties were inherent; and he concluded by assigning them to
the fibrine, almost exclusively.

53   The importance of this constituent is proved by the following experiment:
If an animal be bled until it falls into a state of Syncope, and if fibrineless
blood be injected into the veins, death is nevertheless a consequence of the
hæmorrhage:—But if, when an animal is bled to Syncope, the blood of another animal of the same species be injected into the veins of the one to all appearance dead,[5] we see the inanimate body return to life, gaining accessions

---

1   If. . .149.] *Ellipses marks inserted to replace use of interspersed quotation marks [ed.].*
2   while dispensing] *RO ink insertion over erasure with pencil* 'while dispensing' *interlined
above.*
3   capacities.] *followed by deletion by vertical pencil score of* '{¶}It is not to be supposed,
however, that the venous blood is dead; else we might with reason ask, with Dr. *Davy
{interlined in pencil over deletion of* 'Harvey'}, *if we are to regard the circulation as a
perpetual miracle in which material particles are, without cessation, dying
and reviving?—'.
4   generally] *interlined by RO in pencil without caret.*
5   dead] *with RO pencil comment on facing 54*[v] *without indicated insertion point* 'beaten
blood—it may be deprived of fibrine before it gets cold' *with WC pencil reply* 'How is
this proved by Experiment? because if the blood could *not{interlined with caret}
instantly be deprived of its fibrine on extravasation it would coagulate, and then how
inject it. If dead, by being kept till coagulated, it would kill even if it possessed all its

of vitality with each new quantity of blood that is introduced. Presently it begins to breathe freely, moves with ease, and finally walks as it was wont to do, and recovers completely.

Hunter inferred from the phenomena of transfusion that the blood was of a uniform nature in all animals having the[1] red globules, and he discredited the accounts of the difference of form described by Leeuwenhoeck and Hewson;—[2] but had he extended his experiments in this field, he would doubtless have given due weight to the singular and unexpected phenomena which result from the injection of[3] blood having[4] globules of one form into the vessels of an Animal having blood-globules of a different form.[5]

54   Thus if Sheep's blood be injected into the previously emptied veins of a Bird[6], death suddenly ensues amidst nervous convulsions of extreme [/] violence; and comparable in their rapidity to those which follow the introduction of the most energetic poisons.

Fishes' blood transfused into the veins of a Mammiferous animal produces death as certainly as when introduced into a Bird.

If the blood transfused into the veins of a living Animal differs merely in the size, and not in the form of its globules, a disturbance or derangement of the whole economy supervenes:—[7]an imperfect revival indeed takes place; breathing is not materially affected, but the pulse is quickened, the heat sinks rapidly, the alvine evacuations are slimy and bloody, and death ensues in five or six days.[8]

The fatal effects produced by[9] mammiferous blood when introduced into the vessels of a Bird are as remarkable as they are inexplicable in the present state of our knowledge of the vital properties of the blood-discs. It cannot arise from a mechanical obstruction of the Capillaries because the globules of the mammal are smaller than those of the Bird. These observations tend, however, to show that the blood-discs are of more importance to the Blood than Hunter supposed[10] them to be:—and this conclusion is further supported by a comparative examination of the proportion in which they exist in the blood of different vertebrate animals.

(contd.)
   fibrine. Would it not?—'.
1 the] *interlined with caret.*
2 Hewson;—] *with WC comment on facing 54*[v] 'This killed some of the mangy French Princes it is said—I think sheep's blood was employed—was it not.'.
3 of] *followed by erasure of* 'the'.
4 having] *RO pencil insertion in margin evidently to replace* 'with' *after* 'blood'.
5 different form.] *over erasure with period inserted after* 'form' *[ed.].*
6 Bird] *RO interline in pencil above* 'Goose'.
7 supervenes:—] *with pencil* 'X' *in margin.*
8 bloody. . .days.] *with double pencil* 'XX' *in margin.*
9 produced by] *over erasure of* 'of mammiferous'.
10 supposed] *RO insertion over erasure.*

55   The Table which is[1] exhibited before you, [/] contains the results of numerous careful experiments of M. M. Prevost and Dumas, and presents us with the comparative weight of the solid particles contained in 1000 parts of blood, with the[2] number of pulsations of the heart in a minute;—and the number of inspirations made in the same interval of time [See Cycl: p. 412.]

From these experiments it follows, that in the Cold-blooded Fishes and Batrachia, the blood is poorest in solid particles.[3]

That these considerably increase in their proportion in mammalia;—but that a difference is to be noted between the Carnivorous and Herbivorous tribes in this respect;—the proportion of the globules and fibrine being greater in the former than in the latter:—which[4] in Birds, in which Animal Temperature is highest; respiration most active and extensive; circulation and locomotion most energetic; and the greatest amount of waste to be restored; the Blood is richest, both in red globules and fibrine.[5]

---

1   is] *interlined in ink with caret.*
2   with the] *followed by ink deletion of* 'habitual temperature of different Animals, taken in the rectum:—' *and deleting repeated* 'the'*[ed.].*
3   particles] *With WC pencil comment on facing final cover sheet* 'Qu if anything said of the composition or nature of the red part of the Blood—its chemical properties—its iron—its use—or production.?'.
4   which] *RO interlined in pencil above pencil deletion of* 'but that'.
5   fibrine.] *followed by* 'The End' *in Clift's hand.*

Notes
to
Lecture Seven

3-3   See discussion of the arrangement of
the collection in Note 12-5 to Lecture
One. The blood was treated early in
these displays under the series "Sap and
Blood: Their Different Kinds", which
appears to be the series referred to in this
passage. It is somewhat curious that
Owen treated the sensitive plants in
Lecture Six, and turned to the blood in
Lecture Seven. The arrangement of the
collection would have seemed to dictate
the reverse order of these two lectures.
However, Johannes Müller expounds on
the blood after a similar discussion of
the sensitive plants, and Owen is
following Müller's order of exposition
closely.

3-18   The organs of generation form the
main focus of the second main
subdivision of the gallery collection
(preparations numbered 2224-2851 in
Owen's *DIC* revision of the 1830s).
This series was completed by the
"Products of Generation" (2858-3790).

4-1   Preparations 28-288 (*DIC* I, 1833).

4-3   Owen means the central nervous
system, preparations 1303-1799 (*DIC*
3, 1835).

4-6   The digestive organs form
preparations 289-841 of Series 3 in the
first Division; the muscles are generally
displayed in preparations 33-75 of the
fourth series (*DIC* 1: xiv). Nerves are
covered in preparations 1292A-1385
(*DIC* 3, 1835).

5-3   William Prout, *Chemistry,
Meteorology, and the Function of
Digestion, Considered with Reference to
Natural Theology*, 2nd. Ed. (London,
1831), p. 499. See also *Hunter's Works*
3:47n.

6-4   Müller cites these same figures

(*Handbuch*, p. 93). Owen's passage at 6-12
is essentially a translation of Müller's
text.

7-16   John Hunter, *A Treatise on the
Blood, Inflammation, and Gun-Shot
Wounds* in: *Hunter's Works* 3: 16.
Owen has altered this slightly, inserting
the phrase 'or colourless'.

8-4   See Müller, *Handbuch*, p. 94.

9-1   Müller describes the same technique
in ibid.

9-4   Owen is apparently projecting a
microscopic slide with a version of the
solar microscope.

9-18   See Müller, p. 94.

10-2   See Hunter, *Blood, Works* 3: 21n.

10-17   See ibid., p. 22.

11-1   Ibid., p. 56.

11-13   See Müller, pp. 96ff. Müller there
cites his "Beobachtungender Analyse
der Lymph des Bluts und des Chylus,"
Poggendorfs Annalen **25** (1832): 513-
90.

12-11   See Hunter, *Blood, Works* 3:37.

13-1   Jean Louis Prevost (1790-1850) and
Jean Baptiste Dumas (1800-1884),
"Examen du sang et de son action dans
les divers phénomène de la vie,"
Bibliothèque universelle **17-18** (1821)
218-229; 208-220.

13-6   See Müller, *Handbuch*, p. 109.
Owen is reporting Müller's summary of
Berzelius' analyses, which displayed the
wet/dry ratio of 0.75.

14-5   Owen is drawing on the data reported
by J. J. Berzelius in his "General Views
of the Composition of Animal Fluids,"
Medico-Chirurgical Transactions **3**
(1812), 230; and by Alexander Lecanu in
his "Nouvelles récherches sur le sang,"
Journal de pharmacie **48** (1831), 308-
27. These articles are summarized in

Palmer's edition of the *Blood* in *Hunter's Works* 3: 18n–19.

14–15 note 6] The data is taken directly from Müller, p. 122.

16–4 See Müller, *Handbuch*, pp. 109 110

17–17 Ibid, p.112.

18–9 See Hunter, *Blood, Works* 3:40.

18–12 Ibid., 42n. The reference is to Palmer's note rather than to Hunter himself. Palmer summarizes at length the contemporary discussions of this issue.

19–13 Ibid., p. 48.

20–2 Ibid.

20–8 Müller, *Handbuch*, p. 113 summarizes data on this point from Gay-Lussac, Thenard, Michalis and Prout.

20–14 See Müller, p. 125 on the analyses of Chevreul, LeCanu, Gmelin, and Bourdet.

21–5 Hunter, *Blood, Works* 3:52–3.

21–14 Ibid., p. 48.

21–20 See Palmer's note to ibid., p. 49. Data in the table is directly taken from Müller, p. 125.

25–7 See Palmer's comments in his note to Hunter, *Blood, Works* 3: 50n. Palmer does not identify the source of Brande's claim.

26–6 See Hunter, *Blood, Works* 3:51.

27–2 Ibid., p. 50.

27–13 I have been unsuccessful in locating Owen's original of the diagram being displayed to the audience. Possibly it was an adaptation of Palmer's similar table in ibid., p.20n.

28–11 Ibid., p. 66. Owen has silently deleted a portion of this quotation.

28–18 J. J. Coste and J. M. Delpeck, *Recherches sur la génération des mammifères suivies de recherches sur la formation des embryons* (Paris, 1834). A copy of this is at the RCS.

29–5n See *Hunter's Works* 3: 58n. Owen's reference to his presence in Paris in 1832 would seem to be in error, although it is repeated in his preface to *Works* 4: xiii. Owen visited Paris in July and August of 1831. I have located no other evidence for a trip to the continent in 1832.

29–13 Ibid., p. 59.

30–3 Ibid., p. 61. This reference to Hunter's treatise is to an edition other

than that reprinted in *Hunter's Works*.

30–9 Ibid., p. 62n. Palmer summarizes the data supporting this claim.

30–14 Ibid., p. 59 in Palmer edition.

31–5 Owen is apparently displaying to the audience a new achromatic microscope. The Board of Curators of the Museum had approved Owen's proposal to purchase such a compound microscope in June of 1835 (Minutes of Curators 4: 419).

31–13 See Palmer's summary of the work of Home and Bauer on this issue in *Hunter's Works* 3: 62–3n.

31–20 By use either of the principle of the camera lucida, adapted in the so-called solar microscope, a microscope could be fitted to a window shutter in such a way that the image could be projected in magnified form on a wall in a darkened room. Microscopists commonly utilized this device to obtain accurate size measurements and make large-scale projections of microscopic objects.

33–3 The work of these individuals on this topic is summarized by Palmer in *Hunter's Works* 3: 61-3n.

33–13 note 2 The reference in the deletion is to the so-called "Irish Giant" and the "Sicilian Dwarf." Skeletons of both were on display in the main gallery of the Hunterian Museum.

34–10 William Hewson, *Experiments on the Blood, with Remarks on its Morbid Appearances* (London, 1771). A copy of this is in the RCS library.

35–2 See Palmer's note in *Hunter's Works* 3: 63n.

35–8 This technique is described in Müller, *Handbuch*, p. 96.

35–14 Müller, [ibid.] attributes this claim to studies of Prevost and Dumas. See above, note 13–1.

36–4 Müller, ibid.

36–7 Ibid., p.105.

38ᵛ–6 The reference is to Christian Ehrenberg's doctoral dissertation, *De globularum sanguinis usu* (Berlin, 1833). This is also referred to in the added section to Lecture Four (see end notes to Lecture Four, 100–22). Reference to this treatise does not appear in the holograph draft, and Owen apparently only read this work late in the spring of 1837 and then deleted the prepared material to insert this discussion.

38–3 See Palmer's discussion of this issue in *Hunter's Works* 3: 62n.

39–4 I have been unsuccessful in locating this diagram.

39–12 See "A Report Concerning the Microscope Glasses sent as a Present to the Royal Society by Father de Torre of Naples and Referred to the Examination of Mr. Baker F. R. S.," *Philosophical Transactions of the Royal Society* **56** (1765), 67–71. This describes the remarkably small lenses used in a Wilson screw-barrel single–lens microscope capable, according to the report, of 2,560X. The article mentions a Francis E. Stiles [sic] who conveyed the gift to the Royal Society.

41–1 See Müller, *Handbuch*, p.104.

41–8 Ibid., p. 97.

41–21 Ibid., p.107. Owen has added to Müller's experiments his own determination of the measurements.

42–9 Owen has thus supported Müller's claim that the central area of the blood platelet is not simply a mechanical elevation, but also represents a central nucleus. This would be an important claim for the subsequent cell theory of Schwann and Schleiden.

43–20 See *Hunter's Works* 3:33.

44–7 The reference is to J. A. Prevost and J. A. Dumas, "Examen du Sang et de son action dans les divers phénomènes de la vie," *Annales de chimie et de physique* **23** (1823): 50–68; 90–104, p. 51. This is cited by Palmer in his discussion of coagulation in his notes to *Hunter's Works* 3:20n.

45–2 Ibid., p. 64.

45–12 Müller, *Handbuch*, p. 106.

46–3 Ibid, p. 107.

46–7 Ibid.

46–10 See *Hunter's Works* 3: 24. (1832): 335-60.

47–4 Ibid., p. 33.

47–11 Ibid.

47–18 See note by Palmer to ibid. Palmer refers indirectly to the studies of Marshall Hall. See Marshall Hall, "On Hibernation," *Philosophical Transactions of the Royal Society*

48–6n See Palmer's comments in note to *Works* 3:34n. Palmer cites the experiments on blood coagulation with electricity of C. Scudamore, *Essay on the Blood* (London: Longrave et al., 1824), p. 54.

48–8 William Stevens, *Observations on the Healthy and Diseased Properties of the Blood* (London, 1832). A copy of this is at the RCS.

49–8 Hunter, *Blood*, *Works* 3:34.

51–1 See Palmer's note to ibid., 104.

51–10 Ibid., p. 111. Owen is using a different edition.

51–19 Ibid., p. 112.

53–13 See ibid., p. 13. Palmer comments in his note to this passage that Hunter's claim was refuted by the experiments of John Harwood.

54–3 See Palmer's note to *Hunter's Works* 3:13n. Palmer attributes the experiment to Prevost and Dumas. See above, 44–7 and 13–1.

54–11 Owen here contradicts Palmer's claim (ibid., p. 14n) that transfusion between species, under the proper conditions, is possible and even beneficial.

55–5 The reference is to H. Milne Edwards, "Blood," in R. B. Todd (ed.) *Cyclopedia of Anatomy and Physiology* (London, 1835–6) **1**: 405–14. Milne-Edwards gives on p. 412 a table of the comparative analysis of the blood of various Birds, Mammals, Reptiles and Fishes. I have not been able to locate the enlarged diagram Owen is apparently using at this point of the lecture.

Appendix

Joseph Henry Green's Introductory Hunterian Lecture on the Comparative Anatomy of the Birds, 27 March 1827

Analysis of the Manuscript

The following transcription is the complete text of the first lecture commencing Joseph Henry Green's Hunterian Lectures on the Birds delivered in March and April of 1827. The complete outline of this series is to be found in the Introduction, Table Two, page 25. The manuscript utilized for this transcription is in William Clift's hand, and is found in the Green papers at the College of Surgeons (RCS MSS 67. 6. 11). No holograph manuscripts of any of Green's lectures have been located.

The manuscript is written on folded sheets measuring 44.6 x 36.5 cm on a "W Brookman 1826" paper. The condition of the transcription suggests that it is a careful recopying from rough notes taken down by Clift rather than the notes themselves. Although similar in several aspects to Green's published "Recapitulatory Lecture" of 1828, there are also several important differences in these two texts, and the integration of the 1827 discourse with the prior courses of lectures is more apparent.

Independent notes on this lecture are also found in Green's papers in two locations. There is a rough set of notes taken down by Richard Owen during his attendance at the 1827 lectures (Green Papers, RCS 67. b. 11), and a second more detailed recopy of portions of the opening lecture (Owen Papers, RCS 275. b. 21; see specimen below, p. 309). A second set of notes was taken by T. Egerton Bryant (Green Papers, RCS, 42. 2. 19.). These alternative manuscripts have been compared with Clift's transcription.

Conventions and methods in this transcription have been the same as those used for the Owen lectures. The often clumsy grammatical aspects of these lectures have required more editorial intervention than required in Owen's case. These defects surely reflect Clift's deficient education rather than Green's style. The many misspellings should be assumed to be as given on the manuscript. There is no evidence that this manuscript was subsequently corrected by Green.

Similarities are to be found between the more philosophical aspects of this lecture and Coleridge's manuscript entitled "Volatilia a Day Book for bird-liming, Small Thoughts, impending Stray-Thoughts and holding for trial doubtful thoughts March 1827" located in the first part of his manuscript "Opus Magnum" (Victoria College, Toronto Coleridge MSS S MS 28), fol. 14–15. Coleridge either attended the Green lectures or was provided with notes from them. See S. T. Coleridge, *Aids to Reflection,* "Aphorisms on Spiritual Religion IX: Comment."[1]

---

1   S. T. Coleridge, *Aids to Reflection,* ed. H. N. Coleridge (London: Bell, 1884), p. 160.

Fig. 16. Portrait Of Joseph Henry Green By Thomas Phillips, R. A., Located At The Royal College Of Surgeons. *(Courtesy President and Council, Royal College of Surgeons of England)*

Lecture One
Introduction to the
Natural History of the Birds
Tuesday, 27 March 1827

[1]   I introduced a former Course of these Lectures by distinguishing the objects of a Naturalist and more particularly in that department of Science which regards the organic and animated world into three kinds; and I will connect the present with the foregoing Course by briefly recapitulating thence. The three great divisions into one or other of which all Natural Science resolves itself are, Physiography or description of Nature, Physiology[1] or Theory of Nature. Lastly Physiogony or the History of Nature—The office of the first or *Physiography* is to enumerate and delineate the effects & products of nature as they appear. Its[2] Sphere is that of sensible experience of appearances in contradistinction from[3] truths drawn from immediate facts by inference, the subject matter is not unhappily entitled by elder Naturalists Natura Naturata or Nature considered passively and the result may be compared to an[4] immense family piece the figures of which are all portraits—the office of the Second or *Physiology* is to deduce by inference 1st the rules or principles by which the innumerable facts of Physiography may be reduced into manageable order, either in reference to the convenience of our faculties, which is the principle of all artificial classification or in relation to the objects themselves which should it ever be realized will be the ground of a Natural Classification. 2ndy it is the office of[5] physiology likewise to ascertain the powers which must be inferred from the phenomena and the *laws* under which they act: in other words to ascertain the idea of Life and its constituent forces—as far as it is common to all living bodies—That[6] is in *Kind* without consideration of degree or other difference of the particular Subject—the 3rd or *Physiogony* regards the facts and appearances of the natural world as a series of actions & nature itself as an

---

1   Physiology] 'l' *in darker ink over* 'g'.
2   appear. Its] *period inserted and* 'its' *capitalized [ed.]*
3   from] *deleting repeat of second* 'from' *[ed.]*
4   an] *interlined with caret.*
5   of] *before deleted* 'the'.
6   That] *before deleted* 'S'.

agent acting under the analogy of a will and in pursuit of a purpose,[1] in what sense and whether by a necessary fiction of Science or with some more substantial ground we leave undetermined. Physiogony[2] no less than Physiology investigates the principles of life—but not in the kind principally, but as subsisting in a diversity of degrees even to an unknown minimum— and in that sense, ideal minimum; and this again principally in reference to the original construction of the organs of living bodies and of living bodies contemplated as organisms or systems of organs leaving to the physiologist the affections and disturbances of the vital powers.—considered as functions of the organs and not as productive powers or their formative principle—The far larger portion of my lectures has been and will be physiographic, a delineation of the facts in that order which has appeared to me to unite in the greatest degree of any with which I am acquainted that two objects of classi-

[1ᵛ] fication, the convenience of the mind, and memory, and the correspondence with the order in which we have weighty reasons to beleive[3] that the different classes were really produced but, to have continually before my eyes a multitude of forms, the forms of organic nature which we almost instinctively presume to be significant without even adverting to their actual signification— without a single attempt to discover it, would it seemed to me not alone have argued pusilanimity—but a want of respect for my audience as if their senses and the objects of their senses were alone capable of interesting them. The Historian or Learned traveller cannot behold the undecyphered characters on the Temples & obelisks of Egypt—without yearning for the key—he cannot take up a brick from the vast ruins of Babylon without pondering on its enigmatical characters of 4000 years-antiquity without reflecting,"these assuredly have a meaning, some small fragment in the history of man or of his nature is here contained"—and these yearnings have led from the Very restoration of literature to a succession of efforts each in its turn yielding to some other, and yet never producing despondency but proceding with a hope which in our own times and within the memory of the youngest of my auditors has here justified and rewarded by the discoveries of Dr. Young and Champolion. Shall, then, the characters of *nature* so numerously presented to us in her living forms, yet so evidently prepared for our contemplation by presenting the component lines & traces of her most complex characters—not only each seperately but with a Simplicity and Singleness beyond that which the most perfect human alphabet has ever acheived—Shall these pass unattempted? The Hieroglyphics we can only expect to be mere fragments

---

1   purpose,] *comma inserted [ed.].*
2   Physiogony] *deleting illegible 3 letter word [ed.].*
3   beleive] *sic.*

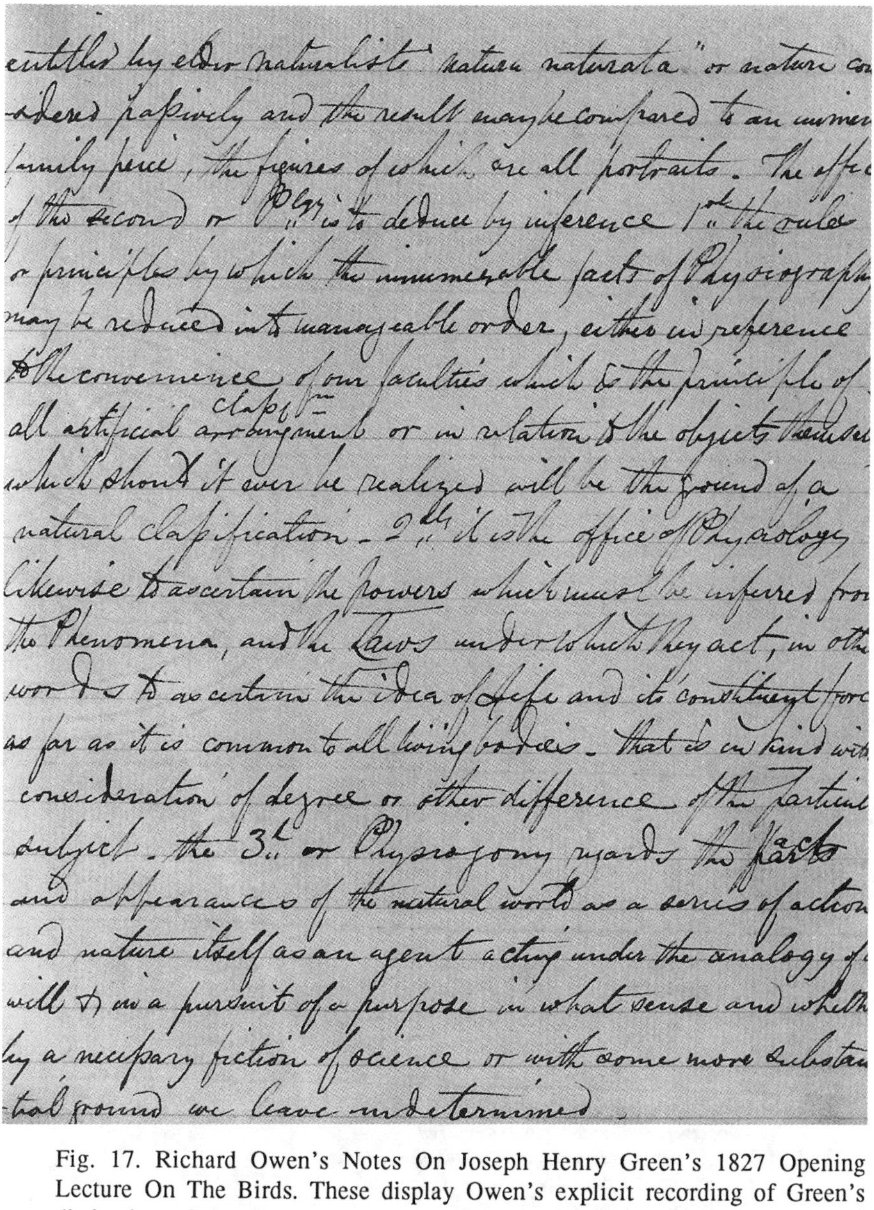

Fig. 17. Richard Owen's Notes On Joseph Henry Green's 1827 Opening Lecture On The Birds. These display Owen's explicit recording of Green's distinction of the three programs for studying nature outlined in this lecture. *(Courtesy President and Council, Royal College of Surgeons of England)*

while in Nature we possess the whole before us as a book not indeed without hiatus and interspaces to be filled up by future discoveries, yet no hiatus of such magnitude or of such importance as to destroy or even obscure the manifest principles of arrangement that pervades the whole. If[1] it be possible in one sentence to convey the sort and the degree of interest which the object of physiogony[2] or the history of nature is calculated to inspire, I might say that its object is by means of evident principles or principles of reason supported in each step by the facts corresponding to exhibit nature, as labouring in birth with *man,* to exhibit every order of living beings, from the Polypi to the Mammalia as so many embryonic states of an organism, to which nature from the beginning attended but which nature alone would not

[2]   realize. It[3] was this Idea which enabled me in former [/] Lectures to present to you Nature's living products—as so many significant *types* of the great process which she is ever tending to complete in the evolution of the organic realm. In each stage of the ascending scale of living beings we see,[4] with evidence increasing directly as the ascent,[5] at once the opposition and the harmony of the two great tendencies which must be regarded as the manufacters or constitutive agents in this great work of nature, namely—that of Nature tending to, *integrate* all into one Comprehensive whole, and consequently retaining each part, and, as in vegetation building upon herself, & on the other hand the *tendency* to *individuality* in the parts and for this purpose the nisus[6] in each to detach itself from the preceeding[7] and more imperfect States, or to supersede them now by building the new edifice out of the Materials of its more rude predecessor and now by distruction as one who by the force of the Vault should crush the platform from which he had taken the Spring. Hence the states which the individual passes through in all the epochs of its embryonic being and which having been, disappear, are preserved in Nature and maintain the rank of external and external[8] forms and thus to present the history of nature as preface & portion of the history of man, the knowledge of nature as a branch of self knowledge—An attempt which can succeed in no degree without either giving or preparing for a deeper insight into those principles of life[9], and reproduction which enters into the interests of practical Science and have within the last half century, nay I might truly assert within the last twenty years given to comparative anatomy a Splendour & equality of rank with its elder sister. Such were the

---

1   whole. If] *period inserted and* 'if' *capitalized [ed.].*
2   physiogony] *reading for* 'phyiogony' *[ed.].*
3   realize. It] *period inserted and* 'it' *capitalized [ed.].*
4   see,] *comma inserted [ed.].*
5   ascent,] *comma inserted [ed.].*
6   nisus] *Clift has added a* '?' *following this.*
7   preceeding] *reading for* 'priceeding' *[ed.].*
8   external] *sic. Clift may have meant to write* 'internal'.
9   life] *in darker ink over illegible word.*

motives that encited me, and the prospects that encouraged me to introduce the physiographic details which form the main body of these Lectures, with an attempt to decypher the forms & characters impressed thereon—not as if I could expect to exhibit a System of natural History or were rash enough to attempt it, but that[1] I might map out the bounds and limits of the Science & endeavour to demonstrate the principles upon which such a Science might be constructed and the main operative powers into which the agency of nature must distinguish itself. And thus perhaps in some mind—for whom leisure & opportunity might supply air[2] & fuel for the spark, while I excited a predeliction for this pursuit I might in some degree to facilitate[3], at least,[4] his labours—, and if I did not share the spoil or the[5] trophy[6] as a fellow combattant I might yet deserve esteem & thanks as a pioneer—

On[7] the other hand I could not hide from myself the difficulties and discouragements of the undertaking under the training &[8] with the discipline and habits of mind general in this Country; and from Causes in the main highly honourable to our character, an almost exclusive Value has been given to pursuits & inventions of intermediate and palpable utility—Our highest aim is to be men of sense; and this is as it should be, were it not that too often the man of the Senses, who resolutely confines his knowledge to the impressions on his Senses, is mistaken for the man of Sense—so often indeed that it is not unfrequently expedient to remind a disputant, that the most certain, and hitherto the most important of all Sciences, the Mathematics I mean, is groun-

[2ᵛ] ded on the intuitions of the Sense, in contradistinction from [/] and exclusive of, the impressions on the Senses.—It is well that we should be men of sense, but not even[9] in the highest import of the term, men of sense exclusively; and I venture to assert that the man who acknowledges no truth and no reality in any Subject, which he cannot reduce in imagination at least to weight, measure or colour, lives in the eclipse of the better half of his intellectual being. He may be[10] a tolerable Mathematician, a philosopher he cannot be. With the diagrams of abstraction he may be familiar & conversant but not with that sublimer geometry & universal arithmatic, the real construction of which forms the history of nature. To the diagrams, such as preserved beneath this roof, formed the Study, fixed and guided the internal constructions of Mr Hunter, those that demonstrate in Succession that individuality and in-

---

1 that] *interlined with caret.*
2 air] *after deletion of illegible word.*
3 facilitate] *altered from* 'fascilitate'.
4 at least] *transposed with line from after* 'his labours'.
5 the] *followed by small superscripted* '3' *without corresponding note.*
6 trophy] *followed by small superscripted* '4' *without corresponding note.*
7 On] *paragraph break inserted to correspond to line gap.*
8 &] *interlined with caret.*
9 even] *interlined with caret.*
10 be] *interlined with caret.*

tegration to a whole are the great polar forces of[1] organic nature, that every, the minutest living creature and every integral part therof acts by a life of its own, and yet that all are permeated & sustained by a common Life—to these he must ever remain a Stranger and too probably will become an enemy. The philosopher, who dissatisfied with lifeless abstract Science seeks after real knowledge, and yet will not confine his enquirey to the impressions received thro' his Senses and generalized under the name of facts, with now & then a theological Make weight, a few religious phrases introduced as substitutes for the ideas that Constitute Religion or inevitably lead to it—he must consent to remain unintelligible for the many, to be represented as a man who had sunk out of the light of common day, and out of the view of common Sense, as if like the ghost of Hamlets Father on the Stage he had suddenly stepped on a trap door and continued his discourse from under ground a vox et preterea Nihil!—And such above all must be the case of every man who undertakes the department of natural history under the full and distinct conception of the words *Nature*[2] *History*. For History has for its subject *actions* and the results & products of *powers* in action. But actions imply or suppose a *Will* a *purpose* and must be interpreted by desires, motives, tendencies by a something at least analogous to purpose, will, desire, and which can only be rendered intelligible by a reference to these as known in ourselves.

But Physiogony or the History of nature has for its peculiar Subject, the activity of *productive* powers or the Sum and Series of those actions of which the facts and phenomena of physiography are the product, under the rule that the product of every given power is to be received as the measure of its forces, and the index of its direction; If Natural *History* be not a misnomer, [3] an erratum in the nomenclature of Science it must be either [/] the history of nature assumed as an *agent*, or the history of a plurality of productive powers considered severally as agents but which taken collectively are called Nature—in the active sense of the term just as the Collective products and results are called nature passively understood—. The same reasoning applies to the immediate subject of these remarks, the investigation of the Significant of organization contemplated as so many *Types* or characters impressed on animal bodies, or into which they are as it were cast. Now Types & characters variously yet significantly combined form a visual language. The Types of nature are a natural language, a language of nature—But a language is as little conceivable without reference to an intelligence, if not inmediately yet ultimately, than a series of determinate actions can be imagined without reference to a will; and a consistant and connected language no less supposes intelligence for its existance than it requires an intillegence for its actual[3] intelligibility; And tho' the Language should not like conventional language stand

---

1   of] *in dark ink over illegible word.*
2   Nature] *followed by ink deletion with deleted interline of* 'and'.
3   actual] *after illegible crossed out word.*

in opposition to the things intended, but be one with them—this would prove nothing more than that it was not a language only.[1] And this I scarcely need say forms but[2] one among very many objects which we recognize in Nature and the number of which acknowledges no other bound but the sphere which comprehends life, enjoyment, protection and perpetuation—

I shall close these prefatory remarks by observing that there are two ways in which a history of nature may be given. The first begins with the highest, sets out from the true and absolute First cause & ground of all; and I scarcely need add that such a history requires an inspired historian for its accomplisment; were it only for this reason, that a familiar knowledge of the great facts & prominent truths of a history so given is indispensible to the well being of man, before it could have been possible for him to have discoverd these truths by the light of his own reason, or the researches of his own industry. The Knowlege I say—would be necessary and indispensible as the condition of man's ever arriving at that degree of moral & intellectual development as at degree of civilization,[3] leisure & security in which individuals could arise capable of commencing such researches—or even of wishing to institute such an enquiry—The other and second way of constructing a history of Nature begins from the lowest, and for its first grounds and materials takes the most general characters and properties of the objects that surround us—together with the active properties inferred from these facts, or known by immediate conciousness—In other words it begins with *nature*—whether the term be understood as no more than a name for the active powers collectively which we construe as one in the calculus of science for the facility of reasoning & the convenience of expression; or whether something more than this verbal or logical unity, should seem to be required by the facts and presumed by the existance and use of the word in every language of civilized man under whatever form of thought, the unity may be conceivable, in either case a *nature* is

[3ᵛ] the first indispensible postulate of Natural [/] history in this second form with which exclusively the naturalist as such is concerned. The first mentioned on the contrary is a branch of Theology, and with all its aid documents & authorities belongs and should be preserved sacred to the Theologician. Nothing can be more injudicious or less equitable than to blend or confound these two forms. The one is complete & immutable and bears immediately on our most important Interests, the other nowise connected with the Motives and conduct of men, the interests of public order, or the grounds of morality, than by the liberalizing & tranquilizing effect of all true knowledge on the mind and heart, is imperfect and a Science yet in its infancy, of rapid growth indeed, but on this very account subject to perpetual

---

1 only.] *period inserted [ed.]*.
2 but] *interlined with caret.*
3 civilization,] *comma inserted [ed.]*.

changes, not only from the correction[1] of errors but from the light which every fresh discovery throws back on the preceding.[2]

It would be equally presumtuous & unreasonable to judge[3] the records of the one with the fluctuating inferences of the other; while on the other hand, to interpret the real or apparent facts of the latter in accomodation to the declaration of the former could only tend by interrupting the progress of Science to prevent it from reaching that higher ground, from which we need not doubt that both will be seen in perfect Harmony—

I have judged it right thus once again to offer to your notice the grounds that led me to adopt the scheme, upon which the several courses of the foregoing seasons have been conducted; and I will now compleat this introduction to the present course, by a brief *recapitulation* of the facts that may serve to justify my arrangement.[4]

Recapitulation (1827)

The[5] subject of the present season is that of Birds, but in thus bringing before you this subject, I feel, *see*, that it would not be doing justice at all to what is past or to what I have to bring before you, without entering into some recapitulation or adverting again to the grounds which have induced me to adopt the Course which I have taken—[6]

We[7] have there set out upon the plan of considering nature as a series of evolutions, from the lower, from that lowest form in which Life manifests it's power in a production of animated being up to its most complex forms— & I presented to you this view not under the idea that the lower had any power of assuming the rank and privileges of the higher nor[4] upon any such fanciful scheme as that which that otherwise most meritorious naturalist Lamarck has proposed for the invertebrated[8] series of animals, but as the lower passing by a series of evolutions to the higher, under the law & influence of a higher power acting in and by nature—In order to facilitate the comprehension & division of the great groups of nature. I stated that I should adopt the arrangement which had been proposed by Cuvier, and let [4] me remind you before I proceed [/] with the primary divisions & classes which that celebrated anatomist has adopted, that they reduce themselves to four primary divisions but of which the first is distinguished from the others

---

1   correction] *altered from* 'corrections'.
2   preceding.] *period inserted [ed.].*
3   judge] *followed by deletion of* 'of'.
4   arrangement.] *at this point horizontal line is drawn across manuscript followed with* 'Diagrams' *before the new heading.*
5   The] *preceded by asterisk probably to indicate point for utilizing first diagram.*
6   taken—] *at this point a short horizontal line to indicate pause.*
7   We] *paragraph break inserted [ed.].*
8   invertebrated] *altered from* 'invettebrated'.

by the very remarkable circumstance, that in the first there is in all a skeleton & they are therefore named from that most important piece of the skeleton the spine—The *Vertebrated* series & consist of 4 classes—two of which I have already considered—viz: Fishes & Reptiles—Birds which are to form the subject of this years course—& the Mammalia—whilst the other 3 are deficient in the skeleton or even in that important piece of it the central column of support—and are the *Invertebrated* series and its divisions which[1] consist as you will see of the soft bodied animals, the Molluscae, the articulated or jointed animals, and those which it is difficult to comprehend under any other name, but which as the lowest grades of organization have been comprehended under the name of Zoophytes—You will recollect then just to[2] bring[3] briefly[4] before your recollection the subjects of which I have to speak, that these molluscae presented to you, these the Cephalopodae as instanced by the cuttle fish, these the Gasteropoda by the snail & slug—these the Acephala by[5] the oyster and mussel[6]—leave these 2 out—5&6—and this 2 —for it is these three to which I wish principally to direct your attention, the others enumerated by these, seem rather to form transitional links—then in the *Articulata*, you have the well known *vermes* the earth worm, the *Crustacea* covered with a hard external tegument as the crab and lobster, the Arachnida or spiders—and lastly the Insecta—of which you have instances in the well known Bee and Beetle; In the Zoophytes or last division you have there, these spiny skinned animals, the Echinodermata as the Echinus and star fish—with the Entozoa, with which you must be familiar as inhabiting other living bodies—the intestinal worms—the Fluke and the Hydatid, the Acalephae—which have been perhaps not very judiciously named, among these are the Actinia or sea nettles,[7] next the Polypi—and lastly the infusoria—or animals of Infusions.[8]

But[9] to present these in an ascending scale has been the business of the foregoing courses and in this point of view, we considered in the very first season the 3 first divisions and you will recollect these presented an almost innumerable variety of shapes of external forms—and internal states presenting as it were rather so many tentative experiments and preparations of

---

1   which] *inserted for grammatical purposes [ed.]*.
2   to] *with small subscripted number '2' underneath possibly to indicate specimens being exhibited to audience.*
3   bring] *with small subscripted number '3' underneath.*
4   briefly] *with small subscripted number '1' underneath.*
5   by] *inserted with caret.*
6   mussel] *altering this from* 'muscle' *[ed.]*.
7   nettles,] *comma inserted [ed.]*.
8   Infusions] Here page divided by horizontal lines with 'To come in here—Table of the Primary divisions with Illustration according to Cuvier—see lect 2—1826—'. *See Figure One, Introductory Essay, p. 22.*
9   But] *paragraph break inserted [ed.]*.

nature, for the higher forms of animated beings rather than the fixed types we observe in the higher Classes, in order to bring this more before your imagination I constructed this scale in order to present as it were at one view the ascending series tho' it included some of the Vertebrated also.[1]

I[2] commenced these from the Infusoria—and while I am speaking you may direct your attention to some of these Diagrams, as for instance here is one of this Class though of a higher order than usually produced under the varying instances of infusions—the Volvox Globator—in the lower of these the general character of structure justifies the place in which they are put—it [4ᵛ] is that of having an uniform gelatinous Body it may be said without distinction of parts at least [/] without distinction of any Specific Organs—there is, as it were, an organismus without Centrality, a life pervades the whole or all the parts equally and if a portion be detached it may become a living animal like the Body from which it is detached—and the mode of generation is very similar,[3] for all these darker points are so many young animals produced by the bursting or splitting asunder of the parent animal but very soon even in these we observe a distinction of parts;[4] there is a cavity hollowed out in this gelatinous body not a membranous cavity but simply a hollow and we have first distinct the corresponding opposition of external and internal surfaces—this central cavity answering the purpose of sexual organs—alimentary cavity—and it might be said likewise of heart;[5] But now in the next and that which is presented here on the left hand line, as in the Polypus—under which the Hydea Viridis ranks—and may serve as an example we find that the organic power begins to exert itself, about this central cavity in the production of Radiated tentacula, feelers, or arms, as you will see represented around the orifice of the cavity for it is a tube with one orifice, and these tentaculae or feelers extending from it, and indeed these sections you may cast your Eye on. These Tentaculae, feelers & arms are in fact the rudiments of locomotive organs & sense, but they are intirely subservient to the wants of the central cavity, are simply & exclusively destined for the supply of that cavity with food for sustaining the Animal; In[6] carrying on the investigation in the line represented by the Radiated[7] or those in which similar parts are disposed round[8] a common centre as in

---

1 also.] *page divided by horizontal lines with* 'To come in here—Table of Ascending Series—of Invertebrates as in Lecture 10—1825 & Lect 2, 1826—& 24.' *See Introductory Essay, Figure 2, p. 22.*
2 I] *paragraph break inserted [ed.].*
3 similar,] *followed by* 'which we may' *and one-half line of blank space in MS. Deleted for grammatical purposes [ed.].*
4 parts;] *semicolon inserted [ed.].*
5 heart] *replacing comma by semicolon [ed.].*
6 In] *altered from* 'in'.
7 Radiated] *altered from* 'radiated'.
8 round] *dubious legibility.*

Acalephae—of which the Actinia or sea nettle is an Instance we find not only an internal and an external surface, not only a cavity with arms or feelers about its margin but you find likewise (this is a section of one) between the central cavity and external surface an organ of reproduction, a[1] sexual organ; not that there is any distinction of sexes but an organ of reproduction and therefore corresponding to what in the higher animals is a sexual organ—nay, even in these we have the first rudiment of a Nervous System at least so it has been discribed by Spix*—and presenting the character, of what afterwards is seen in all the animals of the Invertebrated Series—viz a sort of nervous ring consisting indeed of exceedingly delicate and difficultly to be discovered filiaments in these Animals. Among[2] these the most complex structure in this line of Animals is presented in the Echinodermata of which the Star fish is an example and without at all attempting to describe again its different parts I would notice here that you have not only the distinct opposition between external & internal surfaces but the external formed into a tegument and the internal cavity formed into a membraneous bag or bags for such extend into the different rays of its Star-shaped body, you have here as represented by [5] these red lines, vessels,[3] a Vascular System, and here again you have presented to you the nervous ring which forms the Centre and from each part of the circumference opposite the radiated processes you will find two filiaments which extend into the rays of its Star-shaped body—this seems to be the highest development that is produced in that line, which from the infusoria is represented by all the radiated animals or those in which you have similar parts, placed around a common centre: for the higher forms of development are next to be carried on through a different series—those animals in which length of body preponderates which may be called *nematoidea* or filiform, and of which you will instantly see some in the worms that Inhabit animal bodies;[4] now indeed you will have some of the same grades gone through.[5]

I would remind you of those tubular cavities which are bored as it were in the body of the Fluke and in the different portions of the tape worm. But let us go on to some of the higher, some of the worms which inhabit animals or even into the Vermes;[6] now here we find again in the lengthened body—you will find the external tegument distinct as for instance here represented in the section of the Earth Worm—an external tegument and a distinct membranous Cavity or Alimentary Canal running thro' the Centre of the Animal, the Alimentary canal more developed; since instead of a single orifice it has two orifices that correspond to mouth & anus: But the skin is here still the

---

1   , a] *altered from* 'as' *by WC with comma inserted [ed.].*
2   Among] *period substituted for comma [ed.] and* 'among' *altered to* 'Among' *by WC.*
3   vessels,] *comma inserted [ed.].*
4   bodies] *semi-colon inserted [ed.].*
5   through] *period and paragraph break inserted [ed.].*
6   Vermes;] *semi-colon substituted for comma [ed.].*

respiratory organ so that there is no distinct apparatus for respiration but in understanding the grade of place which they occupy—let us look to the Nervous System—here we find again the nervous ring, but the Nervous System is here indirectly subservient to the alimintary[1] Canal or reproductive Cavity—Therefore this nervous ring is placed about the orifice of the canal,[2] about the mouth as it were,[3] and thus there extend from it filiform threads which extend along the whole length of the animal and in this way animate or influence the different parts so that in these animals you have the opposition of alimentary membrane and cavity but more developed you have the external tegument or skin but still performs the office of respiratory apparatus. There[4] is sexual organ and a further development of the Nervous System in having a ring and also these threads extending thro' the body and placed as may be further noticed on the ventral surface of the animal, and the high place[5] these arms occupy is from our finding in the vessels there is even red Blood which we do not observe in animals higher in the scale but so placed from their more perfectly developed alimentary and locomotive systems—this process of evolution is still first here carried on in the Insecta or Crustacea as we see here, and it is not necessary to distinguish the Arachnida tho' they hold an intermediate place between the Insecta and Crustacea;[6] now if in the Vermes or in those lengthened soft bodied animals we have described you have seen a development such as I have noticed, we see not only in the Insecta a state further perfecting but which is more important to notice, we see a perfecting in a different direction & we find a perfecting of many of those organs which are to form relations with the external world, but without adverting again to the subject of alimentary canal, tho' this is more developed in the Insecta I

[5ᵛ] may notice to you that the [/] respiratory apparatus here becomes distinct, no longer belongs to the skin, but in the Insecta presents itself as stigmatae, tracheae, of which this is in part a representation, but that which shows the ramifications is not here, into tubes which extend thro' so as to aerate the Body. They[7] are always, as it were, with the Vicera in a through draught; & so in the Crustacea you have the respiratory organs appropriated as gills, these which are here represented—But now, in respect to the locomotive organs, as respects their development we find here the External Tegument horny, but divided into several Circular plates or rings, so that it may be moved and form points of attachment to numerous distinct & complex jointed jointed external covering—but there are also legs, wings, maxillae, palpi,

---

1   alimintary] *sic.*
2   canal,] *comma inserted [ed.].*
3   were,] *comma inserted [ed.].*
4   There] *period inserted and* 'There' *capitalized [ed.].*
5   place] *followed by deletion of* 'in'.
6   Crustacea;] *semi-colon substituted for comma [ed.].*
7   They] *period inserted and* 'they' *capitalized [ed.].*

feelers and a variety of complex instruments which answer a variety of purposes—now with respect to the Nervous[1] System we find it also developed
but not so much perhaps as you might expect considering the necessity of
animating such important organs as you have here described, in the Lobster
for instance you have the same nervous collar about the commencement of
the Oesophagus and you will observe that there are here ganglia & I
apprehend these are the first instances where we find developed these relative
centres of the nervous system—now in this nervous collar surrounding the
Oesophagus there is above a bilobed ganglion and the filaments which pass
off here are distributed to the more perfect organs of sense as well as about
the palpi, for in these insects there are eyes as well as the organs of feeling,
but where we use the terms Eyes and feelers do not let us suppose these are
analogous even I would say to the higher,[2] or if there be analogy the difference is so great as hardly, scarcely?, to admit their being considered in the
same point of view. Then this Nervous System or N. Ring repeats itself in
each segment of the articulated body of the Animal so you will see corresponding to each Segment there is a ganglion & between each of these there
are two threads and from the ganglion laterally pass off filaments to be distributed to the segment of the Body, in fact the Nervous System in Insects
seems to be entirely subservient to the irritability which is predominant in
these animals and which is manifest in those organs which I have described,
the locomotive and which perfect the external relation of the animals.—In the
next we return again to the Middle line that represents the development of the
Molluscae and here I only wish[3] to represent to your notice the Acephala
Testacea,[4] the Gasteropoda & the Cephalopoda—Now in these as we observe
in the Insects and Crustacea that there is a perfecting of the external relation
so it would appear in the mollusca there is a perfecting of the internal state,
the important viceral organs and in other respects as in the organs of sense
and locomotion, these may appear far below many of the Insect Tribe and it
[6]　　appears as if nature had [/] again sunk backwards—to concentrate her
organific powers—as a Leaper steps back before he makes his Spring—thus
in nature to acquire force to economise power which may be required for the
perfecting of the development & thus to acquire energy for further evolution[5],
thus we see in the Acephala, the oyster for instance or the the the organs of
circulation become developed and one could not mention any class where the
the organs of circulation become developed and one could not mention any
class where there are more or so many modifications of circulation, as in the

---

1　Nervous] *reading for* 'N.'.
2　analogous. . .higher] *each with small subscripted number from 1–8 below with* 'I
　　would say' *transposed with lead line from after* 'to the higher'.
3　wish] *altered from* 'want'.
4　Testacea,] *comma inserted [ed.]*.
5　evolution] *altered from* 'evolation'.

Molluscae—as[1] if Nature were trying her hand and making so many experiments in the formation of it—The respiratory, consists[2] in gills and a[3] proportion have a highly developed organ for that purpose. The[4] sexual organs are here formed but not advanced—for there is no distinction between the senses—with respect to the nervous system and especially in the Mollusca Acephala—we have a repitition of the Nervous collar or ring. Here[5] you will see above the two ganglia, which here are more fused together, united by threads below to a third surrounding there the entrance to the Oesophagus—there the double thread which extends backwards but here you will observe further a second ganglion which acquires its intelligibility by the Heart which is so situated at the further extremity of the animal in the Gasteropoda and the Slug and Snail—there is a still further development of these organs but the most remarkable is the occurrence of a lung fitted for the breathing in air.[6] Thus you have here the two Sexes tho' both in each individual, and you have added more complete organs of feeling, in the feelers, with the rudiments of eyes at their extremities, tho' even here I should rather regard them as organs of light than any thing for the purpose of vision,[7] but in the Cephalopoda— we have a far higher form of evolution and perhaps these might rather be regarded as transitional steps to the Vertebrated than as comprehended in the Mollusca—We[8] observe this especially in the nervous system and senses, here for instance the whole animal, here the head, now here the bilobed ganglion scarcely appears with its branches to form the Nervous collar *as simple ganglion* but so much increased and united as to appear *as Brain* and here sending out the filaments distributed[9] to these far more important organs of sense assure the complex senses the structure of which you can gain some general notion from this complex structure of the Eye[10], which must here be

---

1   as] *interlined with caret.*
2   consists] *altered from* 'consisting'.
3   a] *followed by deletion of* 'great'.
4   The] *period substituted for comma and* 'The' *capitalized.*
5   Here] *period inserted and* 'Here' *capitalized [ed.].*
6   air.] *period replacing comma [ed.].*
7   vision,] *comma inserted [ed.].*
8   We] *capitalizing* 'we' *[ed.].*
9   distributed] *altered from* 'distinct'.
10 Eye] *deleting dash after* 'Eye' *[ed.]. This sentence is impossibly punctuated. It more intelligibly reads:* ' We observe this especialy in the nervous system and senses. Here, for instance, the whole animal, here the head, now here the bilobed ganglion appears with its branches to form the Nervous collar *as simple ganglion*, but so much increased and united as to appear *as Brain*. [This] sends out the filaments distributed to these far more important organs of sense [to] assure the complex senses the structure of which you can gain some general notion from this complex structure of the Eye, which must here be an organ of vision while there [is] here also a rudiment of an Ear and the first rudiment of a Skeleton. [This] appears for the protection of the noblest organ of all, the Brain. For this is the first proper rudiment of an internal Skeleton, or of [any] Skeleton at all.' *[ed.]*

an organ of vision while there are here also a rudiment of an Ear and the first rudiment of a Skeleton—which appears for the protection of the noblest organ of all, the Brain—for this is the first proper rudiment of an internal Skeleton—or of a Skeleton at all—We have thus seen in the Invertebrates the preparation for that individuality which nature seems ever to be aiming at in the Series of Evolution of animated Beings, and we have noticed this especially in the Nervous System, in this System which of all others & which of course of all others is typical and constitutive of an inner & cerebral unity and we have the proof in this Contemplation of the Invertebrated Animals that the tendency of nature is constantly to the production of individual whole, which at the same time the existence of nature simply consists in the approximation there to—but if I don't tire you with this dis-
[6ᵛ] cription I should go on to [/] observe that in the Class we have to consider, the Vertebrated here we have a far more perfecting of the organisms which is to find its most perfect development in Man—We have seen in the Invertebrated Class all the Component parts with which you are familiar in Man and the higher animals, as for instance you have the Alimentary canal with its appendages—the respiratory organs, a lung for breathing in the air, and gills in the water;[1] you have a perfect circulation or at least with more or less perfection a vascular system containing red Blood & you have the 5 senses, but the perfecting will be principally seen in the nervous system here presenting itself in its greater importance and in its distinct parts of Brain, Spinal Cord, and ganglionic Systems, and it is in these you will first observe there first appears a Skeleton from which doubtless the name as we have seen indicated in the highest of the Invertebrated or the Cephalopoda is as a correspondent to the Nervous System as indicative of its existance & at the same time a protection to this important part. We find indeed that in the Vertebrated the Nervous system first presents itself in its appropriate form & character and I cannot help considering the whole class of the Vertebrated as especially that which is to serve as the evolution of and full development of the Nervous System while in the lower Classes you find it subserviant to the Alimentary or Sexual organs or to the irritability the Muscular and Vascular Systems but in the lowest of the Invertebrated Series you will recollect that.[2]

1　water] *semi-colon inserted [ed.].*
2　that.] *period inserted for terminal dash. At this point manuscript of first lecture ends with one and one-half blank pages before beginning of second lecture. T. Egerton Bryant's notes continue at this point:* 'In the Reptiles we have a still further unfolding of Structure, as we ascend the Scale of beings. In them we have a proper Lung, and external meatus to the Ear, the nervous System more perfect than in the preceeding. In Aves or Birds you find a more perfect vascular System & likewise the Nervous. Indeed organization has become more perfect in the tribe.' *Green papers, RCS 42. 2.1 9.*

Bibliography

Notes to the Lectures contain full references to manuscripts and incidental citations to works cited internally. The following items have been important sources in the preparation of these lectures.

Primary Manuscript Sources

Great Britain

Owen MSS. Archives of the Royal College of Surgeons, London. This archive contains numerous materials, particularly pertaining to Owen's period of residence from 1827 to 1856.

Owen MSS. Archives of the British Museum of Natural History, London. This archive contains Owen's twenty-seven volumes of primary correspondence, five volumes of lecture manuscripts, his pocket notebooks, and other materials related to his career after 1840.

Green MSS. Archives of the Royal College of Surgeons, London.

Whewell MSS. Trinity College, Cambridge.

Darwin MSS. University Library, Cambridge University.

France

Archives of the Museum national d'histoire naturelle.

Procès-verbaux des Assemblées des Professeurs du Muséum national d'histoire naturelle, Paris Archives Nationales of France, AJ15.

Canada

Coleridge MSS. Denderwent Collection, Victoria College, University of Toronto. This has the largest body of Joseph Henry Green material known, with fourteen unpublished letters of Green to Coleridge.

United States

Owen Correspondence, American Philosophical Library, Philadelphia.

The Francis Hirtzel Collection of Richard Owen Correspondence, Temple
  University Archives, Philadelphia. This has three volumes of Owen family
  correspondence, (1, 4, 5) matching the series containing volumes 2 and 3 at
  the Royal College of Surgeons, London.

Unpublished Materials

De Jager, Timothy. "G. R. Treviranus (1776–1837) and the Biology of a World in
  Transition." Unpublished Doctoral Dissertation, University of Toronto,
  1991.
Gruber, Jacob. "Calendar of the Owen Correspondence" (with introductory
  biographical essay). British Museum of Natural History Archives.
Kolb, Daniel. "The Systematic Unity of Kant's Idea of Nature." Unpublished
  Ph.D. Dissertation, Department of Philosophy, University of Notre Dame,
  1983.
Ross, David Lloyd. "A Survey of Some Aspects of the Life and Work of Sir
  Richard Owen, K. C. B. Together with a Working Handlist of the Owen
  Papers at the Royal College of Surgeons of England." Unpublished Ph.D.
  Dissertation, University of London, 1972.
Rupke, Nicolaas. "Richard Owen's Vertebrate Archetype " 36 p. MS (personal
  communication).

Printed Works

Anonymous. United Kingdom. *Report of the Parliamentary Commission on
  Evidence, Oral and Documentary, Taken and Received by the Commissioners
  Appointed . . . for Visiting the Universities of Scotland*, Vol. 1. London:
  Clowes and Sons, 1837.
Abernethy, John. *Introductory Lectures Exhibiting Some of Mr. Hunter's
  Opinions Respecting Life & Disease*. London: Longman et al., 1815.
Appel, Toby. *The Cuvier-Geoffroy Debate: French Biology in the Decades Before
  Darwin*. Oxford: Oxford UP, 1987.
Ashton, Rosemary. *The German Idea*. Cambridge: Cambridge UP, 1980.
Barclay, John. *An Inquiry into the Opinions Ancient and Modern Concerning Life
  and Organization*. Edinburgh, 1822.
Barry, Martin. "On the Unity of Structure in the Animal Kingdom." Pt. 1.
  Edinburgh New Philosophical Journal **22** (January 1837): 116–41; Pt. 2

(April 1837): 345–64.

[Blizard, William.] *Synopsis of the Arrangement of the Preparations, in the Gallery of the Museum, of the Royal College of Surgeons.* London: Carpenter and Son, 1818.

Blumenbach, Johann. *A Short System of Comparative Anatomy.* 2nd. Ed., trans. W. Lawrence. London, 1827.

Bowers, Fredson. "Transcription of Manuscripts: The Record of Variants." *Studies in Bibliography* 29 (1976): 212–64.

[Broderip, William.] "Progress of Comparative Anatomy." *Quarterly Review* 90 (1851–2): 362–413.

Brodie, Benjamin C. *The Hunterian Oration. . .14th February, 1837.* London: Longman et al., 1837.

Carus, Karl Gustav. *Von den Ur-Theilen des Knochen-und Schalengerustes.* Leipzig, 1828.

Charles Darwin's Notebooks, 1836–1844: Geology, Transmutation of Species, Metaphysical Enquiries. Edited by P. Barrett et al. London: BMNH Press, and Ithaca: Cornell UP, 1987.

Clarke, Edwin and L. S. Jacyna. *Nineteenth-Century Origins of Neuroscientific Concepts.* Berkeley and London: U. California Press, 1987.

[Clift, William and Richard Owen.] *Catalogue of the Contents of the Museum of the Royal College of Surgeons in London*, 6 parts. London: Taylor, 1830–31.

Coleridge, Samuel Taylor. *The Collected Letters of Samuel Taylor Coleridge.* Edited by E. L. Griggs. Oxford: Clarendon, 1959.

_____. *Unpublished Letters of Samuel Taylor Coleridge.* Edited by E. L. Griggs. New Haven: Yale University Press, 1933.

_____. *Biographia Literaria*, in: *Complete Works of Samuel Taylor Coleridge.* Edited by W. G. Shedd, Vol. 3. New York: Harper, 1853.

Comrie, J. D. *History of Scottish Medicine.* 2nd. ed., 2 Vols. London: Baillière, Tindall and Cox, 1932.

Cope, Zachary. *The History of the Royal College of Surgeons of England.* London: Blond, 1959.

Corsi, Pietro. *The Age of Lamarck.* Berkeley: U. California Press, 1989.

Cranefield, Paul. *The Way In and the Way Out.* New York: Futura, 1974.

Cunningham, Andrew and Nicholas Jardine, eds. *Romanticism and the Sciences.* Cambridge: Cambridge University Press, 1990.

Cuvier, Georges and A. Valenciennes. *Histoire naturelle des poissons.* 22 Vols. Paris: Levrault, 1828–49.

Cuvier, Georges. "Considérations sur les mollusques, et en particulier sur les céphalopodes." *Annales des sciences naturelles* 19 (1830): 241–59.

_____. *Le Régne animale distribué d'après son organisation, pour servir de base à l'histoire naturelle des animaux et d'introduction à l'anatomie comparée.* 4 vols. Paris: Deterville, 1817.

————. *Leçons d'anatomie comparée.* 2d ed. Paris: Grochard, 1835.

————. *Mémoires pour servir à l'histoire et l'anatomie des mollusques.* Paris: Deterville, 1817.

————. *Rapport historique sur les progrès des sciences naturelles depuis 1789 et sur leur état actuel.* Paris: Imprimerie imperiale, 1810.

Darwin, Charles. *Journal of Researches.* Vol. 3 of *Narrative of the Voyage of the H. M. S. Adventure and Beagle.* Edited by Robert Fitzroy. London, 1839; reprinted New York: AMS Press, 1964.

————. *The Correspondence of Charles Darwin.* 6 Vols. Edited by Frederick Burkhardt, Sydney Smith et al. Cambridge: Cambridge University Press, 1985-91.

Desmond, Adrian. *The Politics of Evolution: Morphology, Medicine, and Reform in Radical London.* Chicago: University of Chicago Press, 1990.

————. "Robert E. Grant: The Social Predicament of a Pre-Darwinian Transmutationist." *Journal of the History of Biology* **17** (1984): 189–223.

Dobson, Jessie. *William Clift.* London: Heinemann, 1954.

————. "John Hunter's Museum." Appendix in Zachary Cope, *The History of the Royal College of Surgeons of England,* 274–306. London: Blond, 1959.

————. "The Architectural History of the Hunterian Museum." *Annals of the Royal College of Surgeons* **29** (1961):113–26.

Douglas, Stair. *The Life and Selections from the Correspondence of William Whewell D.D.* London: Kegan Paul, Trench & Co., 1882.

Ehrenberg, Christian. "On the Magnitude of the Ultimate Particles of Bodies; Infusory Animals Not Formed Immediately from Dead Matter, &c." *Edinburgh New Philosophical Journal* **13** (1832): 319–28.

————. "Vorläufige mittheilung einiger bisher unbekannter Structurverhältnisse bei Acalephen und Echinodermen." *Müllers Archiv* (1834): 562–80.

Farre, Arthur. "Observations on the Minute Structure of Some of the Higher Forms of Polypi, with Views of a More Natural Arrangement of the Class." *Philosophical Transactions of the Royal Society of London* (1837): 387–426.

Galenus, Claudius. *On Anatomical Procedures.* Vol. 1, Translated by C. Singer. London: Oxford UP, 1956.

————. *On Anatomical Procedures.* Vol. 2, Translated by W. L. H. Duckworth. Cambridge: Cambridge UP, 1962.

Girtanner, Christoph. *Ueber das Kantischen Prinzip für Naturgeschichte.* Göttingen: Vandenhoek & Ruprecht, 1796; reprinted Brussels: Culture et Civilization, 1962.

Grant, Robert E. "On the Influence of Light on the Motions of Infusoria." *Edinburgh Journal of Science* **10** (1829): 346–9.

_____. *An Essay on the Study of the Animal Kingdom*. London: Taylor, 1828.

Green, Joseph Henry. *Spiritual Philosophy, Founded on the Teaching of the Late Samuel Taylor Coleridge*. Edited by J. Simon, 2 Vols. London: Macmillan, 1865.

_____. *Vital Dynamics: The Hunterian Oration Delivered. . .14th February, 1840*. London: Pickering,1840.

Gruber, H. E. and P. H. Barrett, *Darwin on Man*. London: Wildwood, 1974.

Harris, W. F. R. *The Heart and Vascular System in Ancient Greek Medicine*. Oxford: Clarendon, 1973.

Harvey, William. *Anatomical Disputation Concerning the Movement of the Heart and Blood in Living Creatures*, Translated by G. Whitteridge. Oxford: Blackwell, 1976.

_____. *Disputations Touching the Generation of Animals*, Translated by G. Whitteridge. Oxford: Blackwell, 1981.

Haupt, H. "Das Homologieprinzip bei Richard Owen." *Sudhoffs Archiv* **28** (1935): 143–228.

Herbert, Sandra. "The Place of Man in the Development of Darwin's Theory of Transmutation, Part I. To July 1837." *Journal of the History of Biology* **7** (1974): 217–58.

Hodge, M. J. S. "Darwin as a Life-long Generation Theorist," In *The Darwinian Heritage*, edited by David Kohn. Princeton: Princeton University Press, 1985.

_____. "Darwin Studies at Work: A Re-examination of Three Decisive Years (1835–7)," In *Nature, Experiment, and the Sciences: Essays on Galileo and the History of Science in Honour of Stillman Drake,* edited by T. H. Levere and W. R. Shea. Dordrecht: Kluwer, 1990, 249–74.

[Home, Everard and W. Blizard.] *Summary of the Arrangement of the Hunterian Collection of the Royal College of Surgeons*. London, 1813.

Hunter, John. *Essays and Observations on Natural History, Anatomy, Physiology, Psychology, and Geology*. Edited by Richard Owen. 2 Vols. London: J. Van Voorst, 1861.

_____. *The Works of John Hunter*. Edited by J. J. Palmer. 4 Vols. London, 1835–7.

Jackson, Heather. "Coleridge's Collaborator, Joseph Henry Green." *Studies in Romanticism* **21** (1982): 161–79.

Jacyna, L. C. "Images of John Hunter in the Nineteenth Century." *British Journal for the History of Science* **21** (1983): 85–108.

_____. "Immanence or Transcendence: Theories of Life and Organization in Britain, 1790–1835." *Isis* **74** (1983): 311–29.

Kant, Immanuel. "Bestimmung des Begriffs einer Menschenrasse." (1785), in *Immanuel Kants Werke*, edited by E. Cassirer. Vol. 4. Berlin: B. Cassirer,

1922, 225–240.

_____. "Ueber den Gebrauch teleologischer Prinzipien in der Philosophie," (1788) in *Immanuel Kants Werke,* edited by E. Cassirer. Vol. 4. Berlin: B. Cassirer, 1922, 489–516.

_____. "Von der Verschiedenheit der Rassen der Menschen." *Kants Werke: Vorkritische Schriften* Vol. 2. Berlin: Reimer, 1912, 429–43.

_____. *Metaphysical Foundations of Natural Science.* Translated by J. Ellington. Indianapolis: Bobbs-Merrill, 1970.

[Keith, Arthur.] *Guide to the Museum of the Royal College of Surgeons.* London: Taylor, 1910.

Lamarck, Jean Baptiste. *Histoire naturelle des animaux sans vertèbres.* 5 Vols. Paris: Verière, 1815–22.

_____. Georges Cuvier, and Bernard de Lacépède. "Rapport des professeurs du Muséum, sur les collections d'histoire naturelle rapportées d'Egypte par E. Geoffroy." *Annales du muséum d'histoire naturelle* 1 (1802): 234–41.

Lawrence, William. *Lectures on Physiology, Zoology, and the Natural History of Man.* 2 Vols. London: Benbow, 1822.

Lenoir, Timothy. "The Göttingen School and the Development of Transcendental Naturphilosophie in the Romantic Era." *Studies in the History of Biology* 5 (1981): 111–205.

_____. "Teleology without Regrets: The Transformation of Physiology in Germany: 1790-1847." *Studies in History and Philosophy of Science* 12 (1982): 293–354.

_____. *The Strategy of Life: Teleology and Mechanics in Nineteenth-Century German Biology.* Dordrecht: Reidel, 1981.

Levere, Trevore. *Poetry Realized in Nature: Samuel Taylor Coleridge and Early Nineteenth-Century Science.* Cambridge: Cambridge UP, 1981.

Linnaeus, Carolus. *Reflections on the Study of Nature.* Translated by J. E. Smith. London: Nichol, 1785.

Lister, J. J. "Some Observations on the Structure and Functions of Tubular and Cellular Polypi, and of Ascidiae." *Philosophical Transactions of the Royal Society* 124 (1834): 365–88.

MacLeay, William Sharp. "Remarks on the Comparative Anatomy of Certain Birds of Cuba, with a View to Their Respective Places in the System of Nature or to Their Relations with Other Animals." *Transactions of the Linnean Society* 16 (1823): 1–46.

_____. *Horae Entomologicae, or Essays on the Annulose Animals.* Vol. 1. London: Bagster, 1819.

_____. "Remarks on the Identity of Certain General Laws Which Have Been Lately Observed to Regulate the Natural Distribution of Insects and Fungi." *Transactions of the Linnean Society of London* 14 (1825): 46–68.

MacLeod, Roy. "Evolutionism and Richard Owen, 1830–1868." *Isis* **56** (1965): 259–80.

Matenko, P., ed. *Tieck and Solger: The Complete Correspondence*. New York: Westermann, 1933.

Milne-Edwards, Henri. "Blood," in *Cyclopedia of Anatomy and Physiology*, edited by Robert B. Todd. Vol. 1. London, 1835–6, 405–14.

Müller, Johannes. *Handbuch der Physiologie des Menschen für vorlesungen*, Vol. 1. Coblenz: Hölscher, 1833–4.

_____. *Elements of Physiology*. Translated by W. Baly. Vol. 1. London: Taylor and Walton, 1837-8.

Ospovat, Dov. "The Influence of Karl Ernst von Baer's Embryology, 1828–1859: A Reappraisal in Light of Richard Owen's and William B. Carpenter's 'Paleontological Application of "Von Baer's Law" '." *Journal of the History of Biology* **9** (1976): 1–28.

Owen, Richard. "Acrita." In *The Cyclopedia of Anatomy and Physiology*, edited by R. B. Todd. London: Sherwood et al., 1836. 1:47–9.

_____. "On the Archetype and Homologies of the Vertebrate Skeleton." *Reports of the British Association for the Advancement of Science* (1846): 169–340.

_____. *Lectures on the Comparative Anatomy and Physiology of the Invertebrate Animals*. London: Longman et al., 1843.

_____. *Memoir on the Pearly Nautilus* (Nautilus Pompilius, *Linn.*). London: Royal College of Surgeons, 1832.

_____. *Odontography, or a Treatise on the Comparative Anatomy of the Teeth*. 2 Vols. London: Baillière, 1840–5.

_____. *Synopsis of the Royal College of Surgeons Physiological Museum*. London: Taylor, 1845.

Owen, Richard Starton. *Life of Richard Owen*. 2 Vols. London: Murray, 1894; reprinted Gregg: Westmead, 1970.

Plarr, V. G. *List of Lecturers and Lectures at the Royal College of Surgeons of England 1810–1900*. London: Taylor and Francis, 1900.

_____. *Catalogue of the Manuscripts of the Royal College of Surgeons of England*. London: Taylor, 1910.

Raine, K. "Thomas Taylor in England." In *Thomas Taylor the Platonist: Selected Writings*, edited by K. Raine and G. M. Harper. Princeton: Princeton University Press, 1969.

Rehbock, Philip H. *The Philosophical Anatomists*. Madison: U. Wisconsin Press, 1983.

Reil, Johann C. "Von Der Lebenskraft." *Reils Archiv für Physiologie* **1** (1796). Reprinted in *Klassiker der Medizin*, edited by Karl Sudhoff. Leipzig: Barth, 1910.

Richards, Eveleen. "A Question of Property Rights: Richard Owen's Evolutionism

Reassessed." *British Journal for the History of Science* **20** (1987): 129–71.

Robinson, Henry Crabb. "Letters on Kant and German Literature." *The Monthly Register and Encyclopedian Magazine,* 1802–3: 6–12; 205–208; 294–8; 397–403; 411–16; 485–8; 492–3.

Rohrer, C. W. G. "Sir Richard Owen: His Life and Works." *Bulletin of the Johns Hopkins Hospital* **22** (1911): 132–40.

Rupke, Nicolaas. "Richard Owen's Hunterian Lectures on Comparative Anatomy and Physiology." *Medical History* **29** (1985): 237–58.

Serres, Etienne. "Anatomie transcendantes.—quatrième mémoire. Loi de symétrie et de conjugaison du système sanguin." *Annales des science naturelles* **21** (Sept. 1830): 5–49.

Shaffer, E. S. "Coleridge and Natural Science: A Review of Recent Literary and Historical Research." *British Journal for the History of Science* **12** (1974): 284–98.

Singer, Charles. *A Short History of Anatomy from the Greeks to Harvey.* New York: Dover, 1957.

Sloan, Phillip R. "The Buffon-Linnaeus Controversy." *Isis* **67** (1976): 356–75.

―――――. "Buffon, German Biology, and the Historical Interpretation of Biological Species." *British Journal for the History of Science* **12** (1979): 109-153.

―――――. "Darwin's Invertebrate Program, 1827–37: Preconditions for Transformism." In *The Darwinian Heritage*, edited by David Kohn. Princeton: Princeton University Press, 1985.

―――――. "Darwin, Vital Matter, and the Transformism of Species." *Journal of the History of Biology* **19** (1986): 369–445.

Stevens, P. F. "Augustin Augier's *Arbre botanique* (1801), a Remarkable Early Representation of the Natural System." *Taxon* **32** (1983): 209–11.

Sulloway, F. J. "Darwin and His Finches: The Evolution of a Legend." *Journal of the History of Biology* **15** (1982): 1–53.

―――――. "Darwin's Conversion: The *Beagle* Voyage and Its Aftermath." *Journal of the History of Biology* **15** (1982): 325–96.

Tanselle, G. Thomas. "The Editing of Historical Documents." *Studies in Bibliography* **31** (1978): 1–56.

Todd, Robert B., ed. *Cyclopedia of Anatomy and Physiology.* 5 Vols. London, 1836–39.

Treviranus, Gottfried Reinhold. *Biologie, oder Philosophie der lebenden Natur für Naturforscher.* 7 Vols. Göttingen: Rower, 1802–22.

[Valenciennes, Achille and John B. Pentland.] *Catalogue des préparations anatomiques laissées dans le cabinet d' anatomie comparée de muséum d' histoire naturelle.* Paris: privately published, [1832].

Von Baer, Karl Ernst. "Ueber das äussere und innere Skelet." *Meckels Archive für Physiologie* **3** (1826): 327–76.

Von Staden, Heinrich. *Herophilus, the Art of Medicine in Early Alexandria.* Cambridge: Cambridge UP, 1989.

[Warburton, Thomas.] "Report from the Select Committee on Medical Education, with the Minutes of Evidence, Appendix and Index." *Reports from the Parliamentary Committees 1834, Part II: Royal College of Surgeons, London.* London, 1834.

Wellek, Réné. *Kant in England.* Rev. ed. Princeton: Princeton UP, 1972.

Index

Abernethy, John; 234; and materialist debates 9; and Owen 10; and electrical theory of life 21; on teeth 98n; on organic matter 211

Absorbent system (*see Lacteal vessels*)

Acalepha; Green on structure of 319

Acrita; defined 45, 263; nervous matter in 255

Aelian; on elephants 120

Air and life 226

Alcmaeon of Croton; place in history of anatomy 86; theory of cranial formation 86; on alleged respiration by ears 86n, 101; on generation 104

Albumen; of egg and blood compared 278–80

Alexandrian anatomy 115; and Erasistratus 116

Amphioxus, Owen's description of 55

Analogy (*see also Homology*); Aristotle on principle of 110; and Harvey 111; distinguished from homology 133; Cuvier on 191

Anatomical plans; Owen on 215

Anatomy; animal and human 142, 154; Mondino da Luzzi on 140 (*see Comparative anatomy*)

Anatomy Act of 1832 140, 195

Anaxagoras; and Socrates 88; and the rationality of nature 88

Animal; definition of 93, 239, 254; laws governing formation of 232

Animal heat 168–9; Hunter on 168; and life 225

Archetype; Owen on 61

Argument from design (*see Natural theology*)

Aristotle 49, 166; place in comparative anatomy 85; reaction to Plato 90; theory of knowledge 90; definition of animal 93; principles of classification 94, 109; blooded animals 94; classification compared to Linnaeus and Cuvier 95n; accuracy of anatomical observations 97; on cephalopods 97, 105n, 106n; on

digestive organs 98; on the elephant 98; on internal organs 99;   on the veins 100; on the teeth 99;   on the absorbent system 99;   on the heart 100; on the nerves 101; on sense of taste 102; on the tongue 102; on generation 104;   on unity of type 106; on hearing in fishes and insects 102; on classification 109–10; on anatomical analogy 112;   editions of works in RCS library 113; recovery by west 141

Arris, Edward: and Gale Surgical Lectures 47,   47n

Articulata 270

Ascaris 108

Asseli, Gaspero, 197; and lacteals 150

Atomists; on chance and design 122

Auroch 103

Azote (*see Nitrogen*)

Babbage, Charles; at Benjamin Brodie's Oration 55

Bacon, Lord Francis; on teeth 99; on natural theology 124; his inductive method 139

Baker, Henry 154

Baly, William;   translates Müller's *Handbuch* 233

Barclay, John; on body-mind separation 8; his proprietary anatomical school 8; and Lawrence-Abernethy debates 9; on lymphatic system 116; Owen's teacher in anatomy 8, 118; on Galen 118

Barnacles; classification of 193

Barry, Sir Charles; and reconstruction of College 44

Barry, Martin; Owen's knowledge of his embryology 62n, 202, 234; and Von Baer on embryology 201;

Bauer, Franz 226, 284, 286; and micrometer 284n;

Beagle voyage; Darwin's development during 3; donation of fossils to College of Surgeons 54n

Bell, Charles 15, 126, 178; as Hunterian Lecturer 47